Bibliotheca Mycologica 61

The Species Concept in Hymenomycetes

Proceedings of a Herbette Symposium
held at the University of Lausanne, Switzerland,
August 16-20, 1976

edited by

H. CLÉMENÇON

1977 · J. CRAMER

In der A.R. Gantner Verlag Kommanditgesellschaft
FL-9490 VADUZ

© 1977 A.R. Gantner Verlag KG., FL-9490 Vaduz
Printed in Germany
by Strauss & Cramer GmbH, 6945 Hirschberg II
ISBN 3-7682-1173-8

Herbette Symposium on Species Concept in Hymenomycetes 1976

CONTENTS

PREFACE 3

LIST OF PARTICIPANTS 5

INTRODUCTORY ADDRESS
    H. Clémençon: Reason and Goals for the symposium. 7

CONTRIBUTIONS
    R. Watling: An Analysis of the taxonomic Characters used in Defining the Species of the Bolbitiaceae. 11
    H.D. Thiers: Species Concepts in the Boletes. 55
    C. Bas: Species Concept in Amaninta Sect. Vaginatae. 79
    H.E. Bigelow: Differentiation of Species in Clitocybe. 105
    A.H. Smith: Speciation in Lactarius. 123
    R. Kühner: A propos de la délimitation des espèces dans les Hygrophorus Fries du sous-genre Hygrocybe Fries. Deux caractéristiques peu ou non utilisées. 157
    K. Esser & P. Hoffmann: Genetic Basis for Speciation in Higher Basidiomycetes with Special Reference to the Genus Polyporus. 189
    R. Blaich: Enzymes as an Aid in Taxonomy of Higher Basidiomycetes. 215
    A. Bresinsky, O. Hilber & H.P. Molitoris: The Genus Pleurotus as an Aid for Understanding the Concept of Species in Basidiomycetes. 229
    R.F.O. Kemp: Oidial Homing and the Taxonomy and Speciation of Basidiomycetes with Special Reference to the Genus Coprinus. 259
    J. Boidin: Intérêt des cultures dans la délimitation des espèces chez les Aphyllophorales et les Auriculariales. 277
    F. Oberwinkler: Species- and Genetic Concepts in the Corticiaceae. 331
    H. Romagnesi: Incidence des caractères non morphologiques sur la notion d'espèce et autres taxa chez les macromycetes. 349

R.H. Petersen: Species Concept in Higher Basidiomycetes: Taxonomy, Biology and Nomenclature. 363
R. Singer: The Species Concept in Agaricales and its Adaptation to Taxonomy. 381

INDIVIDUAL PROPOSITIONS OF A SPECIES CONCEPT PRESENTED BY THE PARTICIPANTS OF THE SYMPOSIUM, AND FINAL DISCUSSION IN ORDER TO ARRIVE AT A COMMON CONCEPT. 393

    INDIVIDUAL DEFINITIONS 393
    FINAL DISCUSSION 395
    COMMON CONCEPT 402

PUBLIC LECTURE (Leçon Herbette)
    R. Kühner: La notion d'espèce chez les Champignons supérieurs. 409

GENERIC INDEX 441
SUBJECT INDEX 443
INDEX OF NEW SPECIES AND NEW COMBINATIONS 444

Herbette Symposium on Species Concept in Hymenomycetes 1976

P R E F A C E

Following recent developments in mycology of hymenomycetes I felt very uneasy about what the current ideas of a fungal species are. I thought that a discussion between experienced mycologists would not only be helpful for me, but that it could be beneficial for all participants. Tentative contacts with taxonomical mycologists met with great enthusiasm. Thus encouraged I also contacted experimental workers and again found interested collaborators.

With this academic support I set out to obtain financial aids. Several commercial enterprises favored the symposium academically, and some even contributed to the costs in a very early phase of the preparations. These contributions were extremely useful and even of crucial importance and also allowed some participants to continue the meeting in the form of a workshop, actually collecting fungi and discussing them. It is not only my duty, but a pleasure to thank the following companies for this essential support:

   Migros Genossenschafts Bund, Zürich
   Nestlé Alimenta SA, Vevey
   Zyma SA, Nyon
   La Suisse, Assurances, Lausanne

The symposium itself and the public lecture were made possible by the generous gift from the HERBETTE FOUNDATION. Our meeting thus became a HERBETTE SYMPOSIUM. The Herbette Foundation encourages scientific research and communication at the Faculty of Sciences of the University of Lausanne. The Foundation also sponsers symposia and public lectures taking its name.

JEAN HERBETTE, former ambassador of France in Moscow and Madrid, was a passionate scienctis working in his private laboratory in his home at Montreux, Switzerland. He bequeathed his possessions to the Faculty of Sciences of the University of Lausanne. The Foundation dates from 1974.

Our meeting in August 1976, discussing the Species Concept in Hymenomycetes is the first "symposium Herbette" and Dr. Kühner's public lecture became the first "leçon Herbette".

All participants thank the Herbette Foundation for its ample support.

H. Clémençon

Herbette Symposium on Species Concept in Hymenomycetes 1976

LIST OF PARTICIPANTS

Bas, Cornelius, Rijksherbarium, Schelpenkade 6, Leiden, Holland

Bigelow, H.E., Department of Botany, University of Massachusetts, Amherst, 01002 Mass. USA

Blaich, R., Bundesforschungsanstalt für Rebenzüchtung, D-6741 Siebeldingen, Germany

Boidin, J., Université Claude Bernard, Laboratoire de Mycologie F-69100 Villeurbanne, France

Bresinsky, A., Universität Regensburg, Institut für Botanik II, D-84 Regensburg, Germany

Clémençon, H., Institut de Botanique systématique et de Géobotanique, Université de Lausanne, CH-1007 Lausanne, Switzerland

Esser, K., Lehrstuhl für allgemeine Botanik, Ruhr-Universität Bochum, Postfach 2148, D-463 Bochum, Germany

Horak, E., Institut für Spezielle Botanik, ETH, Universitätstrasse 2, CH-8092 Zürich, Switzerland

Kemp, R., Department of Botany, Mayfield Road, Edinburgh EH9 3JH, Scotland

Kühner, R., Université Claude Bernard, Laboratoire de Mycologie F-69100 Villeurbanne, France

Oberwinkler, F., Lehrstuhl für Spezielle Botanik, Institut für Biologie, Universität Tübingen, Auf Morgenstelle 1, D-7500 Tübingen, Germany

Petersen, R.H., Department of Botany, University of Tennessee, Knoxville Tenn. 37916, USA

Romagnesi, H., Av. Daumesnil, F-75012 Paris, France

Singer, R., Field Museum of Natural History, Roosvelt Road at Lake Shore Drive, Chicago, Illinois 60605, USA
Actual address: INPA, Caixa postal 478, 69.000 Manaos, Brasilia

Smith, Alexander H., University of Michigan Herbarium, Ann Arbor Michigan 48104, USA

Thiers, H.D., Department of Ecology & Systematic Biology, San Francisco State University, San Francisco California 94132, USA

Watling, R., Royal Botanic Gardens, Edinburgh EH3 5LR, Scotland

Herbette Symposium on Species Concept in Hymenomycetes 1976

REASON AND GOALS FOR THE SYMPOSIUM

H. Clémençon

Singer (1936) divides the History of Mycology into three major periods:
- (I) from Persoon and Fries to Patouillard and Fayod, which is the period of artificial systems,
- (II) Patouillard, Fayod to Lohwag and Atkinson, where first attemps and some results of a natural classification became visible,
- (III) the third, and present, period which is "characterized by an immense accumulation of additional descriptive facts and new theories to explain them" (Singer 1975).

Those who added to the mass of descriptive data also built a revised and well-reasoned classification of Hymenomycetes reflected in excellent Monographs and such outstanding works as Kühner and Romagnesi's "Flore Analytique", Dennis, Orton and Hora's "New Check List", and Singer's "Agaricales in Modern Taxonomy". Most of these are the result of a nearly life-long quest for the understanding of fungal species, genera and families, and in some way represent a crop of the third period in Mycology. I do not pretend, of course, that the harvest is done by now, since the third period is still very much alive and continues to stimulate competent researchers, but I think time has come to complement the crop by an effort to try to define and illustrate the taxonomic concepts that have evolved during this period by those who gave it to us.

In order to simplify our task, and in the hope to be more efficient, one level of taxonomic thinking has been selected for main discussion: the species concept. It is inevitable that lower and higher taxonomic levels will be invoked, but this is by no means an inconvenience, quite to the contrary.

To anyone working in fungal taxonomy it is obvious that the species concept of different authors, even contemporary ones, do not match each other. It must be very interesting indeed to compare the concepts that evolved and matured in different mycological schools during the third period. The participation at our Symposium of researchers who helped to shape present day mycology ensures a true and authentic record of the different ideas and gives us much needed stability for our discussions. Perhaps it will be possible to find enough points in common to come to some conclusions, however modest they may be, that express present day taxonomic concepts.

There is another main reason for this Symposium. Gradually at first, but then greatly activated by Raper and his school, fungal genetic information became available. A new dimension was introduced in the 1930's and has never stopped growing since. In the 1950's it became so important that no taxonomist since can afford to ignore it. There are two main though related phenomena that are of high interest and direct bearing for taxonomy: the incompatibility systems and parasexuality. Although the latter is still predominately a domain of Ascomycete research, it probably plays a very active and efficient role in genetic exchange in Hymenomycetes, perhaps via Buller phenomena.

Up to now all species in Biology have explicitly or implicitly been based on sexuality. Mycology tried to apply these concepts to fungi, but here things are complicated by incompatibility systems which caused Raper to speak of "sexes by the thousands" (1966). To make things worse, a system totally unknown in classical Biology, the parasexuality, introduces a new dimension defying all attempts to transfer classical species concepts in Mycology. Since fungi have so much more than mere sexuality, they are so much more complex and challenging. We should try to take this into account in our struggle for a modern species concept in Hymenomycetes.

Another point to discuss is the relationship between Mendelian genetics, incompatibility and parasexuality. It is known, that a single mutation can cause enormous (in the taxonomist's mind) morphological changes, e.g. Panus tigrinus and its Lentodium squamulosum state (Rosinksi and Robinson 1968). On the other hand, only very minor morphological differences can be detected between two incompatible strains, e.g. in Psathyrella (Romagnesi 1975). Although academically easy to handle, such cases are nightmares for the practical taxonomist. What types of morphological differences are essential?

But there is still another nightmare, horrifying some taxonomists, tantalizing others with promise and simultaneous frustration: the possibility of "vegetative hybridization", perhaps by parasexual mechanisms. Alexander Smith's "Pluteus explosion" (Smith 1971) may be an excellent illustration of what kind of difficulties arise for the taxonomist. If this is true, could a species of Hymenomycetes be created (almost mathematically out of given gene-pools) several times at different locations and at different times?

I do not think our Symposium can answer all these questions, but it certainly should work them out neatly to bring them to wider attention. If we were to have some success, we may even forecast the coming of a fourth period of mycological taxonomy.

Literature.

Raper, J.R. 1966: Genetics of Sexuality in Higher Fungi. Ronald Press Company, New York.
Romagnesi, H. 1975: Description de quelques espèces de Drosophila Quél. (Psathyrella ss. dilat.). Bull. Soc. Myc. France 91: 137-224.
Rosinksi, M.A. & A.D. Robinson, 1968: Hybridization of Panus tigrinus and Lentodium squamulosum. Amer.J.Bot. 55: 242-246.
Singer, R. 1936: Das System der Agaricales. Annales Mycologici 34: 286-378.
Singer, R. 1975: The Agaricales in Modern Taxonomy. Cramer, Vaduz.
Smith, A.H. 1971: The Origin and Evolution of the Agaricales. <u>in</u> Evolution in the Higher Basidiomycetes (R.H. Petersen edit.), University of Tennessee Press, Knoxville.

Herbette Symposium on Species Concept in Hymenomycetes 1976

AN ANALYSIS OF THE TAXONOMIC CHARACTERS USED
IN DEFINING THE SPECIES OF THE BOLBITIACEAE *)

R o y   W a t l i n g

Summary

The characters used in delimiting taxa of the Bolbitiaceae are critically reappraised, particularly those used in the laboratory. It is suggested that less emphasis than in the past should be placed on basidiospore size and the position of velar remains on mature basidiocarps; caution is expressed in the use of microchemical reactions. Cystidial characters are considered of utmost importance and a return to the greater use of macromorphology supported; cultural analysis so far carried out supports the latter. It is requested that a sympathetic hearing should be given to the adoption of the concept of micro- and macrospecies when dealing with members of the Bolbitiaceae.

Introduction

The family Bolbitiaceae forms a well-defined group which has long been neglected by both formal and experimental taxonomists. In fact from a review of the literature not only have very few of the members of the family been experimentally investigated but few have been obtained in pure culture.

Although the constituent genera of the family have been examined by classical taxonomic methods of agaricology, the family has never been studied as a unit, until comparatively recently. Descriptions of members of the main genera are scattered throughout the literature and although some are in considerable detail, the information has neither been brought together nor the characters used in identification reassessed.

---

*) This paper expands some of the results originally submitted as the subject of a doctorate thesis at the University of Edinburgh, 1964.

The *Bolbitiaceae* was first recognised in the literature as a distinct group of fungi by Heim (1931); Singer (1936) gave it subfamily rank and later (1948) he erected the family name now used. He obviously considered it a very natural group of agarics and if, as I also beleive this is the case it is quite unlike some other families currently accepted in the Agaricales. The family name has been derived from that of the genus selected as type, i.e. *Bolbitius*, and Singer included in the family seven genera. There is little doubt that Singer's ideas on the *Bolbitiaceae* as in many other groups of the agarics were moulded around the original principles outlined by Fayod (1889), who applied anatomical and developmental information to an understanding of the taxonomy and affinities of the agarics.

Indeed Fayod originally not only described the two most prominent genera in the family, *Agrocybe* with hazel-brown spore-print, and *Conocybe* with rust-brown spore-print but also the genus *Pholiotina*. There is much discussion about the latter genus as to its placing within the family and whether it is in fact distinct enough to merit generic status.

In compilation of notes and keys Singer (1951) again was an important forerunner, particularly in the genus *Agrocybe*. Of the other genera of the family *Bolbitiaceae* Kühner (1935) produced a beautiful annotated account "Le genre Galera" based primarily on fresh collections. Kühner's treatment of the old Friesian tribe *Galera* laid down the foundation of many modern studies on the Bolbitiaceae, both directly and indirectly. He recognised that the genus *Galera*, a generic combination made by Kummer embracing Fries's original tribe *Galera*, was in fact polygeneric and separated out the two important components *Conocybe* and *Galerina*. Although members of the latter genus had always stood in earlier works beside true members of the family *Bolbitiaceae* Fayod, Atkinson and Kühner found them to be only superficially similar. The variation, however, in the genus *Galera* was well known to Fries himself, who recognised (1838) two major groups and although he placed them both in a single tribe (and later subgenus)of *Agaricus* he distributed one, i.e. *Conocybe*, in the section *Conocephalae* and the other i.e. *Galerina*, in the section *Bryogenae*.

Kühner's and Singer's papers in conjunction with the introductory chapter of Heim's "Le genre Inocybe" (1931) are the starting points in the modern analysis of the Bolbitiaceae. Heim in the same publication described a new genus for the family, i.e. Tubariopsis, and a new subgenus of Conocybe i.e. Cyttarophyllum, both from Madagascar. He recognised their affinities to the temperate species of Conocybe and Bolbitius and in a key to the brown-spored groups refers to them all as the "Groupe des Bolbitius". Singer (1947, 1951, 1962) in his later studies, although raising many of Kühner's sections to higher rank, does in fact retain the major groupings first established by this meticulously careful French mycologist.

Since Kühner's publication and Singer's earliest systematic work on the genus Agrocybe many new species have been described and attributed to the family. However, there has been no recent revision of the family and both the taxonomy and nomenclature, especially the latter, are sliding into chaos. In Leiden in 1972 an attempt was made to discuss characters of the family in open forum and this apparently was a very useful exercise, (Watling, in Kits van Waveren, 1972).

Although the genera of the Bolbitiaceae are well defined the species now placed in them have been attributed to a whole host of unrelated brown-spored and even purple-spored agaric genera. The majority have been distributed in either the very artificial genus Pholiota or in the genus Galera, depending on whether the gills before becoming exposed to the atmosphere were protected by an annulus (Pholiota) or by a fugacious veil (Galera). The time is ripe for an assessment of the species in the family and the species concept to be adopted.

The Bolbitiaceae as here understood is as defined in an earlier paper (Watling 1965) consisting of three agaricoid genera, Bolbitius and Conocybe with rust-coloured spore-prints and Agrocybe with hazel-brown spore-print , and two gasteromycetoid genera Gastrocybe parallel to Bolbitius and Cyttarophyllum parallel to Conocybe; inclusion of Tubariopsis will only be resolved until fresh material is found. A third gasteromycetoid genus parallel to Agrocybe is not only predicted but in fact known, although not as yet formally described. I am of the opinion that Descolea should be excluded from the family and support Horak (1971) in his retention of this genus in the Cortinariaceae along with the gasteromycetoid Setchelliogaster which I believe is closely related.

I am aware that I am reintroducing Cyttarophyllum although both Heim and Singer have synonimised it with Galeropsis Vel. but after examination of very small primordia of several galeropsoid fungi and many of the type collections of so-called species it has been concluded that Galeropsis desertorum and its allies are not conspecific with the North American G. besseyi; the latter has a "cellular" cuticle. If all the species of Galeropsis belong to one genus then one can derive the agaricoid families Strophariaceae and Bolbitiaceae from it; the converse is true also,if they are considered derivatives of agaricoid families different gastroid genera will be found with similar and parallel morphology to those genera in the agarics.

The defining characters of the family Bolbitiaceae are
1. basidiocarps with monomitic structure with or without laticiferous hyphae,
2) lack of indefinite growing margin,
3) swollen units yet general economy in use of cell-wall materiel,
4) a "cellular" cuticle, and
5) brown spores.

The "cellular cuticle" is composed of spheropedunculate units arranged in a palisade with their longest axes at right angles to the pileus surface e.g., a palisadoderm. This palisade has been described as hymeniform because the cells are arranged in a similar way to those of the hymenium. Beneath this single-celled layer is a zone in which further hymeniform elements are formed and which ultimately push up to increase the surface area of the pileus. This lower layer is composed of irregularly swollen cells and would be termed the subpellis. Thus the palisade is composed of cells with differing length.

The cells of the suprapellis vary in the length of their shortest axis between 10 and 30 microns. In some species they are at the smallest end of the range, e.g. C. rickeniana, whereas other species always have large cells, e.g. C. silignea. They are variable even in the same pileus, those at the margin being slightly smaller than those at the disc; this may be related more to the age of the cell and pileus and the availability of water, than to the taxonomic position of the fungus. The size does not appear to be related to the size of the basidiocarp. The variation in the cell size is too great to assist in distinguishing taxa. In Bolbitius the palisadoderm is

Bolbitiaceae

covered in a gluten composed of a compound which stains blue green with alcian blue and becomes pigmented after treatment with the periodic acid Schiff's staining method. This layer is quite thick often one and a half or even twice as thick as the hymeniform suprapellis itself. It is easily separable and has been called a viscid pellicle, frequently including loosened cells from the palisadoderm. In Conocybe and Agrocybe a similar compound may be present although in the majority of species it is of limited development; C. coprophila is probably the most important exception.

In Conocybe the cells of the palisadoderm are very well defined with the pedicel either thickened, especially in s.g. Pholiotina and/or coloured in aqueous solutions of alkali. In Agrocybe these same cells although difficult to see in the primordia are very evident in young material but soon become intermixed with filamentous cells with age or they may separate to expose the pileus-trama beneath; it must be borne in mind that the basidiocarp of many species of Agrocybe is much more persistent than those of Conocybe, and Bolbitius is very ephemeral being parallel to Coprinus in general facies yet not autodigesting. Thus the "cellular" aspect can be lost with age in species of Agrocybe or in wet weather, a feature which is accentuated in Cyttarophyllum were the suprapellis soon disorganises through attrition from sand-grains.

Pileocystidia are frequent in Conocybe s.g. Conocybe, and some species of Agrocybe, pushing up between the pyriform cells of the suprapellis; pileocystidia are formed in s.g. Piliferae and in Conocybe laricina by modification of the hymeniform elements. The pilocystidia and pileocystidia may be accompanied by hyphal strands as in Bolbitius aleuriatus which can be interpreted as the remnants of the former blematogen.

The structural organization described above gives to the pileus a characteristic appearance when seen in the field as on drying it glistens and minutely wrinkles, unlike the pileus of a wide range of other brown-spored agarics. This allows many members of the family to be recognised at once, except perhaps certain Agrocybe spp. which because of their large size and thicker flesh approach members of the Strophariaceae.

The characters of the pileus of several veiled species of Conocybe as seen in the field have been discussed by van Waveren (1970) and by Watling (1971) and the observations correlated with anatomical details. The reader is referred to these papers as little more can be added except to report that in large specimens the flesh is frequently slightly thicker and as a result the suprapellis puckers to give a wrinkled appearance; this appears for instance to be the main and possibly only difference between Bolbitius reticulatus and B. aleuriatus.

Most Conocybe spp. are hygrophanous, only the thicker fleshed species being expallent; a few members of the genus Agrocybe are neither hygrophanous nor expallent, e.g. A. dura. Bolbitius ssp. are slightly hygrophanous in the areas where the pellicle has pushed away and exposed the subpellis.

Species of Descolea possess an epithelium-like pileipellis which may be covered and/or have incorporated into it, more or less ornamented, clamp-connected filamentous units. The majority of the filamentous units are from the original enveloping veil whilst the rest are from elongation of similar hyphae to those which terminate in the spheropedunculate end-cells. Thus although in mature specimens a cursory glance would suggest a relationship with the Bolbitiaceae in detail this is not so.

The colour of the spore-print is taxonomically paramount. The spore-print colour found in the Bolbitiaceae ranges in the main from amber brown to hazel brown although small deviations from the central patterns can be demonstrated, e.g. Conocybe michiganensis, Agrocybe leechii and a Bolbitius sp. from Virginia U.S.A. (Watling 1975 b). However, although these might be expected a careful look-out should also be kept out for white-spored taxa equivalent to Leucocortinarius bulbiger and Hebelomina. None have yet been found but pale coloured spore-prints are found in C. spiculoides and forms of C. lactea, but this is not a lack of pigment but a reduction in wall thickness in these species. The lack of vinaceous tints helps to distinguish the family from the central pattern in Strophariaceae and the lack of a rufous flush from members of certain sections of Cortinariaceae; they are more strongly coloured in mass, however, than those found in the clay coloured series i.e. Inocybe and Hebeloma.

Bolbitiaceae

The spores forming on the basidium are at first hyaline but rapidly become golden-yellow on approaching maturity. Anomalies are rare, the only case out of over 1,000 fresh collections being that of _Conocybe lactea_ which had half its basidiospores colourless.

The pigment of the basidiospore in all the members examined is localised in the exosporium of the spore-wall and is darkened in the presence of even weak aqueous solutions of alkali. It is a resistant pigment destroyed neither when the basidiospores are mounted in concentrated aqueous solutions of caustic alkalis nor when they are mounted in concentrated mineral acids. Even after a twelve hour treatment with an acidic solution of sodium sulphite a hint of colour is retained in the basidiospores of members of the _Conocybe pubescens_ group and _Bolbitius vitellinus_. In the same time and under similar conditions the pigmentation is lost in the basidiospores of _Conocybe aporos_ and _C. rickeniana_ i.e. those with a slightly thinner wall than in _Bolbitius_. Ammoniacal solutions of hydrogen peroxide take a similar time to bring about discolouration but freshly prepared solutions of eau de javelle have no measurable results. A 5% aqueous solution of potassium permanganate oxidises the brown pigment in the wall of the basidiospore in a twentieth of the time taken by ammoniacal peroxide.

The characteristic spore type is one with an optically smooth our face, generally distinct and often large germ-pore, and ellipsoid out-line. Generally under the electron microscope the surface is apparently smooth or at most minutely undulate (Pegler & Young, 1971) but the latter phenomenon may be caused by desiccation during the preparation of material. Deviations from this basic pattern, however are to be found; ornamented spores are rare but not absent from the family. Thus in several major sections roughened spores are found, _C. subverrucispora_ in _Piliferae_, _C. subnuda_ in _Pholiotina_, and _Ochromarasmius_ e.g. _C. laricina_. The range of ornamentation found in the Bolbitiaceae has been the subject of an earlier paper (Watling 1977), and apparently is because of localised thickening and deeper pigmentation of the exosporium that of the episporium below.

What, however, is clear is that in all the species so far studied the wall is rather complex consisting of a thick exosporium and a thinner episporium; the former is uniformly electron dense although some striation can be seen running parallel to the spore-surface and the latter is stratified although not significantly so. Apparently there is a full complement of wall layers, including an exosporium, both in species with smooth and ornamented spores. The one example, C. oculispora in which a separating perispore was thought to exist has been explained as abberrant development of the epispore (Watling 1975a).

The hilar appendix is small in all species studied; Pegler and Young (1971) record that an open-pore hilum is present in all the spores they examined. A slight suprahilar depression is occasionally found in Bolbitius aleuriatus according to Pegler and Young; an obvious suprahilar plage is lacking in all species although a slight one has been found in some of those members of the genus Conocybe with spores slightly amygdaliform shape in side-view. There is no doubt that the size of the hilar appendix is related to that of the basidiospore itself for all those species with larger basidiospores have larger hilar appendices and this in turn is related to the size and number of sterigmata per basidum. In the gastroid members of the family the hilar appendix is irregular in shape and often appears as if torn.

Other variations from the basic pattern are found in the size, presence of a germ-pore etc. Thus C. aporos in the annulate group lacks a germ-pore as does Agrocybe vervacti and C. vestita. Generally in small basidiospores the germ-pore is not very distinct whereas in contrast in large basidiospores its presence is very obvious and sometimes extremely prominent. However, even when the basidiospore is large the germ-pore is not always of the same shape, nor in the same position; thus in Agrocybe leechii the germ-pore is placed excentrically (Watling 1975b). A whole morphological series from aporate to distinctly porate basidiospores is found in both the major genera Agrocybe and Conocybe. The character of the germ-pore although dependent on the size of the basidiospore can be correlated in many species with habitat preferences and does not appear in this group of agarics to be a simple indicator of phylogeny. A large germ-pore is invariably found in coprophiles e.g. Bolbitius vitellinus, Conocybe pubescens, C. coprophila, and a small germ-pore in those species inhabiting places of a less ephemeral nature such as pathsides, scree slopes e.g. C. pygmaeoaffinis, C. arrhenii and Agrocybe erebia; in the last species the spore has an attenuated apex. Double germ-pores although aberrant are not infrequent in members of the family.

The presence or absence of a germ-pore is taxonomically significant and if present the morphology and position in relation to the spore axis and hilar appendix is important and constant for a single taxon.

The shape of the spores is basically ellipsoid to ovoid in face-view tending to be slightly flattened in side-view; they are more or less terete in section except for some degree of applanation on the adaxial surface. They are rarely lentiform although exceptions are found e.g. C. intrusa, but more frequently an amygdaliform shape is exhibited. In the gastromycetoid members the spores are more symmetric about the hilar appendage even in profile. Some species in the family are adaxially-abaxially compressed converting the amygdaliform or limoniform outline to a more or less hexagonal pattern which in some species may even be angular e.g. C. antipus. An undescribed species from Queensland possesses mitriform spores further extension of the familiar pattern is the presence of nodulose spores resembling those in Inocybe Section Clypeus and this appears to be a feature of a whole group of species in SE Asia.

The basidiospore shape varies very little within a given taxon excluding Conocybe Section Candidae where the basidiospores are broadly ellipsoid to ovoid to almost subglobose in optical section although the more typical ellipsoid basidiospore might also be found. Shape varies very little with development of the basidiospores although during periods of initial spore shedding and again in senescence of the basidiocarp, basidiospores might be slightly contracted, significantly so in some collections.

The basidia on which the spores borne are typical holobasidia and thickening and septation, primary or secondary have not been seen. The presence of such sporogenous cells is a fundamental character in placing the Bolbitiaceae in the (Homo-) Basidiomycetes; the structure and morphology of the basidium, except for a few examples is fairly constant throughout the family. There is only one particular type of basidium exhibited in the family; they are relatively short and wide, thus contrasting with those found in other families of the Agaricales. Although they are of the same general shape as those in the Coprinaceae, they are never as long; the basidia of Agaricus spp., although frequently of similar length, are narrower. The sterigmata in the Bolbitiaceae are also usually shorter than those found in other families particularly in the heterogeneous Hygrophoraceae, where they are frequently extremely long.

At specific level in the Bolbitiaceae the differences in the basidial size are usually unimportant, except in a few special cases e.g. Conocybe laricina, C. mairei and C. pilosella. C. laricina differs from members of Conocybe s.g. Conocybe in that the basidia are more cylindric than claviform and therefore approach those of s.g. Pholiotina.

The basidiospore is a single unit in its own right, whereas the basidum throughout its life-history is part of a complex tissue. Basidia are difficult to measure for as soon as the hymenium is disturbed the lateral pressures exerted on both the mature and developing basidia are redistributed and changes of shape ensue. This probably explains why the range for the basidia of a single species is often very wide. In the gasteromycetoid forms the basidia collapse after the spores, along with part of the sterigmata, have fallen off.

Basidia should be examined whenever possible from sections of gills of live material mounted in water. When this is impossible, as with herbarium material where the natural structure of the hymenium has already been disorganised, parallel measurements can only be taken from squash-mounts.

With experience small differences can be observed especially by direct comparison of camera lucida drawings of the basidia but these differences are difficult to express in formal diagnoses. Thus Conocybe magnicapitata has a rather shorter but much wider basidum than related species. The presence or absence of clamp-connections at the base of the basidia is undoubtedly of great value.

Frequently the number of sterigmata produced per basidum has been used as a criterion for separation, at varietal (e.g. Agrocybe semiorbicularis) or at specific level e.g. (Conocybe rickenii). During the present study in the species which were induced to produce basidiocarps in pure culture i.e. Agrocybe dura, A. praecox, A. paludosa, Conocybe farinacea and C. coprophila, the number of sterigmata per basidum did not alter. In cultures of Conocybe rickenii under semi-sterile conditions two-spored basida were always produced in the

C. intrusa.

The volume of the basidiospore has been shown to be interrelated with that of the basidium (Corner 1947) and thus the dimensions of both spore and basidia reflect a single character. In the present study large basidia always were found to produce large spores and small basidia small spores; never-the-less the shape of the basidiospore is also reflected in its dimension but as the shape varies from one species to another the size of the basidiospore cannot be substituted as a measure of the basidium.

The size of the basidiospore has since the early 1930's been used to separate species and the problems experienced in judging the significance of differences in spore-size in members of the Bolbitiaceae is common to other groups of agarics. However, as spore-size is apparently not only dependent on the species involved but also the environmental conditions great care must be exercised in the use of such a character.

Shape and size are interconnected and the proportion of one reflected in the other. The "Q" value, which is the numerical value for the ratio of average length of the basidiospore to its average width, is calculated for all specimens as a matter of course. This value simply expresses numerically the position and the relationship of the basidiospore dimensions in a sample from a given spore-print on a length-width graph. The length-width graph and the "Q" value were used as methods of quickly ascertaining affinities between large numbers of spore-prints of different collections of the same and of different species. The "E" value is the same as the "Q" value but applied to any one basidiospore which makes up the sample of basidiospores.

Although shape may be significant, great care must always be adopted when interpreting numerical data of this kind as one can show that significant changes take place between the first spores shed and those formed towards the end of the life of the basidiocarp. The calculation of the fiducial limits for a given spore sample have been found most useful in helping to determine the significance of differences in spore-data from one collection or species to another.

It is suggested that the application of an asymptotic concept should basidiocarps in agreement with the specific description; even though a four-spored race is known this has never been induced to develop in culture from the two-spored race.

In the two-spored forms and species of *Conocybe* and *Agrocybe* the basidia are hardly, or not noticeably, larger (not significant at a given probability level of 1 in 20) than in corresponding or very closely related four-spored species. The sterigmata, which are produced by a two-spored basidium, are both more prominent and thicker at the base than those of a four-spored basidium e.g. species-pairs *Conocybe crispa* and *C. lactea*, *C. rickenii* and *C. ochracea*, two-spored and four-spored respectively. This is probably because of the mechanics of supporting and discharging larger basidiospores. The number of sterigmata produced by each basidium was therefore genetically fixed; the operation of such two-spored basidium has been discussed by Kühner.

In *Conocybe* other correlatable morphological characters usually separate the two-spored "races" from their four-spored counterparts, characters which could hardly be explained by the presence of a pleotropic gene-system e.g. *Conocybe rickenii* is coprophilous, has a white stipe and mat, non-striate, non-hygrophanous pileus, while its counterpart *C. ochracea* grows amongst herbs and agrees with but few of the characters listed. However, we know as yet nothing about the action of such gene-systems in *Conocybe*, or if they in fact exist at all in the agarics.

At present this study suggests that it would be prudent to keep the different spored races quite separate, even at specific rank, if su table correlatory evidence from other characters is available. Two spored basidia are said to have little taxonomic importance in cer tain agaric groups e.g. *Mycena galericulata* which has two and four spored races, but some experimental evidence suggests that this i not strictly true. However, a character which may have little tax mic significance in one group may be important in another. Not ir quently basidia have been observed lacking sterigmata or producir one or three sterigmata but they were always a very small percen of the total number of basidia present; they probably reflect a tain nuclear inbalance. In the present study dimorphic basidial tems were not observed and dermatobasidia are recorded only for be adopted in the interpretation of spore measurements obtained from the rather ephemeral agarics characteristic of the Bolbitiaceae to overcome the environmental parameter in the spore data.

The cystidia are of rather uniform structure in the Bolbitiaceae, marginal cystidia (cheilo-) are common to all members whereas facial cystidia (pleuro-) absent in Conocybe are found only in certain sections of Agrocybe and less frequently in Bolbitius. Brachycystidia (pseudoparaphyses or pavement cells) are characteristic of Bolbitius, are found in a few species of Conocybe particularly those in section Candidae and in Gastrocybe. Metuloids, lampro-cystidia and chryso-cystidia are not found in any members of the family.

The cheilocystidia are leptocystidia developing from similar cells as the basidia, in fact it can been seen that they develop in a parallel way to basidia but are sterile, hence they have been referred to as cystidioles. They are thinwalled infrequently contain even faintly coloured contents and are rarely encrusted with crystalline material, although they frequently have drop-like exudates at their apex. They are always relatively simple in morphology. Many species of Conocybe are characterised by precisely symmetric lecythiform cheilocystidia. The dimensions of the various parts of the cystidia are paramount in separating species. The actual shape of the cheilocystidium assists in placing the fungus in a section or even subgenus of Conocybe or Agrocybe and their morphology is generally correlated with other characters.

In section Pholiotina the cheilocystidia are very firmly fixed to the subhymenium and are difficult to extract from a section except after vigorous taping quite unlike those found in section Conocybe where they are easily separated one from another.

The pleurocystidia are also leptocystidia and are also formed from similar cells to the basidia. When present they are usually found in those species with widely spaced gills. It is suggested that the pleurocystidia in these fungi have evolved to act as baffles to reduce water loss between the gill-faces and from around the basidia, so allowing the spores to develop in a uniform humid atmosphere. Those species with closely arranged gills do not require modification as the gills by their proximity help to retain the humidity between them.

<u>Conocybe</u> <u>farinacea</u> was grown for several generations under several different treatments and in all cases the cheilocystidia remained constant in morphological detail and distribution pattern. There seems to be a naturally induced cline from slightly larger cheilocystidia nearest to the stipe to smaller headed cheilocystidia nearest to the pileus margin.

This may result from the cells at different places on the gill-edge having had different developmental times. When the measurements of cheilocystidia for both small and large gills from a single basidiocarp were compared similar cystidial size ranges from each gill were obtained, reflecting a constancy in the two types of gill. When the measurements were repeated using basidiocarps produced under artificial conditions, the results did not differ significantly.

The caulocystidia on the same basidiocarp, however, were more variable in size depending on the treatment. Between individuals grown on dung, paper-pulp and sand-mixture there is a slight variation in the percentage of subcapitate cells and their distribution on the stipe. In the basidiocarps grown under high humidity the heads of the caulocystidia were invariably absent, or reduced in size.

Specimens of <u>Conocybe</u> <u>rickeniana</u> kept for long periods in a damp chamber, produced caulocystidia whose necks elongated and whose heads were 1-2 microns smaller than those of specimens collected in the field. However, in this species they have rarely been induced to become subcapitate or non-capitate. The shape was therefore constant under these conditions although the relationship between the different parts of the cystidium varied.

Cheilocystidial characters therefore remained constant whereas caulocystidia were much more variable. Pileocystidia were not examined quantitively other than to observe that under no experimental treatments was their development in <u>Conocybe</u> <u>coprophila</u> inhibited. In the field it was observed that these same pileocystidia in <u>C</u>. <u>coprophila</u> could easily be lost by abrasion with dung particles etc. Similar observations were made for <u>Conocybe</u> <u>brunnea</u> under normal conditions of pileus expansion.

When young specimens of typical members of Conocybe e.g. Conocybe rickeniana and C. silignea were allowed to develop in a very damp atmosphere, hairs were induced to form on both stipe and pileus. In the specimens examined hairs were few in number on the pileus even after 48 hours but many flexuous hairs developed on the stipe during this same time. Specimens kept in tubes as controls without a pad of damp pulp or wet sphagnum gave no such results.

These results indicated that Fries (1821) and J. Schaeffer (1930) had some justification for their opinion, based only on field observations, that pilose forms in Conocybe were "atmospheric" forms of other more familiar glabrous species. Kühner (1935) did not entirely accept this view but observations by Herregods (1952) again supported the fact that species attributable to Kühner's sections Capitatae can develop hairs.

When incisions were made in the stipe and an area of the cortex removed hairs were induced to form in numbers similar to those under humid conditions. The hairs so produced were flexuous and similar in many characteristics to the hairs of C. farinacea, but were frequently branched, clamp-connected and multi-septate. It is considered that an atmosphere of relative humidity approaching 90%, as in the above experiments, induces further hyphal activity in the basidiocarps. The high humidity of the chamber, however, is rarely realised in the habitats in which the original specimens were collected and so the importance of these observations must be considerably reduced. The main factor in the taxonomy of the group is not the ability to produce hairs but the ability to produce hairs at the initial stages of growth, under all conditions, and to retain them e.g. Conocybe farinacea.

The validity of Conocybe s.g. Conocybe section Mixtae, as originally described by Kühner, based on these experiments and on the examintion of both fresh and dried specimens is challenged on two points; the variability of caulocystidia in shape and number on the stipe and the ability of many Conocybe ssp. to produce hairs.

Watling (1971) has shown that in Conocybe s.g. Conocybe the cheilocystidia are connected in development to the caulocystidia. This is also apparently true in the genus Agrocybe and Bolbitius and annulate

species of _Conocybe_. Only in the last example does one have to make
some modification to the hypothesis because of the presence of the
ring. The cheilocystidia and caulocystidia are developmentally interrelated only above the ring; the ring joins the pileus margin to the
stipe and so the gills do not during development come into contact
with the lower areas of the stipe. This area is covered in a dense
net-work of filamentous hyphae the morphology of which does not show
sufficient variation to be helpful in separating species but nevertheless it is an assistance in species grouping. The loose, usually
clamp-connected hyphae are broader towards their ends and are remnants of the original blematogen. By careless handling or with maturity this silvery silky white covering is lost exposing the stipe
cortex below, the cells of which particularly with age may be slightly or even strongly coloured.

The stipe in members of the Bolbitiaceae has been variously described
as floccose, scaly, roughened or silky striate, fibrillose etc.
There is no doubt these field characters reflect the presence of remnants of the blematogen covering and may either terminate in a chain
of up to five short broad cells or may themselves proliferate to form
such chains.

In some exannulate species and in _C_. teneroides they are adpressed to
the stipe surface or may ruffle up under adverse weather conditions;
in the _C_. vexans group the chains tend to clump together particularly
in age. They are very rarely gelatinised; indeed the stipe is only
viscid in one member of the family i.e. the poorly documented
_B_. gloiocyaneus from N. America. In _Conocybe_ s.g. _Conocybe_ similar
cells are only found at the very base of the stipe almost indistinguishable from the vegetative hyphae. In the annulate species of
Agrocybe although similar hyphae are present they are far less differentiated and are only found towards the stipe-base where they adhere
together to form white rhizoids; these rhizoids are quite characteristic.

The presence of brachycystidia can be used taxonomically with significant results. In _Bolbitius_ they are prominent and well developed,
but are absent in _Agrocybe_, although pleurocystidia here may be present. Both structures act as spacing agents but the latter in addition also as baffles to air movement; they are absent from most

species of _Conocybe_. _C. lactea_ in many ways resembles _Bolbitius_, indeed it was originally described as _Bolbitius tener_. It and its relatives possess brachycystidia also; Donk (1964) considered these structures as a means of expanding the sporeproducing layer. The subhymenium during the production of hymenial elements increases in size and number; thus the age of the gill at any one point will dictate the thickness of the subhymenium and the number or size of brachycystidia.

Apparently by a study of the inter-relationships of cystidia we arrive at a developmental hypothesis which would indicate that the Bolbitiaceae are fundamentally paravelangiocarpic. This is the case for all the species so far studied except where the veil material is almost completely absent and then they are termed gymnangiocarpic, and in _Agrocybe cylindracea_ where the basidiocarp possesses a blematogen of considerable development making the fungus bivelangiocarpic. Thus as might be expected a theme is found in the family with minor deviations to one side and to the other. From the discussion above it is clear that a morphological series exists depending on the nature of the lipsanenchyma and on the structure to which it adheres, i.e. stipe or pileus. This series cuts across the classification of both Fayod's genera _Conocybe_ and _Pholiotina_ as emended by Singer (1951) e.g. _Conocybe mairei_ and _C. coprophila_, both by some placed in the latter genus, approach in veil characters _C. tenera_, the type of _Conocybe_; _C. subvelata_ on the other hand is a true member of _Conocybe_ s.g. _Conocybe_ from the literature appears to approach in velar characters _C. vestita_, which is placed in _Pholiotina_. In _Agrocybe_ both annulate and non-annulate species exist side by side forming otherwise a fairly homogeneous group. The lack of veil is regarded by some authors as indicative of primitiveness but _Conocybe_ ssp. and _Bolbitius_ spp. are far from primitive in their ecological and life-history characteristics. It is suggested that in this brown-spored group of agarics members can be primitive in one way and very advanced otherwise; this conforms with Corner's suggestions for the three species of _Oudemansiella_ (1934) he studied.

The constancy of the veil in species grown in culture indicates that it is a useful specific character. The presence or absence of a veil appears to be related to ecological preferences; it is absent or poorly developed in coprophilous and praticolous _Conocybe_ spp. but present in those which have a longer developmental period and those which inhabit fairly stable often exposed communities e.g. <u>Conocybe arrhenii</u>, <u>C</u>. <u>aporos</u> and <u>C</u>. <u>vexans</u>. The veil is linked intimately with the development of the basidiocarp and should be considered in this light. More careful observations are required on fresh collections as the more detailed characteristics of many species are inadequately known.

Although the veil characters of a single species or group of species are constant, from evidence of basidiospore morphology, hymenophoral trama arrangement and pileus "cortex" structure, it is considered that too much taxonomic emphasis has been previously put on the veil.

If the development of the basidiocarp is followed in the laboratory the critical stages of velar formation can be more easily seen but it is also apparent that many species in the Bolbitiaceae possess conidia. These are thallic arthroconidia although there is some evidence to indicate that they might form actually within the parent hyphae as a specialised type of thallic conidium. It is too soon to say whether these are useful in identification or circumscription of species as they are in certain sections of <u>Coprinus</u> but what is worth exploring, however, is the way in which these conidia fuse with compatible hyphae, with hyphae of related species, their reaction to unrelated "species" and whether they themselves actually germinate at all and so act as vegetative propagules,or spermatia.

The basidia and cystidia are connected to the hymenophoral trama through a subhymenium composed of swollen units, and experience has shown that the distribution of the hyphal strands in the trama are important in circumscribing the higher taxa of the Agaricales.

Fayod (1889) on first suggesting the autonomy of the genera <u>Pholiotina</u> and <u>Conocybe</u> recognised similarities not only in the "cellular" cuticle of the pileus and in basidiospore structure, but he also separated them on differences of structure of the hymenophoral trama which he considered distinct. Singer (1951) accepted this view and extended it to embrace many extra - European species although originally Fayod had based his observations on only two species of <u>Pholiotina</u> and five species of <u>Conocybe</u>. When considered separately the <u>Conocybe</u> <u>arrhenii</u> (<u>as</u> <u>togularis</u>) - <u>blattaria</u> group, referred to the genus *Pholiotina* by Fayod is sufficiently different from *Conocybe tenera* and its allies to place it in a different genus. Although this is perfectly true for the few species named above it was considered necessary in the present study to examine in detail many more fungi attributed to both *Conocybe* and *Pholiotina*. In this way the hiatus between them, if one existed, could be found. This was particularly important for those species morphologically intermediate between the extremes of each genus.

Thus the species,"*Galera*" *plicatella*, often placed in the separate genus *Galerella* by virtue of the cystidial type and plicate striate pileus, is equidistant in structure of the hymenophoral trama between Fayod's two genera.

A group of *Conocybe* spp. were examined i.e. *C. pygmaeoaffinis*, *C. mairei*,*C. coprophila* and *C. appendiculata*, which had a fairly well developed broad mediostratum approaching in structure that found in the *C. blattaria* group, but which did not possess the reduced subhymenium and hymenopodium. One of these *C. mairei* had a hymnophoral trama which approached the structure observed in some gills of *C. farinacea* although in view of the presence of subulate dermatocystidia and caulocystidia it was clearly unrelated.
The presence of dermatocystidia and non-lecythiform cheilocystidia was shared by all members of the group described above.

Thus the two types of hymenophoral trama were not completely separate and this was confirmed after the examination of sections of secondary and primary gills of single basidiocarps of several unrelated species of *Conocybe*. In many *Conocybe* spp. attributable to *Pholiotina* Fayod on macromorphology e.g. *Conocybe* vexans, *C. filaris* the sections of the secondary gills indicated that even here the mediostratum was reduced, as it was in the tips of the primary gills. In *Conocybe farinacea* the sections of the primary cells showed that at the base they had a structure resembling that found in *Conocybe coprophila*, i.e. well developed mediostratum, subhymenium and hymenopodium. In culture *Conocybe coprophila*, *C. farinacea*, *C. rickenii*, *Agrocybe dura*, *A. paludosa*, *A. praecox* and *Bolbitius vitellinus* retained the characteristics of their hymenophoral trama, the only differences between gills of the same basidiocarp being those already noted above. These differences were constant. The development of the gills indicates the close connection between the hymenophoral trama and the flesh of both pileus and stipe.

The evidence does not support separate lines of development in the hymenophoral trama in the pholiotoid and galeroid members of the genus Conocybe. Many intermediates have been examined which connect the two extreme forms of hymenophoral trama. However, correlation might be sought between hymenophoral trama structure and ecological preferences and this would cut across the divisions erected by Fayod and more rigorously applied by Singer. Thus the thin-fleshed species, and coprophilous or praticolous species of Conocybe, on the whole have a hymenophoral trama resembling that of Fayod's Conocybe type, whereas those Conocybe spp. inhabiting rocky places, clay banks, pathsides etc., have a hymenophoral trama resembling that of the Pholiotina type. Amongst the former would be grouped nearly all Conocybe s.g. Conocybe but also C. plicatella, C. mairei, C. coprophila, some collections of C. appendiculata, and C. aberrans judging from the literature. In the second group would be placed the Conocybe arrhenii - blattaria group, C. brunnea and some collections of Conocybe appendiculata. In parallel to this Bondartzev (1963) has showed that differences in trama structure of the polypores are indicative of ecological requirements.

In the young primordium of Conocybe farinacea the tissue at the junction of the pileus and stipe resembles that of the same region in a much older basidiocarp of Conocybe aporos. In this species the expansion of the primordium is not as rapid as in C. farinacea and because of this, primordia are frequently located in the field. On the other hand the primordia of members of Conocybe s.g. Conocybe are rarely found except when buried in dung or chaff. In C. farinacea the cells are unable to retain their regularity at the junction of pileus and stipe because of the rapid growth once the basidiocarp has reached 5 millimetres or so in size. In C. arrhenii with its slower development the balance between the cells of the pileus and of the stipe-flesh is retained. It is because of this that in C. arrhenii the cells which connect stipe to pileus are more regular in shape, less entangled and more dense than in C. farinacea. In all Conocybe spp. of s.g. Conocybe the actual passage from the stipe of the pileus is more or less abrupt yet less so in some members of the s.g. Pholiotina e.g. C. appendiculata. No distinct colour differences across the

In _Bolbitius_ _vitellinus_ the transition from stipe to pileus is also abrupt. In this same species the development is much more rapid and the cells of both the stipe and the pileus become distorted and disarranged and frequently separate to produce an alveolate structure. The flesh is also reduced considerably and in mature and fully expanded specimens of _B_. _vitellinus_ the pileus flesh is less than one millimetre thick.

In _Agrocybe_ spp. primordia are frequently seen in the field and this reflects their much slower growth as compared to that of _Conocybe_ _farinacea_; they take approximately two to three times as long to develop. In many Agrocybe spp. e.g. _A_. _dura_, the flesh is quite thick but in others, e.g. _A_. _paludosa_ it is almost as thin as in some _Conocybe_ spp. These facts indicate that it is extremely difficult to describe flesh textures of the Bolbitiaceae without including the characters of the cortical zones.

Recent work on the agarics often includes the results of treatment of the flesh with a selection of chemical reagents. In the _Bolbitiaceae_ such results have rarely been available, except for those obtained by the chemical use of an aqueous ammoniacal solution.

Unlike the results from treating the flesh of _Russula_ spp. and _Lactarius_ spp. or _Cortinarius_ spp. with chemicals it appears that in _Conocybe_ such results are of little importance in species separation. Moreover, there is reason to suppose, from the isolated tests carried out on fresh collections of other members of the genera _Conocybe_ and _Agrocybe_ that this is general for most members of the _Bolbitiaceae_.

Colour reactions are dependent on the presence of particular metabolic constituents in the tissue, bye-products, enzymes etc. Tests can be made to locate the presence of a compound not only by a colour change but by the evolution of gas. Such a test is the reaction between the basidiocarp tissue and hydrogen peroxide. Fresh and dried material of _Conocybe_ _farinacea_, _Agrocybe_ _dura_ and _A_. _paludosa_ gave vigorous bubbling with a solution of hydrogen peroxide because of the presence of an active catalase system. The mycelium in culture of the same collections did not exhibit similar properties. The presence of a catalase was confirmed by electrophoretic methods using a starch gel and it appeared to be a true haem-enzyme. It has further been shown from the few bolbitiaceous species tested that it would be a fruitful approach to the higher fungi if a comparison of the specific characters of these enzymes was made.

In agaricology the taste and odour of the basidiocarp is frequently utilised in the separation of closely related groups or species even though many myologists cannot distinguish them or detect them. In the Bolbitiaceae there seems only to be few distinctive odours and tastes; most species of the genus Conocybe lack a distinct taste or odour. One of the reasons for the separation of Conocybe farinacea and C. pubescens is the exceedingly strong mealy taste and odour, absent in any other member of the genus studied in the present work. Many basidiocarps of C. farinacea have been grown on many different substrate mixtures and in all treatments the mealy odour was constantly present and only rarely was it not easily detected.
The taste and odour are not produced by the dicaryotic or monocaryotic mycelium in culture. Basidiocarps of the Conocybe arrhenii agg. often have a rather acidulous taste and faintly pungent disagreeable odour, resembling that found in Lepiota cristata. In an unnamed collection close to C. appendiculata, the acidulous odour was frequently accompanied by the odour of Pelargonium, and in a collection from Lundell's herbarium one unnamed species with an aromatic smell was recorded.

A few of the larger Agrocybe spp. have a slightly bitter taste coupled with an odour of new meal e.g. A. praecox. On the other hand Agrocybe dura lacks the odour of meal but has a similarly unpleasant taste. The smell of A. dura resembles that of uncut earth-balls or newly ploughed soil, earthy, as is found on cutting potatoes. Both the monocaryotic and dicaryotic cultures of A. dura possess the odour of the fresh basidiocarp whereas those of A. praecox do not; in the latter the mycelium is not bitter as is the basidiocarp tissue.

Kühner (1935) observed that long acicular crystals formed when gills of Conocybe spp. were treated with an aqueous solution of ammonia. They had been deposited in the aqueous phase about the fungal tissue and at the interface of air and liquid at the edge of the coverslip. The group which produced these crystals were always members of the

Bolbitiaceae

Conocybe tenera complex, the phenonemon rarely being observed in other species although slight positive results were occasionally obtained in members of the C. spicula complex. Crystal formation was never found by Kühner outside his group Capitatae. Kühner recognised the significance of his observations and in his monograph used the presence or absence of the ability to form crystals with ammonia as a means of differentiating species; this was extended by Singer to extra European species.

Until the present study, the crystals have been neither isolated in a pure form nor identified, even though their formation has been used as a taxonomic character.

Experiments were designed to identify the crystals and their precurser, and to determine whether the formation of such crystals was as important taxonomically as was indicated in the literature. The part played by the precurser in the basidiocarp tissue was not investigated.

Different areas of the basidiocarp were tested and it appeared that the precurser was only to be located in the hymenial tissue in sufficient amounts as to be detectable by assaying with an ammoniacal solution. It was absent from the stipe tissue, although in thin-fleshed specimens this tissue was very difficult to separate completely from all the hymenial tissue. Spore-prints, did not give a positive reaction even after the basidiospore walls had been ruptured by crushing. The precurser was easily leached out of the fragments of gill-tissue by water; after leaching fragments of gill tissue produced no crystals on the application of ammonia. If the leachate was similarly treated with ammonia, however, crystalline material was formed. Positive results will certainly assist in the placing of a collection in the macro-species C. tenera but negative results do not exclude a specimen from the same group. There is every likelihood of this character being found elsewhere in the family and in other agarics.

The crystals were finally identified with the help of Dr Brian Caddy Glasgow, Scotland as a polysaccharide probably an aminopolysaccharide such as an oligosaccharide which may occur naturally, or arise as a result of treatment with the ammonia the -CH-O-CH- unit of a constituent being broken. Mucopolysaccharides, which are also oligosaccharides, are found in most Bolbitiaceae.

Pileus shape, striation and colour, stipe shape and colour and the degree of hygrophanity of the basidiocarp all appeared within small limits to be constant for a particular collection. The formation of a pseudorhiza, the grouping of the base of stipes, the presence of a bulb at the base of the stipe and its shape, and the presence of rhizoids were all indicative of a certain species or group of species and were very constant in their appearance. Basidiocarp size on the other hand was not constant and care had to be exercised when dimensions of stipe and pileus were considered. Stature and colour have deliberately been omitted in this survey as they are in the main problems experienced by all agaricologists. Restrictions to the use of colour of the pileus have been explored by van Waveren (1970) and myself, and stature can be varied considerably in culture. The latter may be in part because of limitations to our culture technique where the correct time control on the different stages of development have not been achieved. There is no doubt that field agaricologists appreciate the variation found in the macroscopic characters of the basidiocarp and some details have been indicated for the Bolbitiaceae in Watling (1975).

The colour in the Bolbitiaceae is usually membranal, only in species of Bolbitius is an intracellular pigmentation found. The expression of colour in those lacking intracellular pigmentation is usually a function of the context and its ability to loose water rapidly or retain it. The flesh is composed of swollen units in the pileus forming an open structure which on drying may trap air; the stipe is composed of parallel hyphae in the context with more irregularly arranged cells towards the centre and in the usually thickened base. The latter cells frequently break down during the maturation of the fungus although apparently water (and food) conducting hyphae on the inner surface of the cortex remain intact.

Pigmentation is almost lacking in the C. lactea group hence the erection of the section Candidae, but other species can possess pale caps! In Agrocybe erebia the membranal pigment is strongly developed.

From field observations it was seen that the family Bolbitia typified soils of base rich areas, usually those containing the free alkaline metals calcium and magnesium. This view was confirmed by the analysis of pH data taken from several hundred fresh collections of various members of the family.

The careful examination of the habitat of all collections yielded interesting results as to the possible association of some members of the family with particular plant-communities and especially with soils of higher nitrogenous content than average. However, the latter observation could also be explained by a possible change in the carbon-nitrogen ratio of the soil colonised, a parameter demonstrated as being of importance to agaric fructification by Grainger (1957). In addition to the large number of soil pH readings from the base of the basidiocarp, % water and % organic content have been determined. Substrates of high field capacity and organic content were always indicated except in some Agrocybe spp. which invariably produced rhizoids and often colonised dry soils of poor structure. Only a very few species characterised soils with gleyed horizons and, even in a rich alluvial area of grass, members of Conocybe s.g. Conocybe were nearly always associated with worm-casts and occasionally with sides of mole-hills. Both phenonema bring about the formation of an aerated area free from grass and forbs. Under suitable conditions this would be colonised rapidly by the mycelium of the agaric and the life-history completed before bryophyte or phanerogamic colonists of the "micro-area" established themselves. This samo pattern was found with dung for it established a similar "micro-area" for colonization; such areas were colonized by C. coprophila or C. pubescens both of which can be considered more or less strong coprophiles. By the careful noting of the plant-communities in which basidiocarps were found growing it was demonstrated that members of the Bolbitiaceae were fairly restricted in their habitat preferences. Thus in general terms members of s.g. Conocybe, as noted above, frequented meadows, grasslands, lawns, alluvial flats, glades in woods, whilst members of Conocybe s.g. Pholiotina (sect. Pholiotina) were found in more open habitats, e.g. clayey banks with sparse covering (Conocybe aporos), amongst road scrapings (C. vexans), or on pathsides (C. appendiculata). Some Agrocybe spp. had similar preferences and were frequently encounted near farmyards or in stubble fields e.g. A. dura, or in gardens e.g. A. praecox; other species were an integral part of the fixed sand-dunes and downland communities e.g. A. semiorbicularis and A. vervacti. Conocybe laricina was rather unique in that it occurred in grykes in limestone pavement or amongst limestone rubble often attached to pieces of woody debris or old stems of Mercurialis perennis; C. plicatella grows in grassland and has similar preferences to the members of s.g. Conocybe.

Little taxonomic value has been given to the accurate description of the ecological characteristics of a species. The early mycologists were often very careful in noting basidiocarp habitats but with the introduction of microscopic, chemical and cultural techniques their observations were never extended. The present study shows how little we know of the biology of members of even this one small family of the agarics. Many more seasons of field notes throughout the world are required to fill in many gaps for even our commonest species.

Dung has a high nitrogenous content and this may indicate preference by members of the family for substrates of high nitrogenous nature. Parle (1963) has shown that total, and mineral, nitrogen of wormcasts is higher than that of the surrounding soil. The presence of nitrophiles such as Melandrium rubrum, Urtica dioica and Mercurialis perennis suggests that the nitrogen and phosphate content of the soil on which many collections are found is higher than normal, e.g. Conocybe filaris.

From the present study the basidiospore-structure has been found in most cases to run parallel to the habitat preferences of the particular species. Thus coprophilous members of the genus Conocybe inevitably have thick-walled basidiospores with both a prominent apiculus and large germ-pore. The members attributed to Conocybe s.g. Pholitoina (sect. Pholiotina) on the other hand have thinner-walled basidiospores and are nearly always to be found on pathsides or in open patches in woods and copses. Even in Conocybe s.g. Conocybe closely related species have differently shaped basidiospores. Thus C. rickeniana growing amongst herbs in copses has thin-walled basidiospores whilst C. magnicapitate growing in grassland has larger and thick-walled basidiospores. Some exceptions have been found e.g. Conocybe filaris (Watling herb. 154 C), collected on a pathside on a cliff-top had thin-walled basidiospores.

Singer (1959) describes alpine forms of Conocybe spp. as always having larger basidiospores than those of lowland collections. The experimental and ecological data obtained in our studies suggest that this can be interpreted as an effort on the part of the agaric to ensure enough food material is available for initial germination under rather adverse conditions. The alpine environment would tend to reduce the basidiospore size by a drying influence greater than that acting on specimens of the same species in lowland pasture. Size of basidiospores would therefore be expected to decrease more rapidly and thus approach the minimum size for the normal function of a basidiospore more quickly than in a habitat free from winds. Under these conditions therefore phenotypes would be selected for their ability to produce large basidiospores. These alpine forms can therefore be considered more specialised than the type form, some might prove by cultural techniques to be specifically different a suggestion supported even by some of the descriptive data.

Habit and structural characteristics of the basidiocarps e.g. structure of the hymenophoral trama and veil characters, appear to be correlatable with habitat also. Thus members of Conocybe s.g. Conocybe found in pastures and in turf have fairly long, often slender stipes thus enabling them to grow up amongst the grass and to raise the pileus above the boundary layer in order to favour dispersal of the basidiospores. These species are protected for much of their life-history by the buffering nature of the vegetation and lack a veil. Woodland species on the other hand are favoured by the microclimate developed under the trees. Those silvicolous species which grow at woodland margins, in open places, on pathsides and amongst road scrapings, are slightly sturdier in the structure of their basidiocarps, usually thick-fleshed, and frequently veiled e.g. Conocybe arrhenii agg.

Conclusion

After this lengthy analysis of the characters found in the Bolbitiaceae it is concluded that although there is a general homogeneity found in the family with all members very similar in general facies and without bright colours or unique basidiocarp outlines there are sufficient characters available to separate taxa at the specific level. It is suggested on the basis of this analysis that although in the laboratory it will be possible to circumscribe biological species based on the action of the conidia and/or detailed microscopic analysis in correlation with field data this may not be possible to do from field data alone, or if so only as an experienced guess. These taxa I would suggest should be treated as microspecies which can be clustered into a fairly homogeneous group which should be designated as a macrospecies. These latter species agree in the main with Kühner's concept adopted in 1935 and serve a useful service in that skilled amateurs and non-mycological professionals without cultural facilities or sophisticated equipment can identify and use specific epithets. If the species concept is too narrow then we (1) will not acquire the help from ecologists and the like in ascertaining the true distribution of taxa and (2) will not persuade our taxonomy to be adopted by non-mycological scientists. Mycologists are few and far between and although still few in number natural historians are never-the-less in greater proportion and do collect in all parts of the world. Even with the field notes which are possible to make under expedition conditions, there will always be a difficulty of applying a narrow species concept, although biologically I would agree such a concept is correct.

The complexities of the Agrocybe semiorbicularis group have been demonstrated by Singer (1950) and in various groups of Conocybe by Kühner. Singer has indicated complexes around A. cylindracea and certainly one can see similar parameters working in the Bolbitius vitellinus group.

If one does not decide on the relative importance of characters in the better known European arena, understanding the fungi from unexplored areas will be almost impossible. It is not advisable at this stage to describe new species of single collections unless we have good field data and this probably applies perhaps even more so to the gastroid forms.

Bolbitiaceae 39

Interpretations and authorities of species discussed in text

Agrocybe cylindracea (DC. ex Fr.) Maire, 1938
   dura (Bolt. ex Fr.) Singer, 1936
   erebia (Fr.) Kühner apud Singer, 1939
   leechii (A.H. Smith) Watling, 1975
   paludosa (J. Lange) Kühner & Romagn., 1953
   praecox (Pers. ex Fr.) Fayod, 1889
   semiorbicularis (Bull. ex St Amans) Fayod, 1889
   vervacti (Fr.) Maire, 1938

Bolbitius aleuriatus (Fr. ex Fr.) Singer, 1951 (including B. reticulatus (Pers. ex Fr.) Ricken, 1915)
   gloiocyaneus Atkinson, 1908
   tener Berkeley (= Conocybe lactea (J. Lange) Métrod)
   vitellinus (Pers. ex Fr.) Fries, 1838

Conocybe aberrans (Kühner) Kühner, 1935
   antipus (Lasch) Kühner, 1935
   aporos Kits van Waveren, 1971
   appendiculata (J. Lange&Kühner apud Kühner) ex Watling, 1971
   arrhenii (Fr.) Kits van Waveren, 1971
   blattaria (Fr.) Kühner, 1935 (non Kühner, 1935)
   brunnea (J. Lange&Kühner apud Kühner) ex Watling, 1971
   coprophila (Kühner) Kühner, 1935
   crispa (Longyear) Singer, 1942
   farinacea Watling, 1964
   filaris (Fr.) Kühner, 1935
   intrusa (Peck) Singer, 1950
   lactea (J. Lange) Métrod, 1940
   laricina (Kühner) Kühner, 1935
   magnicapitata Orton, 1960 (C. spicula f. macrospora Kühner)
   mairei (Kühner) Watling - see below
   michiganense (A.H. Smith) Watling, 1975
   oculispora Locquin ex Watling, 1974
   ochracea (Kühner) Singer, 1950

Conocybe plicatella (Peck) Kühner, 1935
   pubescens (Gillet) Kühner, 1935 (non Kühner, 1935)

pygmaeoaffinis (Fr.) Kühner, 1935

rickeniana Singer ex P.D. Orton, 1960

rickenii (J. Schaeffer) Kühner, 1935

silignea (Fr. ex Fr.) Kühner, 1935 (non Kühner, Singer, Moser etc.)

spicula (Lasch) Kühner 1935 (= C. rickeniana Singer ex P.D. Orton, 1960)

spicula var. spiculoides Kühner, 1935 nomen nudum

subnuda Kühner & Maire, apud Kühner, 1935

subverrucispora Veselský & Watling, 1972

subvelata Singer, 1950

tenera (Schaeffer ex Fr.) Kühner, 1935

teneroides J. Lange, 1921 (= C. blattaria see above)

vestita (Fr.) Kühner, 1935

vexans P.D. Orton, 1960

Galeropsis besseyii (Peck) Heim, 1950 (= Cyttarrophyllum)
 desertorum Vel., 1930

Leucocortinarius bulbiger (A. & S. ex Fr.) Singer, 1945

Mycena galericulata (Scop. ex Fr.) S.F. Gray, 1821

N o m e n c l a t u r e   o f   K ü h n e r ' s   C o n o c y b e   s p e c i e s

Kühner in "Le genre Galera" (1935) described under the genus Conocybe seven species, nine varieties and seven forms, of which two species and one variety were in conjunction with R. Maire and one species and one variety in conjunction with J. Lange. The last group of taxa with Lange have since been validated (Watling, 1971) but there still remains several nomina nuda, because although Kühner's descriptions are excellent and well documented Article 36 of the International Code of Botanical Nomenclature requires latin descriptions to be supplied for epithets published after Jan. 1, 1935. Kühner's book was published after this date.

In the list above C. subnuda, C. mairei and C. spicula var. spiculoides all require validation. I decline to supply latin descriptions

Bolbitiaceae

to the first and the last as only N. American material of what is probably C. subnuda has been examined; C. spicula has been replaced by the more precise name C. rickeniana, and it must be ascertained whether var. spiculoides is genetically connected to Orton's fungus. However, through the kindness of Professor Kühner it has been possible to study authentic material of the third species which I here designate as type. The material agrees in all details with Kühner's description and it needs only a latin diagnosis.

C o n o c y b e   m a i r e i   (K ü h n e r)   e x
W a t l i n g   s p .   n o v .

Pileus 4-8 (-9) mm latus, e convexo vel expanso-convexo, ochraceus vel ochraceo-brunneus, pruinosus vel canus, striatus, hygrophanus. Stipes 10-25 x 0.7 - 1 mm aequalis vel sub-bulbillosus, fragilis, albidus vel pallido bubalinus, pruinoso-pubescentibus. Velum nullum. Lamellae ex pallido-mellinae dein ochraceae, adnatae.

Basidiosporae ellipsoideae leviter amygdaliformes 6-8 x 3-4 μm. Pleurocystidiae absentia; cheilocystidiae lageniformia vel cucurbiformia ad apicem obtusum, 17-28 x 4.2 - 7 μm ad apicem 1.5 - 2.2 μm.

Typus France, Samoëns 8.IX.1953: Herb. Kühner, Lyon. Slide in E.

References

Bondartzev, M.A. 1963: On the anatomical criterion in the taxonomy of Aphyllophorales. Bot. Zh. SSSR. 48(3): 362-372.

Corner, E.J.H. 1934: An evolutionary study in agarics: Collybia apalosarca and the veils. Trans. Br. Mycol. Soc. 19: 39-88.

Corner, E.J.H. 1947: Variation in the size and shape of spores, basidia and cystidia in Basidiomycetes. New Phytol. 46: 195-228

Donk, M.A. 1964: Conspectus of the families of the Aphyllophorales. Persoonia 3(2): 199-324.

Fayod, V. 1889: Prodrome d'une Histoire Naturelles des Agaricinés. Ann. Sci. nat (Botanique) 7th series, 9: 181-411.

Fries, E.M. 1838: Systematis Mycologici scu. Synopsis Hymenomycetum. Upsala.

Grainger, J. 1957: From Field to Laboratory and Back. Presidential address, Naturalist, July-Sept., London.

Heim, R. 1931: Le Genre Inocybe, Encyclopédie Mycologique 1, Paris.

Herregods, M. 1952: La villosité chez les Conocybes de la Section Capitatae (Kühner), Bull. Soc. Mycol. Fr., 68: 258-262.

Horak, E. 1971: Studies on the Genus Descolea Sing., Persoonia 6(2): 231-248.

Kühner, R. 1935: Le Genre Galera, Encyclopédie Mycologique, VII, Paris.

Parle, J.N. 1963: A microbiological study of earth-worm casts, Journ. Gen. Microbiol. 31(1): 13-22

Pegler, D.N. & Young, T.W.K. 1971: Basidiospore morphology in the Agaricales. Beih. Nova Hedwigia. 35: 1-210.

Schaeffer, J. 1930: Die Sammethäubchen (Galera). Zeits. für Pilzk. 14(9): 16.

Singer, R. 1936: Das System der Agaricales. Ann. mycol. 34: 286-378.

Singer, R. 1948: New and Interesting species of Basidiomycetes II, Pap. Mich. Acad. Sci. 32(1946): 103-150

Singer, R. 1947: New Genera of Fungi III. Mycologia 39: 77-89.

Singer, R. 1950: Naucoria Fr. in the U.S.S.R. Acta. Inst. Bot. Acad. Sci. Ser II 6: 403-493.

Singer, R. 1951: The Agaricales in Modern Taxonomy. Lilloa 22.(1949).

Singer, R. 1959: New and Interesting species of Basidiomycetes.
 VI Mycologia 51(3): 375-400.
Singer, R. 1967: The Agaricales in Modern Taxonomy. 2nd Edition Weinheim.
Singer, R. 1975: The Agaricales in Modern Taxonomy. 3rd Edition Weinheim.
van Wavern, Kits, E. 1970: The Genus Conocybe Subgen. Pholiotina.
 I. The European annulate species, Persoonia 6(1): 119-165.
Watling, R. 1965: Observations on the Bolbitiaceae 2. A conspectus of the family. Notes. Roy. Bot. Gdn., Edin. 26: 289-323.
Watling, R. 1971: Observations on the Bolbitiaceae IV. Developmental studies on Conocybe with particular reference to the annulate species. Persoonia 6(2): 281-289.
Watling, R. 1972: In Kits van Wavern: account o/ meeting on Bolbitiaceae. Coolia 15: 101-110.
Watling, R. 1975a: Observations on the Bolbitiaceae 10. The enigma of the perispore. Notes Roy. Bot. Gdn., Edin. 34(1): 131-134.
Watling, R. 1975b: Observations on the Bolbitiaceae 12. The affinities of two anomalous species. Notes Roy. Bot. Gdn., Edin. 34(2): 245-251.
Watling, R. 1977: Observations on the Bolbitiaceae 15. The taxonomic position of those species o/ Conocybe possessing ornamented basidiospores. Rev. Mycol. 40 (1976): 31-37.

Legend to Plates

Plate 1. <u>Bolbitius vitellinus</u> (Pers. ex Fr.) Fr. Basidiospore data.
Graph a. length/width graph (in microns): 1-5. selected collections
to show range; 6.G420 <u>B</u>. <u>lacteus</u> Lange?; 7. Lundell & Nannfeld exsiccata No   ; xx material produced in growth chamber, Graph b.
"Q" value for samples/length. Graph c. "Q" value for samples/width.

Plate 2a. Hymenophoral trama of sections in <u>Bolbitiaceae</u> accompanied
by characteristic cheilocystidium;
1. <u>Conocybe</u> <u>pubescens</u> (Gillet) Kühn. (s.g. <u>Conocybe</u> sect. <u>Mixtae</u>)
2. <u>C</u>. <u>rickenii</u> (J. Schaeff.) Kühn. (s.g. <u>Conocybe</u> sect. Pilosellae)
3. <u>C</u>. <u>magnicapitata</u> Orton (s.g. <u>Conocybe</u> sect. <u>Conocybe</u> (Capitatae)
4. <u>C</u>. <u>plicatella</u> (Peck) Kühn. (s.g. <u>Galerella</u>)
5. <u>C</u>. <u>intrusa</u> (Peck) Singer (s.g. <u>Conocybe</u> sect. <u>Gigantae</u>)
6. <u>C</u>. <u>coprophila</u> (Kühn.) Kühn. (s.g. <u>Pholiotina</u> sect. <u>Piliferae</u>)
7. <u>C</u>. <u>lactea</u> (Lange) Métrod (s.g. <u>Conocybe</u> sect. <u>Candidae</u>)
8. <u>Bolbitius</u> <u>aleuriatus</u> (Fr. ex Fr.) Singer (s.g. <u>Pluteolus</u>)
Plate 2b, as 2a.
9. <u>Conocybe</u> <u>brunnea</u> (J. Lange & Kühn.) ex Watling (s.g. <u>Pholiotina</u>
              sect. <u>Intermediae</u>)
10. <u>C</u>. <u>laricina</u> (Kühn.) Kühn. (s.g. <u>Ochromarasmius</u>)
11. <u>Agrocybe</u> <u>vervacti</u> (Fr.) Maire (s.g. <u>Agrocybe</u> sect. <u>Microsporae</u>)
12. <u>A</u>. <u>erebia</u> (Fr.) Kühn. (s.g. <u>Aporus</u> sect. <u>Velatae</u>)
13. <u>A</u>. <u>semiorbicularis</u> (Bull. ex St Amans) Fayod. (s.g. <u>Agrocybe</u>
              sect. <u>Pediadeae</u>)

Plate 3. Graph a. Mycelial weight/pH of culture media for dicaryotic
<u>Conocybe</u> <u>farinacea</u> Watling grown in pure culture. Graph b. Expression
of pH of substrate for members of the Bolbitiaceae; comparison with
<u>Galerina</u> (<u>Cortinariaceae</u>).pH of substrate about base of basidiocarp-
number of collections on specimens examined. Magnification shown where
necessary.

Plate 4 a & b. Conidial development in Bolbitiaceae: Plate 4a, A-I
production of thallic arthroconidia in <u>Conocybe</u> <u>farinacea</u> Watling.
Plate 4b. J-M continuation of above with a.<u>C</u>. <u>percincta</u> Orton and
b. <u>Agrocybe</u> <u>semiorbicularis</u> (Bull. ex St Amans) Fayod.
Plate 5a. <u>Conocybe</u> <u>farinacea</u> Watling cultured on sterile dung.
Plate 5b. <u>Agrocybe</u> <u>acericola</u> (Peck) Singer on paper-pulp.

Dr Roy Watling introduced his paper as follows:

Colleagues,

It is something of the order of seventy years which separates Linnaeus' Species Plantarum from Fries' Systema Mycologicum. There is little wonder therefore that agaricologists are behind the phanerogamist in their nomenclature and taxonomy, and that a symposium such as this is necessary.

We are entering an exciting period, however, with a rapidly expanding future and with a whole spectrum of techniques available, some of which have been very well tested over the years in other groups of organisms. Techniques in ecology, chemotaxonomy, serology, physiology and genetics.

But alas the fungi often have been only dealt with as simply an extrapolation of the flowering plant flora and perhaps this has been our down fall, certainly a very great disadvantage, in that we have adopted the concept of species based on the so called developed plants without question. We must ask ourselves what our species concept is to achieve, and what it is to be used for. Is it to be recognised in the field, after limited, but critical microscopic analysis, or only after application of experimental techniques. Is it a "morphological" species or a "biological" species; and is there a place for both? In such a discussion we can hardly omit to mention subspecies, variety and form ranks for not knowing what our species concept is, it is more difficult to define the lower categories.

As many of you know I have for some time now looked at various aspects of the Bolbitiaceae, the nomenclature of its members, and taxonomy from both experimental and classical stand points. The paper which has been distributed as I indicated to our host Heinz Clémençon was much more complete than I intended, as you will be pleased to hear, to deliver to the symposium. In fact it was in effect a proto-pre-publication in order to give delegates a background to the family and the thoughts which the studies provoke. Through the hard work of Heinz Clémençon and his staff, delegates have now had an opportunity to read and digest the various contributions and judge the parameters by which the individual speakers have worked. The subject matter of these papers range from the new characters which can be used to define species to more philosophical thoughts, or to the genetical approach. I therefore wish in my paper to select and talk around a particular aspect which has been given less emphasis than others, ecology because of its importance as an isolating factor.

Dr Watling then followed his introductory remarks with that part of the printed text dealing with ecological parameters and concluded:

We have found ecology of great use in our cultural studies but little value has often been given in formal taxonomy to the <u>accurate</u> descriptions of the ecological characteristics of a species. The early mycologists were frequently very careful in noting basidiocarp habitats but with the introduction of microscopic, chemical and cultural techniques their observations have not been systematically extended. My present study shows how little we know of the biology of members even of this one small family of the agarics; no doubt it applies to other agarics groups too. Many more seasons of field notes throughout the world are required to fill in many gaps for even our commonest species. I can only plead therefore that in the future it is essential to attempt to study the entire biology of the species in our quest to understand the species concept in agarics and within this framework particularly the ecology. The problems involved appear to be the difficulties in distinguishing between the individual and the population, the degree of substrate specialization and the limited range of morphological characters which are available for defining taxa. There is no doubt in my mind that we are describing in out morphological analysis a reflection of ecology both phenotypic modification and the result of numerous generations of evolutionary adaptation. Evidence suggests that we require a greater effort in the future to ascertain which morphological characters can be used as taxonomic indicators and which cannot, and which can be correlated with information from other spheres of study and what these correlations mean.

DISCUSSION FOLLOWING DR. WATLING'S PAPER

CLEMENCON: What is the difference between a variety and a microspecies?

WATLING: My concept of a microspecies applies to collections which can be best separated on cultural characteristics and, where one finds incompatibility between two strains the morphology of which is very close, such that in the field, with the known variability of the characters, very difficult to separate. Nevertheless, in the laboratory one can separate them quite easily. One gets compatibility between a variety and its species, whereas microspecies, I would say, are totally incompatible one with the other although they may show some morphological similarities.

SINGER: It seems to me that bringing together what we observe experimentally and in nature within the hierarchy of accepted taxa is a very difficult task, and especially the word variety is used in very different ways. I would not think this is so awfully bad for the time being. What I would like to see is that every author of taxonomic papers define what he thinks a variety is. I wish everybody here would give us an idea of how he uses this term. Whether or not we reach an agreement on definitions does not matter at first.

WATLING: In the world there are an incredible number of species which at the moment are not documented. If we adopt a very narrow species concept, which may be biological, we are not going to obtain the assistance of other botanists going on expeditions. I think we must perhaps accept that there is a biological species and understand what it means, and appreciate that ecological isolation is a very important factor, but nevertheless base our taxonomy very much on the shoulders of macromorphology. In this way we can easily obtain definable information from our expeditions. Maybe in another 20 years we can have another conference like this with all the information that has come in by then and we can start to discuss again. I think we have based so much work on fungi from temperate countries that finds from tropical areas will shatter a lot of our ideas.

THIERS: When _Conocybes_ fruit in culture, they are in a very fine, ideal environment. Do you find differences, let us say in spore size, between those produced on carpophores developed in culture and those from carpophores produced in nature, and do you find other micromorphological differences between these basidiocarps?

WATLING: Generally, fruiting bodies that you obtain in culture give you a whole range of what you would call aberrant forms through to normal forms. Now if you go in fact and look carefully in the place where you collected your so-called normal form in nature, you will see these aberrant forms as well. In culture you do in fact get perfectly normal fruiting bodies. I have collected them in the field, described them, both macroscopically and microscopically, transferred them vegetatively and through spores, grown them in culture, and induced the formation of fruiting bodies. Between these natural and cultured fruit-bodies I cannot see any differences.

THIERS: Spore size?

WATLING: Spore size varies anyway! Of course you can modify spore size depending on the amount of relative humidity of the culture, but they do not vary above an asymptote. I suggest when we use basidiospore size in Bolbitiaceae we should not say "spores 12 to 14 microns", but we should say "spores never greater than 14 microns".

THIERS: Would you use the word "never"?

WATLING: Yes! We must have keys that work, so people can use them.

BRESINSKY: I have some problems in seeing a theoretical background to distinguish between biological species on one side and morphological species on the other side. Morphology is part of biology and we should talk about other alternatives. One alternative would be genetic species versus phenetic species, or good versus bad species. I think what you would like to express is the possibility of having biological species on one side and on the other side a species concept which I would like to call a "bona fide" species concept which is governed by practical reasons without information or very little information about barriers, population genetics, etc. I think there is no possibility to make a difference between biological and morphological species.

WATLING: I would agree as far as the biological species in the Bolbitiaceae are concerned the morphological differences are so small that I think it would be too narrow a species concept to use at this stage of our mycology.

SINGER: I sympathize with Dr. Bresinsky's point of view. We should concentrate our efforts to obtain information on biological species, and I have the feeling that the possibilities, on other continents than Europe, to present the data necessary are perhaps somewhat underestimated. If we need for our taxonomic work to obtain the necessary data on cultured species I believe it is not too difficult to obtain such data from other continents. Europe has no particular advantage over other continents, as far as cultural work is concerned.

WATLING: I would agree with you. It is possible to do mycology with a pressure cooker and Petri dishes, even amateurs can do it, but I think we have to tread very carefully because there seems to be a resistance to adopting cultural approaches to the higher fungi. After all, the microscope was a barrier at the beginning of the century, and there are now enzymological and chemotaxonomic techniques to which I think morphological taxonomists may also have a slight barrier. At each technological stage we have to take the hurdles as they come along.

SINGER: When I talked, two days ago, to Alex Smith he said to me that the greatest mistakes we ever made (and I think he used the word "we"), we made on insufficient information given by other people. I sort of agree with this statement, so why not rather try to influence the collectors to get more and better data .
My last point is that I am really impressed by what you said about the ecological approach which is something that always interested me. I merely would like to say that what is observed in England or Europe is not necessarily a law for all continents. E.g. I can see that in tropical South America the number of species that grow on rotten wood in the rain forest is relatively large.

WATLING: Yes, but rotten wood in a rain forest is not a truly lignicolous habitat, it is simply a substrate which is above the highly reducing conditions one gets on the floor of such rain forests. I would accept in tropical rain forests on such substrates we find spe-

cies which would normally be called terricolous. I think it is our
approach really that is wrong, not the ecology of the fungus. In the
slashings of Michigan you have the Conocybe septentrionalis group
growing on the large old Fagus trunks, and they are above the ground
in an aerated situation. In Britain we vacuum-clean our woods so there is no such trash, but the same species grow on the ground, ie,
equally aerated conditions.

SINGER: Comparing the tropical and temperate zones we have clear evidence that in the tropics you have fewer humicolous species and more
species growing on wood. The main reason for that is that in the tropics you have practically no ectotrophic mycorrhiza. So I think the
tropics are a bit separate; we have special ecological observations
on that.

PETERSEN: (to Watling) When we talk about biological versus taxonomic species, don't we run into some nomenclatural problems. When we
deposit a dead type specimen we can no longer mate it, nor can we
perform biochemical procedures.

WATLING: Well, there is a method. If we describe a species today
there is no reason why we should not keep a living, "sleeping" specimen; liquid nitrogen allows us to do this, and even if we cannot germinate its spores now, 20 years hence we might be able to do so. One
can use tissue, vegetative mycelium and spores. We have had members
of the Bolbitiaceae, Strophariaceae and Coprinaceae in liquid nitrogen now for 7 years; one can take them out, and off they go as if
they were collected only yesterday. So, in fact, we should not only
deposit a dead specimen to agree with the Rules, but perhaps also we
should consider very carefully having living, "sleeping" specimens in
an international collection. If culture work is going to play a very
important part in our understanding of the species in the future,
then I think we should keep living specimens.

BRESINSKY: What about the danger to have a contaminant as the type?

OBERWINKLER: Can you give an example where you can differentiate
species the biological way?

WATLING: Yes, I have very good examples. They are in the genus

Bolbitiaceae

Coprinus from Kemp's work. But I have similar examples, although not nearly, as much experimental evidence as in Coprinus. In Conocybe, in the pubescens group, e.g. Conocybe farinacea, which has typical conical cap, down the stem caulocystidia which are mixed, some lecythiform, some filamentous, and spores which are very large equipped with a germpore; most members of the group grow on dung and are uniform tawny or cinnamon-brown. Now, in the field you can distinguish four morphological entities in the group. They differ certainly in the size of the spores, which you cannot alter in culture; also they differ in the number of lecythiform cysitidia which are produced on the stem. There are in addition within these four groups collections with slight differences in cap colour etc; however, only with slight morphological differences which in the field and in the laboratory I would have great difficulty in saying this is how you distinguish these strains. I would call such collections microspecies. In the laboratory you have a barrier produced between such cultures, on taking the oidia of one very little attraction to the hyphae of the other is shown, hyphae do not anastomose in culture etc. I would therefore consider that these are microspecies. If you go back into the field and ask, which group of these gave such a reaction and look carefully one will find that in fact there are very slight supporting morphological differences. But I would have great difficulties in taking you now into the field and say that is Conocybe pubescens A, C. pubescens B, C. pubescens C. It may be a problem only in the Bolbitiaceae; it may not apply to other mushrooms! I would suggest that we carefully think of this; what might apply to one genus or family may not necessarily apply to another family or genus. Septate spores in Ascomycetes is a good example. (So I do have examples).

BAS: Why not call your microspecies subspecies? What would be the problem?

WATLING: Subspecies to the zoologist and to the flowering plant taxonomist have some geographical connotation and at the moment I have no evidence of such geographical influences in Conocybe. My results are parallel to the Auricularia judae work of Duncan and McDonald where you get two strains of Auricularia judae in North America. It should be noted, however, A. judae in Europe will not cross with the ones from North America.

BAS: I know, of course, that subspecies are generally connected with differences in geographical distribution, but why should we stick to it? Because I think in this case you would avoid a great problem. I mean, although it is generally done, why should we stick to this concept of subspecies. I also ask you this because you did not explain what your concept of subspecies was.

WATLING: I do not have any concept. The one variety that I have ever published was a variety of <u>Boletus calopus</u>, and the only difference between that and the type is the fact, that the "cuticle" of the cap tore apart as it expanded. It seemed not to have enough material available to produce a fully expanded cap, and so when the cap developed it pulled apart and produced cracks. That seemed to be the only difference coupled with the distribution only in N. America. I would think, coming from a Botanic Garden, a variety should have, maybe, just one gene or two genes difference which gives one just a slight colour difference, a slight size difference.

BLAICH: If you have two species which do not intercross, and you find a third species which intercrosses with both of them, what would you say?

WATLING: (Well, what I would say first is something rather rude). I have not yet had that. I have had great difficulty in bringing together my morphological and my cultural information. The concepts are still very fluid in my mind.

THIERS: Is there an example of such a bridging species?

WATLING: I have not found any in the Bolbitiaceae.

ESSER: Well, I think, Dr. Blaich is up to what we call heterogenic incompatibility, and there are indeed some examples. The first examples were given by Ascomycetes, but there are also examples found, if you look carefully, in the higher Basidiomycetes, what we call the ABC-System. A does not with B, B does with C, and A does with C, and the conclusion is that you can exchange genetic information from A to B via C. So this has to be, from my point of view, seriously to be considered if one is talking about species. The more we look into this, the more we do experimental taxonomy, I think we will find further examples of this phenomenon.

BRESINSKY: A concrete example in Basidiomycetes: in _Fomitopsis pinicola_ there are two strains in America which are intersterile, but both could be crossed with the European strain. That would be the ABC-System.

Herbette Symposium on Species Concept in Hymenomycetes 1976

SPECIES CONCEPTS IN THE BOLETES

Harry D. Thiers

In beginning this discussion I am reminded of a remark a colleague of mine made when he learned I would be discussing species concepts of the boletes at this conference. His cryptic reply was that "no such a concept exists and it defies formulation". As I worked on the preparation of this paper I became somewhat sympathetic with his conviction. In the first place there is no precise definition of a species. Each of us, to be sure, has his own concept of this taxonomic unit, but can we measure its validity, and do we always remain objective and consistent in formulating such concepts? Furthermore, do we, in general, agree in the evaluation and ranking of the characters which are used at the various taxonomic levels? When is a character of sufficient importance, for example, to be of generic significance and what is it that reduces a character to such a position that it is valid only at the species level? Certainly intuition is involved in formulating species concepts, and I have often wondered if, perhaps, the keenness of this perception might not determine whether we are classified as "lumpers" or "splitters". Obviously the answers to the questions posed above are not readily apparent, otherwise we would be in essential agreement concerning the classification of the higher fungi, which, unfortunately, is not the case.

We who are working in the field of taxonomy of the higher fungi have certain handicaps with which I know you are fully aware. Nevertheless in self defense I find it necessary to dwell upon this topic for a brief moment. The most important handicap is that, even in this advanced sophisticated age, we are still confined largely to the utilization of characters supplied by only one phase of the life cycle of these organisms with which to make taxonomic decisions and that, of course, is the basidiocarp. This rather unusual predicament has resulted in the placement of heavy, even undue, emphasis on characters which in other groups of organisms would often be neglected or given only slight attention. For example, what flowering plant taxonomist, in making species distinctions, measures the width and length of the seeds of an ovary or measures the diameter of the cells across the blade of a leaf or counts the number of pollen grains per pollen sac in an anther? In a high percentage of the species with which we must work, we are unable to determine the presence or absence of hybrids or to obtain any reliable information regarding when and how hybridization might occur. Is there anyone in this audience who has not given serious thought to the possibility that some of the species reported in the literature as known "only from the type locality" might represent the product of some segregating generation and which may never be seen again?

We are severely handicapped by the lack of knowledge concerning the chromosomes of the higher fungi, including data on the numbers of chromosomes and their morphology. We hardly know what a chromosome of a bolete looks like much less being able to determine what the N or 2N number might equal at some stage in the life cycle. In contrast, as all of you know, there are almost monthly reports giving such information on the various groups of vascular plants.

Let us now turn our attention to the boletes. By way of review I will, at the outset, present only a brief circumscription of this group since all of you, I am sure, are familiar with them. These fungi constitute a rather closely related group of Homobasidiomycetes that produce macroscopic, putrescent basidiocarps in which the hymenophore is developed in the form of numerous, vertically arranged, relatively short, narrow tubes with angular to elongate pores; the hymenophoral trama is typically divergent to a greater or lesser extent, and hymenial cystidia are usually present. These cystidia may occur solitarily or, less commonly, in bundles; caulocystidia may be present. The spores are highly distinctive in shape and are often relatively large. There is, however, considerable variation in many of the characteristics of the spores, and as we shall see, they form one of the most useful sources of characters for use at various taxonomic levels. Another feature of some significance is the fact that, with only a very few exceptions, the boletes are all ectomycorrhizal associates and sometimes show a very high degree of specificity in their mycorrhizal associates.

With this background we can now explore the group at the generic level and determine, if possible, any trend in the ranking of characters at this level. A brief, capsule description of the general classification of the boletes will be presented, but before doing so, I

Boletes

should like to point out that in dealing with these fungi, as in other groups, there is not complete agreement as the generic disposition of some of the species and, in addition, we also do not agree on the number of genera which should be recognized. The fact that I might include a genus in this presentation does not necessarily mean that I accept it.

Suillus - Veil and annulus sometimes present; spores brown or olive brown in mass; glandulae may be present on stipe surface; pileus viscid or, if dry, then squamulose; fascicled cystidia often present and typically incrusted and stain dark brown in potassium hydroxide; mycorrhizal associations with conifers.

Segregates or Closely Related Genera:

Fuscoboletinus - As in Suillus, but with dark purple brown to dark reddish brown spore deposit.

Boletinus - Dry, squamulose pileus; no glandulae on stipe surface; pores radiately arranged; clamp connections present.

Paragyrodon - As in Suillus except for ovoid spores; heavy glutinous veil; hymenophore decurrent.

Gyrodon - Pores more or less radiately arranged; hymenophore arcuate-decurrent; small, ellipsoid spores; no fascicled cystidia; mycorrhizal association with alder.

Boletinellus - Pores radiately arranged; hymenophore arcuate-decurrent; stipe usually eccentric; associated with Fraxinus.

Psiloboletinus - Pileus tomentose but not scaly; boletinioid; no veil; spores elongate, smooth; incrusted cystidia present; clamp connections present.

Boletus - Veil and annulus absent; spores brown to olive brown or yellow brown in deposit; stipe lacking scabrosities as in Leccinum and glandulae as in Suillus; cystidia present but not clustered; cuticle typically dry, rarely viscid; spores smooth, or rarely, roughened.

Segregates or Closely Related Genera:

Aureoboletus - As in Boletus except pores brilliant yellow and cystidia with golden yellow content.

Xanthoconium - As in Boletus except spores yellow-brown in deposit.

Pulveroboletus - As in Boletus except for the bright yellow, pulverulent veil.

Chalciporus - As in Boletus except that the hymenophore is colored red to reddish brown; cuticle sometimes subviscid.

Xerocomus - As in Boletus except basidiocarps with dry, velvety to tomentose cuticle and often with large, irregular pores; dissepiments not readily separable from each other.

Boletochaete - As in Boletus except for thick-walled, strongly colored cystidia; short spores; no clamp connections.

Gyroporus - Spores yellow in mass; hymenophore typically white when young then yellow; spores small, ellipsoid; stipe frequently hollow maturity; cortical hyphae of stipe transversely oriented.

Segregates or Closely Related Genera:

Phaeogyroporus - Spore print yellow brown to more or less olive brown; hymenophore adnexed to depressed; spores relatively short; clamp connections present.

Leccinum - Boletoid; hymenophore typically white to pallid at first; stipe ornamented with small scales which usually darken with age; spores brown in mass, smooth.

Tyolopilus - Boletoid; hymenophore whitish to gray or pallid when
young, often pinkish vinaceous to dark reddish brown with
age; spores vinaceous to vinaceous brown to purplish brown
in mass, smooth or slightly roughened.

Strobilomyces - Boletoid; pileus dark colored, squamulose; hymenophore gray to black; spores black to blackish in deposit,
ovoid to globose, walls reticulate or echinulate.

Boletellus - Boletoid; hymenophore yellow or vinaceous; spores brown
in deposit; walls roughened by wings, ridges or striations.

Porphyrellus - Boletoid; hymenophore white to grayish when young;
spored not olivaceous, smooth or slightly roughened.

Heimiella - As in Boletellus but spores ovoid, with reticulate surface.

Meiorganum - Merulioid; dimidiate; stipe lateral; pores boletinioid,
hexagonal; spores ovoid; clamp connections present.

In a brief and inadequate analysis of the major or key characters
used in delimiting these different genera, the following were found
to be of some significance.

Veil and/or annulus - Suillus, Pulverulentus, Boletinus, Paragyrodon.
Stipe ornamentation - Leccinum, Suillus.
Cuticle - Suillus, Leccinum.
Spore color in mass - Emphasized in practically all genera.
Spores (microscopic characters) - Strobilomyces, Paragyrodon, Gyrodon
Porphyrellus, Boletellus, Heimiella
Surface - Suillus, Leccinum.
Cystidia - Suillus, Leccinum, Aureoboletus, Boletochaete.
Hymenophore - Chalciporus, Boletinellus, Boletinus.
Clamp connections - Gyroporus, Phaeogyroporus, Boletinus, Meiorganum.
Mycorrhizal associate - Suillus, Gyrodon, Boletinellus.

Obviously no single character typically segregates one genus from
another, instead, there may be several major characters as in Suillus, for example, or there may be only one major character which is

supplemented by numerous satellite characters. The overwhelming use, however, of spore characters, whether it be their color in mass or their microscopic characters is strongly apparent at the generic level. It is interesting to speculate as to what would happen if we devalued the spore color character, as is being done in some of the groups of gill fungi. Would the genera as we now know them survive? I think many of them would. Suillus and its related group would remain largely unchanged, with but a few exceptions. Leccinum would survive because of its characteristic stipe ornamentation supplemented by other characters. What would happen, though, to such genera as Tylopilus, Xanthoconium, Porphyrellus, Boletellus, Gyroporus and Fuscoboletinus?

How do the characters used at the generic and higher taxonomic levels compare with those used in delimiting species? As would be expected the species characters are of less significance and of greater variety that those previously mentioned. It must be emphasized that, as at the generic level, most species are distinguished by a set or combination of characters rather than a single factor. If we focus our attention first on the macroscopic features of the basidiocarp that are used in species separation we find that there are two major features and a few of lesser significance. These two major characteristics are pigmentation and the surface characters of the basidiocarp. These are supplemented to a lesser extent by the size and shape of the various parts of the basidiocarp and, to even lesser extent, by taste and odor. Substrates and mycorrhizal associates are sparingly used in most genera. Let us now discuss in more detail some of the characters enumerated above.

Pigmentation

There is little doubt but that coloration of the basidiocarp is the most popular source of characters for species distinction in the boletes, and, for that matter, apparently for most of the Hymenomycetes. This feature is heavily utilized in the large genera because of the wide range of colors and the variation in the disposition of these pigmented compounds. Not only do we utilize the color of the pileus, stipe or hymenophore, but we are careful to note color changes which might occur as the basidiocarp ages or becomes more exposed. In addition, we commonly consider as distinctive, various color combinations of the stipe and pileus or the combinations of colors in the pore and body of the tube of the hymenophore. We also make full use of the color changes which occur as a result of some mechanical injury such as bruising or exposure. Frequently the ornamentations of some parts of the basidiocarp are differently colored from the pileus surface. All such characteristics are fully utilized in the separation of species. Finally, we should not overlook the use of color and color changes of the context in species delimitations. This feature is used in most genera of boletes. Another clear indication of the heavy reliance we put upon the use of color are the rather detailed and highly sophisticated color guides which we now employ in describing the different taxa of mushrooms. As already noted we utilize color at the generic level in relation to the spore mass color in separating many of the common genera of boletes. It seems apparent that the basis for the color use at a higher taxonomic level is the rather restricted range of color variations in spore color in mass which makes it possible to group several entities together. On the other hand, the wide range of pigmentation in the basidiocarp precludes its use at a higher taxonomic level. We frequently employ minor color variations in delimiting subspecific categories. As will be pointed out later we also use pigmentation of such units as cystidia, hyphae and hymenium to assist in distinguishing various species.

Surface Characters

Let us next consider the use made of the surface features of the basidiocarp in formulating species concepts in the boletes. There are, in my opinion, three different aspects of the surface which furnish characters for making species distinctions. It is somewhat difficult to delineate between these three different aspects but here is what I have in mind. The first is the nature of the surface, that is, is it smooth, rugulose, pitted or in some other way characteristic? This feature is not a source of species delineation in most instances, but there are a few species in Boletus and Leccinum where there are distinctive features in this area. The second aspect also deals with the nature of the surface, but this time I have in mind whether the surface is viscid, lubricous, moist or dry. These features, surprisingly, do not supply as many characters for making species distinctions as might be thought. As a matter of fact many features from this source are used at the generic level as, for example, the viscid nature of many of the species of Suillus or the squamulose surface in Strobilomyces and other genera. There are, however, at least a few species in some of the larger genera such as Boletus, Leccinum and Suillus in which these characters are used in species separation. The third category is concerned with the nature of the ornamentations on the surface. We are familiar with the extensive use that is made of these features, mostly, but not exclusively, at the species level. On the pileus surface we eagerly search for appressed fibrils, fibrillose scales, squamules or any other distinctive type of ornamentation. On the stipe no only do we search for the same kind of features, but we also look for reticulations as in Boletus or Tylopilus, or glandulae as in Suillus, as well as for other distinctive features. These surface features are not always restricted to use at the species level, and are sometimes used at the generic level as exemplified by the scabrous nature of the stipe in Leccinum or at the section level as exemplified in the genus Boletus.

Size and Shape

The third most popular source of characters is, as I have already indicated, much less frequently used than the two previously mentioned. This category I have designated as Size and Shape and it applies to all parts of the basidiocarp either singly or in ratios between the different parts. The combination of the size and shape of the basidiocarp results in the overall aspect of the species and, whereas we do not often find a description of such in our species circumscription, it is nevertheless, very important in formulating our own mental image of a given species and is one which we all use in species recognition. There are instances where the shape of the pileus is specific such as in S. umbonatus or T. conicus. In other cases the shape of the stipe is distinctive as in B. satanas. In the hymenophore about the only place where these characters are at all extensively used is in the size of the pores. This variation in pore size and shape is used in several different genera, but is particularly useful in Suillus and Boletus. Generally the length of the tubes is very consistent and is of little value in distinguishing most species. The changes in shape or, less frequently, size resulting from age or maturity are sometimes incorporated in species descriptions.

Taste and Odor

There are a few instances where the basidiocarp is characterized by a distinctive taste. Perhaps the best example is B. piperatus or one of its closely related species. However, the presence of a bitter taste in also used as a species character in B. calopus, T. felleus and related species. Odor is less frequently used but has been employed in a few cases in distinguishing some species in Leccinum and Boletus.

Other macroscopic features which are occasionaly used in species delimitations include the following. The presence or absence of a veil and or annulus. This character is frequently used for generic designations such as in Suillus and Pulveroboletus, but within the genus Suillus it is also utilized sometimes at either the species or sectional level. The presence of a veil is also characteristic of one species of Leccinum. Since practically all species of boletes are terrestrial the type of substrate is of no great value in distinguishing species. There are a few instances, however, where the nature of the substrate is helpful, perhaps best exemplified by B. parasiticus which is parasitic upon Scleroderma. In addition there are a few species which appear to be at least facultatively lignicolous and others which rather consistently occur in the close vicinity of rotten wood. The type of mycorrhizal associate, at least in some instances, is in my opinion, a highly justifiable character that can be used in species separations. This character is, I think, of value in the genus Suillus where there appears to be a very high degree of specificity of mycorrhizal hosts in several species. In other genera there may be less host restriction, but there are several instances where the mycorrhizal host can be used as a supplemental character. Last among the macroscopic characters which might be mentioned are the chemical reactions of various parts of the basidiocarp with different compounds and reagents. I think we all more or less agree that we are still in the process of acquiring and evaluating data in this field and are not, at this time, prepared to make extensive use of this information and perhaps, do not fully comprehend its significance in distinguishing species. Nevertheless, there are a few instances, such as the blue green color reaction of B. spadiceus to ammonia, where these data are of value in species separation.

Let us turn our attention now to the microscopic characters which are utilized in formulating species concepts in the boletes. I am sure we all agree that even though, as has been indicated, extensive use is made of basidiospore characters at the generic level, a comparable use is also made of the spore at the species level. In addition, as we shall indicate, characters of the cuticle are commonly used in establishing species concepts as are those of the cystidia and other microscopic characters.

First consider the spore. It is not very often that we can clearly distinguish a species by spore size along, but it is apparent that perhaps this is the most commonly used feature. This character, as you know, is also frequently used to group related species together in stirpes or complexes. In addition to spore size, use is also made of spore ornamentation. There is not general agreement as to the amount of emphasis that thould be placed on this character since some of us use the ornamentation of the spores at the generic level while others include both smooth and ornamented spore-forming species in the same genus. The type of ornamentation and the degree of development of the ornamentation are of value in formulating species concepts. Other features of the spores which are used when available are the spore apex such as the truncate spores of B. truncatus, the presence of a germ pore, or the amyloid reaction of the spore wall that has now been noted in several different species.

The second significant microscopic feature at the species level in the boletes is the cuticle, particularly that of the pileus, but also, in some instances the stipe cuticle as well. Most boletes have either a trichodermium or an ixotrichodermium so the type of cuticle is of no real value in distinguishing species. However, considerable emphasis, at least in some genera, is placed upon the arrangement of the cells of the cuticular hyphae, the shape and size of these cells and, in Leccinum, particularly whether they disarticulate when mechanically disturbed. Also, the incrustations on the walls of the cuticular hyphae are often helpful, especially in Boletus and Leccinum. In addition, the contents of the cuticular cells are of importance not only because of the presence of differently colored cells in the cuticle, but also because of the distinctive reactions, as in Leccinum, of these contents when mounted in Melzer's Reagent. Color changes sometimes occur in the incrustations when mounted in various reagents and such data are of value in species distinctions. There is need for much more data on the nature of the cellular contents of these cells and on the reactions pathways which the various reagents follow resulting in these different color changes. The size of the hyphal cells, particularly the terminal or subterminal ones, are sometimes of value in species concepts. These cells are often differently shaped or even differentiated into pileocystidia which increase their value in species delimitations.

Mention should also be made of the cystidia and the use made of their characteristics in formulating species concepts. In most instances only minor use is made of these structures at this level. They are used, however, more often at the generic level, as for example in Suillus and Leccinum and to a lesser degree in Boletus. Some use is made, however, at the species level in Leccinum where special emphasis is placed on the caulocystidia and where significant variation exists. In other genera they are of much less value.

Other microscopic structures such as the nature of the hymenophoral trama, presence of clamp connections and features of the basidium are sometimes employed in distinguishing species, but are generally of little value at this level.

Before finishing the discussion of the different characters used for distinguishing species in boletes, mention should be made of the contributions derived from the culture of these fungi on artificial media. The work of Pantidou, particularly, has shown that this is an area where there is some potential for species separation at least for supplemental data to support data from other sources. Admittedly much more data are needed and as we are now learning it is a very slow, tedious and of late, rather expensive process, nevertheless, it is a field that should be pursued with enthusiasm, and the results will undoubtedly be highly gratifying.

The difficulty encountered in making species determinations in some bolete genera has stimulated us to seek additional characters which will assist in species separations. One area of high potential is chemotaxonomy and data being supplied from chromatographic analysis of various extracts from the basidiocarps. We are all familiar with the numerous contributions of the French workers at Lyon and of Prof. A. Bresinsky of Regensburg and their pioneering work in this field and certainly we owe them a considerable debt of gratitude.

If I may, I would like to present at this time as an example of the help we might expect from this source, the results of a chemotaxonomic study involving thin layer chromatography of the species of _Suillus_ which occur in California that was carried out by one of my graduate students. He made absolute methanol extracts of twenty different species and spotted them on TLC plates which were coated with MN 300 Cellulose. He ran the plates three times with the first and third runs in the same direction and the second run at right angles to those runs. The solvent system for the first and third run was acetic acid:water (3:17), and for the second run it was n-amyl alcohol:acetic acid:water (2:1:1). The usual procedure was followed of examining each plate under long range UV light with subsequent exposure to ammonia vapors. After all spots that were visible were traced the plates were then sprayed with a 2.5% aqueous solution of potassium ferricyanide and followed immediately by a 2.5% ferric chloride solution in methanol and then examined under normal light.

His results showed over seventy chromatographic spots from the twenty species. Of these spots, he found ten which he thought to be of considerable importance as taxonomic merkers. He was able to divide the California species into two groups. Group A had only three species and all remaining species belonged to Group B.

Group A: _S. acerbus_, _S. fuscotomentosus_, _S. tomentosus_.
Group B:
   Subgroup B-1: _S. albidipes_, _S. brevipes_, _S. borealis_, _S. brunnescens_, _S. glandulosipes_, _S. granulatus_, _S. megaporinus_, _S. pungens_, _S. pseudobrevipes_.
   Subgroup B-2: _S. monticolus_, _S. riparius_, _S. umbonatus_, _S. volcanalis_, _S. punctatipes_, _S. caerulescens_, _S. ponderosus_, _S. lakei_.

These results provoke some interesting thoughts. Species in Group A all belong to the Section _Suillus_ and, most importantly, all have a highly ornamented pileus cuticle in the form of fibrils or fibrillose scales. It would be interesting to compare _S. variegatus_ and _S. hirtellus_ with this group. I suspect that they would also belong in this group.

Group B is a bit more interesting and certainly a great deal more complex. Subgroup B-1 brings together species all placed in Section

Suillus with the exception of S. pseudobrevipes. As the name implies
S. pseudobrevipes has similarities with at least some members of this
group, and, as a matter of fact, it is my opinion that it has closer
affinities with members of the Section Suillus instead of Section Boletinus where it is presently placed. It is encouraging to note that
these results still group together certain species which were thought
to be similar based upon their morphology. Such a group is seen in
S. granulatus, S. albidipes, S. glandulosipes and S. pungens.

Those species belonging to Subgroup B-2 do not form such a homogeneous group. S. volcanalis, for example, from a morphological point
of view seems quite close to S. brevipes, and S. riparius is quite
similar in appearance to S. megaporinus. However, S. caerulescens and
S. ponderosus which look quite similar morphologically also appear
quite similar in these results. S. lakei is quite distinct chromatographically from all of the others and possibly should have been placed in a separate group. It is interesting to note that it is the only species in with a dry, brightly colored, fibrillose pileus.

Presently we are carrying out a similar type of project with the species of Leccinum, but at this time it is much too early to draw any
conclusions, however, the results are rather encouraging. The use of
gas chromatography might well contribute additional data of value for
distinguishing species in these fungi.

The potential for additional contributions from the artificial culture of these organisms toward solving taxonomic problems is rather
great. There is, at this time, in the literature several reports of
the successful formation of basidiocarps in culture and more can be
expected. And should the difficulty of obtaining spore germination in
the boletes be overcome, the door would then be opened for investigations into hybridization and genetic studies of these fungi.

I view the potential contributions from scanning electron microscopy
with mixed emotions. Like you, I am grateful for all information that
can be made available from any source regarding any of the fungi,
much less the boletes, and the fact that the spores of some species
of boletes appear roughened when viewed under very high magnification, or in some other way appear distinctive, is of great interest.
But what role should these data assume in formulating species con-

cepts? If we incorporate this information into our routine descriptions and taxonomic keys, we have rather effectively placed the field of taxonomy out of reach of many and, in addition, have made it a much more tedious process than it is at present.

I would now like to take a few minutes and present a brief survey of some of the larger genera of boletes and in so doing attempt to demonstrate how some of the species characters previously discussed have been employed in species distinctions. A brief, incomplete synopsis of the classification of each genus is included to facilitate understanding of the species grouping.

Suillus

I should like to begin with the genus Suillus. In the sense of Smith and Thiers there are three sections of the genus.

Sec. Paragyrodon: Spores ovoid to subglobose, yellow to yellowish in deposit; hymenophore typically decurrent at maturity; heavy glutinous veil.

Sec. Boletinus: Pileus surface dry, fibrillose to squamulose, or if viscid, then stipe annulate and with glandulae.

Sec. Suillus: Pileus surface viscid, glabrous or fibrillose, if annulate, glandulae present or absent.

Pigmentation and surface characters are the most frequently used for species separation in this genus, and often are used in conjunction with each other. The predominant colors are either yellow or yellowish or red to reddish brown, but certainly other colors are present. Most of the surface characters are derived from the variations in the type of ornamentation on the surface of the pileus and stipe. Microscopic characters play a reduced role and do not appear to be of great value at the species level, but are given some emphasis at the generic level. The bluing reaction of the context is given more emphasis in this genus than in any other group of boletes. It is a bit surprising to note that size and even shape of the pileus and stipe are frequently employed in species separations and that taste is used as an important character in several instances. It is my opinion that the genus Suillus is a very old genus and possibly represents the ancestral stock from which other boletes might have arisen. The bases for this assumption are, among other things, the large, often lamel-

Graph a
Width
8
7
6
5
9
10
Length
11
12
1
2
3
4
5
6
7
Graph b
Q
2
1.5
9
10
Length
11
12
7
6
Graph c
Q
2
1.5
6
7
Width
8
9
7
6

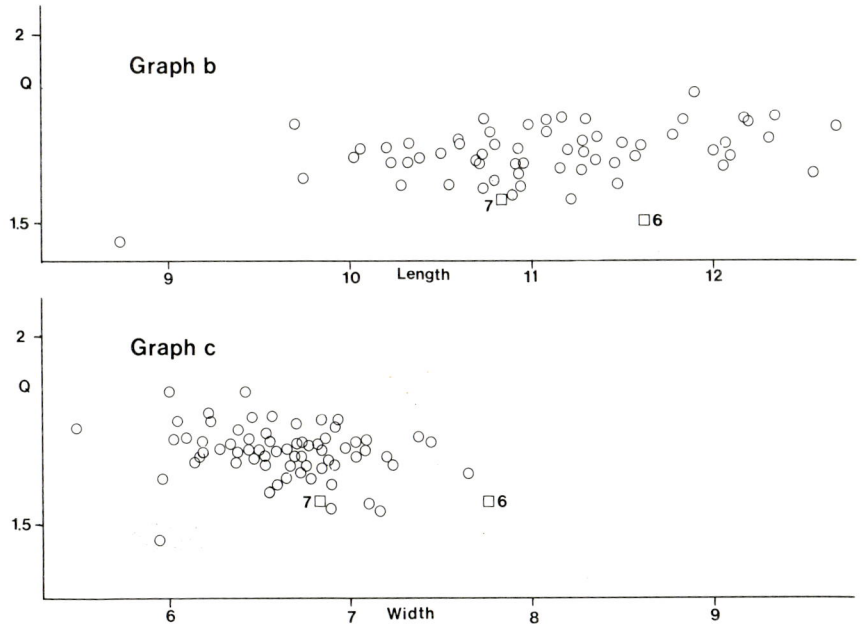

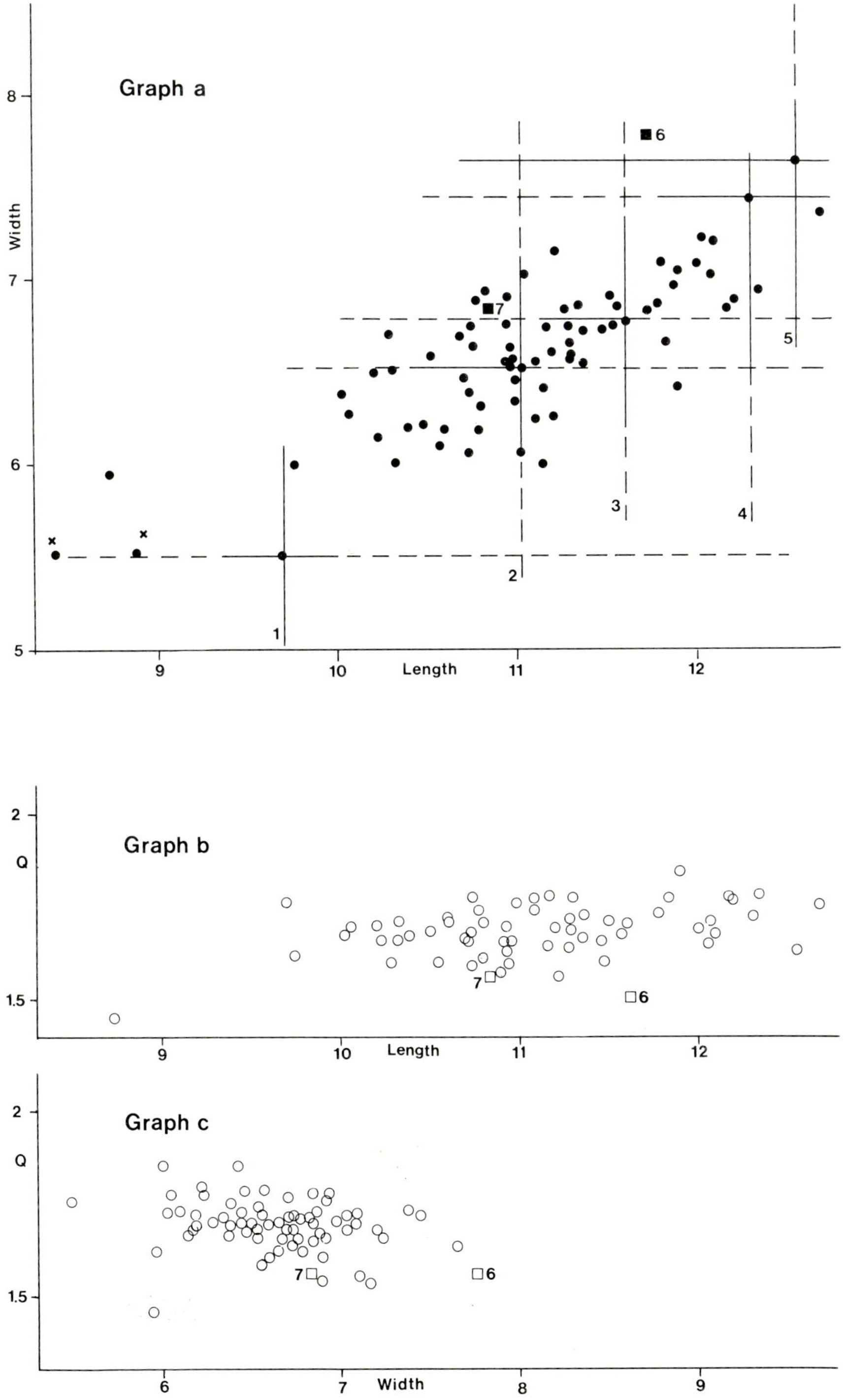

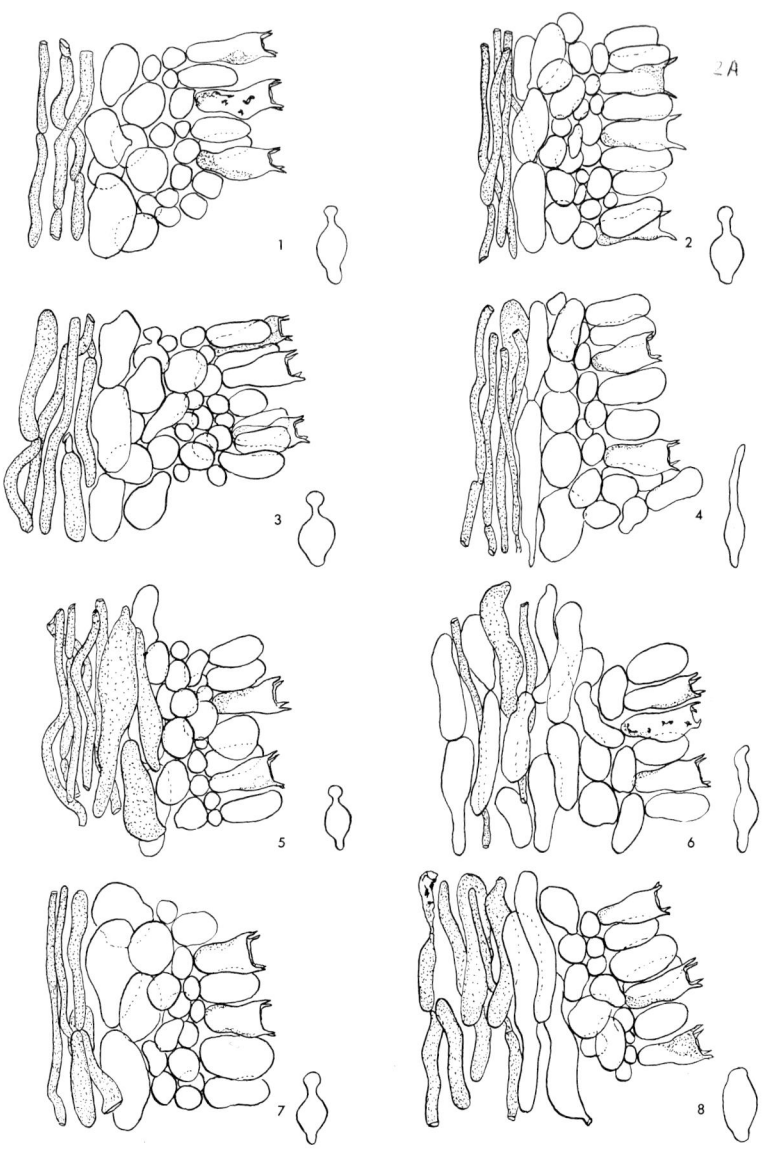

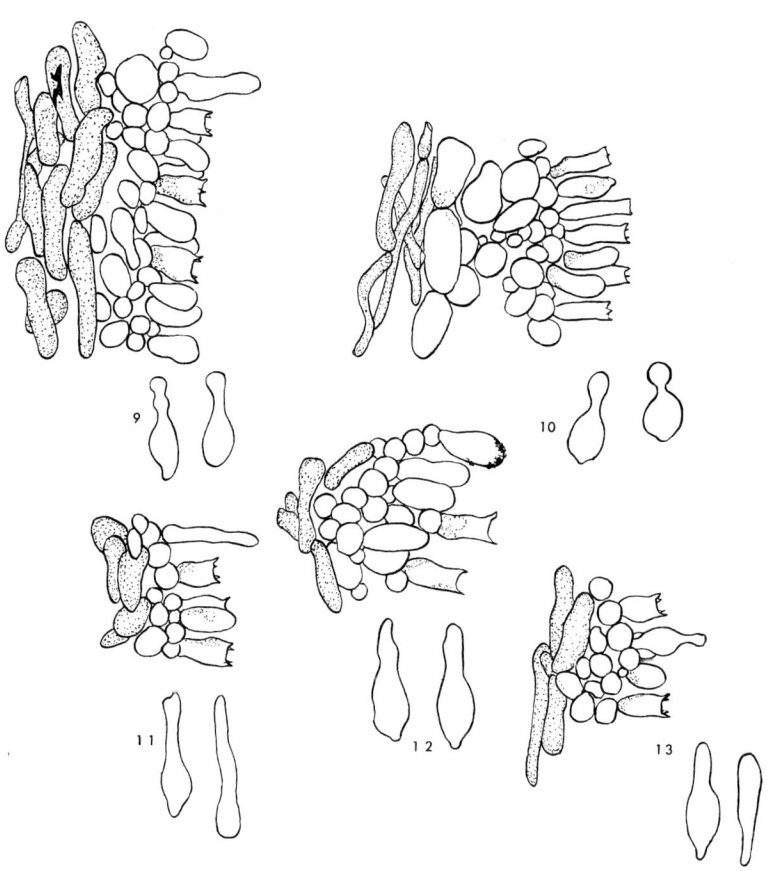

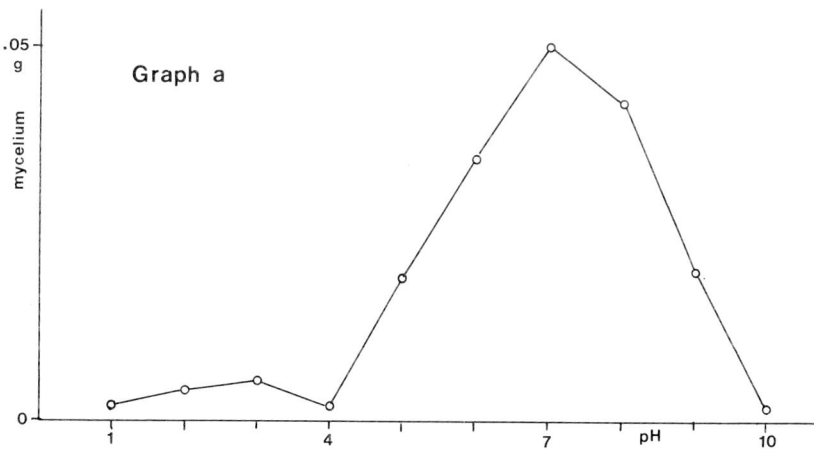

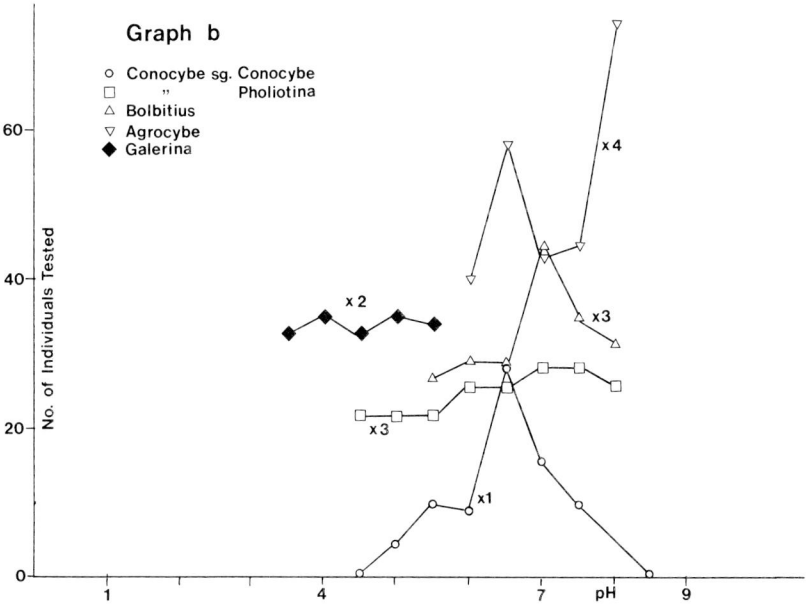

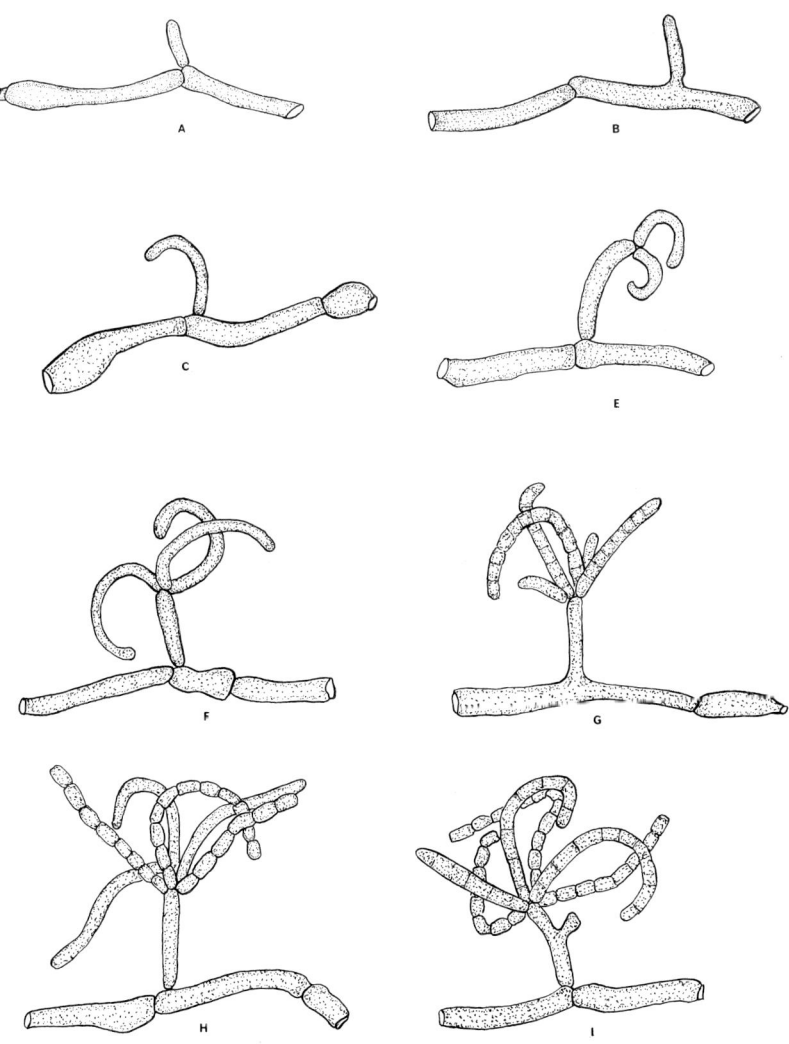

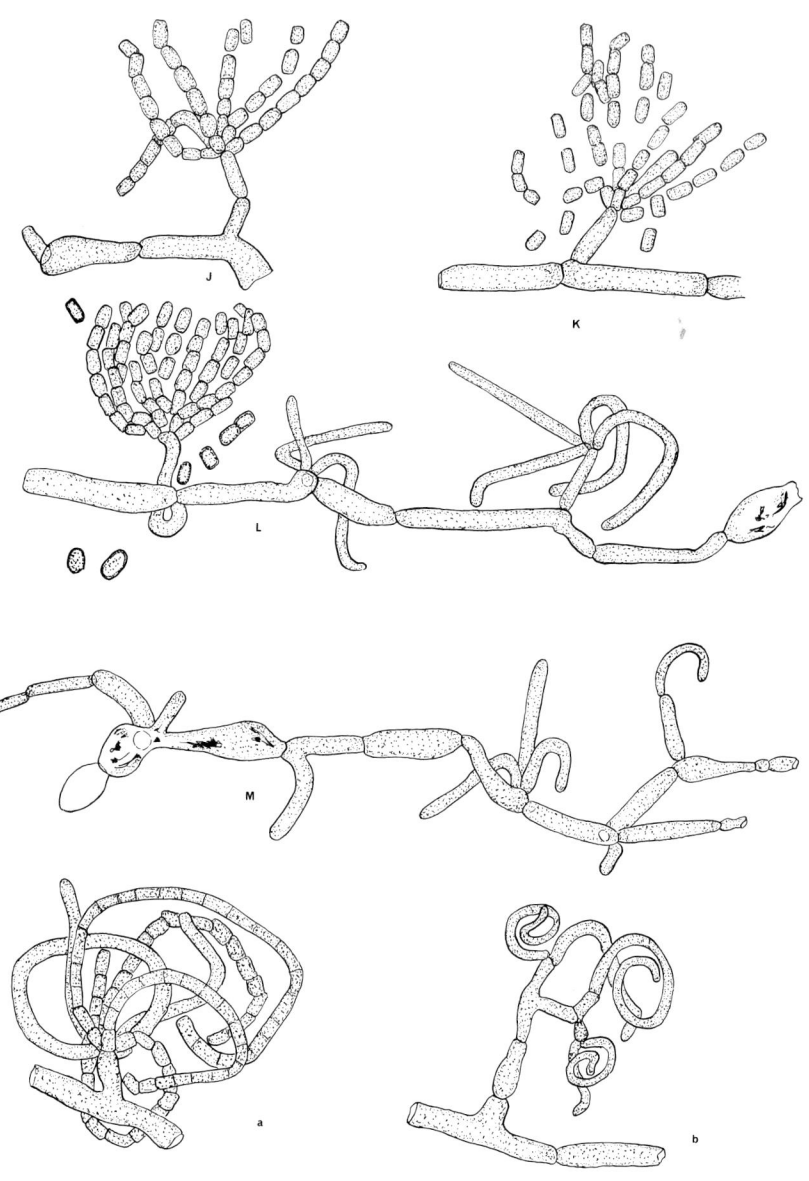

5A

loid pores seen in several species, the coniferous mycorrhizal associates and the rather clear distinctions between the different species. There seem to be relatively few intermediates.

Boletus

Smith and Thiers have divided this large genus into the following sections.

Sec. <u>Allospori</u>: Spores roughened; stipe alveolate to shaggy reticulate.
Sec. <u>Truncati</u>: Spores smooth, often with flattened or notched apex.
Sec. <u>Piperati</u>: Hymenophore reddish to reddish brown throughout; pileus surface viscid to moist.
Sec. <u>Boleti</u>, Subsect. <u>Luridi</u>: Pores differently colored from the tubes, usually orange, pink to reddish or vinaceous.
Sec. <u>Boleti</u>, Subsect. <u>Boletus</u>: Stipe reticulate; pores concolorous with tubes; pores "stuffed" when young.
Sec. <u>Subtomentosi</u>: Stipe not reticulate; pores not stuffed; pileus cuticle dry, dull, velvety to tomentose.
Sec. <u>Pseudoleccinum</u>: Stipe not reticulate but is roughened to granulose or fibrillose scaly; scales not darkening as in <u>Leccinum</u>.
Sec. <u>Pseudoboleti</u>: Stipe glabrous to more or less pruinose; not reticulate.

Pigmentation, as in <u>Suillus</u>, plays a very important role in forming species concepts in <u>Boletus</u>. However, microscopic characters are given more emphasis than in <u>Suillus</u>. Spore characters assume a very important role in <u>Boletus</u> involving not only size, but other features as well. Likewise cuticular characters and cystidial characters are commonly employed. Size, chemical reactions, taste and characteristics of the context are also used relatively often. Surface characters of the pileus and stipe are used but not as much as in other genera. I regard the genus Boletus as the central genus, possibly derived from a suilloid type of ancestor and which has given rise to several of the boletoid genera such as <u>Tylopilus</u> and <u>Leccinum</u>.

Tylopilus

The following divisions of the genus are presented. I should like to point out that this does not represent a complete breakdown of the genus.

Subg. Roseoscabra: Stipe scabrous-roughened, scales white to yellowish, not darkening as in Leccinum.

Subg. Porphyrellus: Spores smooth or slightly roughened, some very dark reddish brown in mass; hymenophore gray to brown or dark reddish brown.

Subg. Tylopilus: Spores smooth, vinaceous to vinaceous brown in mass; hymenophore usually vinaceous in age.

In Tylopilus surface and color characters plus a combination of these two are those most commonly employed in formulating species concepts. Spore characters, however, are frequently used and cystidial characters are sometimes employed. The cuticle appears to be of little value since there is little variation within the genus. Taste is certainly important but often it is used at sectional levels. Size and shape as well as mycorrhizal associates appear to be of little importance. As indicated I believe this genus was derived from a Boletus-like ancestor and I see no evidence that it has been involved in the ancestry of any other group of boletes.

Leccinum

At least three major subdivisions of the genus can be recognized.

Sec. Leccinum: Sterile "flaps" or extensions of the cuticle forming a membrane around the pileus margin.

Sec. Luteoscabra: Sterile margin absent; cuticle composed of short globose to ellipsoid cells, cellular, or with filaments with inflated cells.

Sec. Scabra: Sterile margin absent; cuticle of pileus filamentous; lacking noticeably inflated cells.

I have saved the genus Leccinum until last in this restricted coverage of the genera of boletes. I have done so because I believe that this genus is very actively evolving at present and that it is one of possibly relatively recent vintage. Those of us who have worked with the species of this genus are painfully aware of the difficulties encountered in making species separations and of the numerous intermediates that are always present. An analysis of the monographic studies by Smith, Thiers and Watling reveals that they, too, used color characters most frequently in formulating species concepts. Spore characters were frequently used as were color changes of the context. The surface features were used to a greatly reduced extent. Cystidia particularly the caulocystidia, provided a source of characters that were frequently used. Other features such as mycorrhizal associates, cuticle characters, shape and size of the basidiocarp were also used but less frequently than the ones mentioned above.

In summary we can say that morphological features continue to provide the largest source of characters employed in formulating species concepts in the boletes. Those characters such as pigmentation and surface are most frequently employed in all groups because of the wide range of variation in their expression. Those of lesser variation such as spore color in mass, fascicled cystidia and others are often used at the generic level. In the future it is to be expected that contributions to the taxonomy of the boletes will be made from such fields as chemotaxonomy, genetics, culture work and scanning electron microscopy. It is important here not to overlook the fact that more intensive collecting of boletes will undoubtedly also reveal additional characters that can be used in delimiting species as shown by the recent discovery of the apparent mycorrhizal associations between Leccinum and members of the Ericaceae in western North America. Before closing I would like to express my sincere appreciation to Prof. Clémençon for arranging this conference, for making us give serious consideration to an often neglected taxonomic problem and for giving me an opportunity to discus the boletes with you.

Bibliography

Corner, E.J.H. 1972: Boletus in Malaysia. The Botanic Gardens, Singapore. 263 Pp. 23 Pls.

Disbrey, B.D. & R. Watling 1966: Histological and Histochemical Techniques Applied to the Agaricales. Mycopathol. Mycol. Appl. 32: 81-114.

Edwards, R.J. 1976: Chromatographic Investigations of California Species of the Genus Suillus. Masters Thesis. San Francisco State University, San Francisco, Calif. 105 Pp.

Fries, E. 1821: Systema Mycologicum. 1: 1-508. Lund.

Horak, E. 1968: Synopsis Generum Agaricalium. Beiträge zur Kryptogamenflora der Schweiz. Band XIII. 741.Pp.

Kallenbach, F. 1926-1938: Die Rohrlinge in die Pilze Mitteleuropas 1: 1-158. 55Pls. Leipzig.

Murrill, W.A. 1910: Boletaceae. North American Flora 9: 133-161.

Pantidou, M.E. 1961: Cultural Studies of Boletaceae. Can. Jour. Bot. 36: 1149-1162.

Pantidou, M.E. & J.W. Groves 1966: Cultural Studies of Boletaceae. Can Jour. Bot. 44: 1371-1392. 6 Pls.

Pantidou, M.E. & R. Watling 1970: A Contribution to the Study of the Boletaceae - Suilloideae. Notes from the Royal Botanic Garden, Edinburgh, 30:207-237.

Pomerleau, R. & A.H. Smith 1962: Fuscoboletinus, a New Genus of the Boletales. Brittonia 14: 156-172.

Singer, R. 1945: The Boletineae of Florida with Notes on Extralimital Species I. The Strobilomycetaceae. Farlowia 2: 97-141.

Singer, R. 1945: The Boletineae of Florida with Notes on Extralimital Species II. The Boletaceae (Gyroporideae). Farlowia 2: 223-303.

Singer, R. 1947: The Boletoideae of Florida. The Boletineae of Florida with Notes on Extralimital Species III. American Midland Naturalist 37: 1-135.

Singer, R. 1965: Die Röhrlinge Teil 1: 131 Pp. 21 Pls. Jules Klinkhardt, Bad Heilbrunn

Singer, R. 1967: Die Rohrlinge Teil II: 151 Pp., 26 Pls. Jules Klinkhardt, Bad Heilbrunn

Singer, R. 1975: The Agaricales in Modern Taxonomy. Cramer, Vaduz. 912 Pp. 84 Pls.

Smith, A.H. & H.D. Thiers. 1964: A Contribution toward a Monograph of the North American Species of Suillus. Ann Arbor, Mich. 116 Pp. 46 Pls.

Smith, A.H. & H.D. Thiers. 1971: The Boletes of Michigan. The University of Michigan Press. Ann Arbor, Mich., 428 Pp. 157 Pls.

Smith, A.H., H.D. Thiers & R. Watling 1966: A Preliminary Account of the North American Species of Leccinum, Section Leccinum. Mich. Bot. 5: 131-178.

Smith, A.H., H.D. Thiers & R. Watling 1967: A Preliminary Account of the North American Species of Leccinum, Sections Luteoscabra and Scabra. Mich. Bot. 6: 107-154.

Snell, W.H. 1941: The Genera of the Boletaceae. Mycologia 33: 415-423.

Snell, W.H. & E.A. Dick 1970: The Boleti of Northeastern North America. Cramer, Lehre. 115 Pp. 87 Pls.

Thiers, H.D. 1975: California Mushrooms. A Field Guide to the Boletes. Hafner Press, New York. 261 Pp. Microfiche.

Thiers, H.D. 1975: The Status of the Genus Suillus in the United States. Beiheft 51 zur Nova Hedwigia. Cramer.

Thiers, H.D. & J.M. Trappe 1969: Studies in the genus Gastroboletus. Brittonia 21: 244-254.

Trappe, J.M. 1962: Fungus Associates of Ectotrophic Mycorrhizae. Bot. Rev. 28: 538-606.

Trappe, J.M. 1963: Some Probable Mycorrhizal Associations in the Pacific Northwest. Northwest Sci. 37: 39-43.

Watling, R. 1970: British Fungus Flora. I. Boletaceae, Gomphidiaceae, Paxillaceae. Royal Botanic Garden, Edinburgh. 124 Pp.

DISCUSSION FOLLOWING DR. THIERS' PAPER

BAS: You have mentioned nearly only species in your lecture; I have heard you speak of only one variety. Does this mean that you recognize very few infraspecific taxa in the boletes?

THIERS: There are several, e.g., in <u>Boletus edulis</u> many varieties have been described.

BAS: They have been described, but do you recognize them?

THIERS: I have no aversion to recognizing varieties. I do, in fact, recognize varieties and have described some myself.

BAS: Do you extend your concept of subspecies also to mycorrhizal types? I mean in the case of slight morphological differences between taxa with different mycorrhizal hosts?

THIERS: Did you say subspecies?

BAS: Yes.

THIERS: I do not use subspecies very much at all. I have not had an occasion to do so, and if I did, I would follow the usual tendency of mycorrhizal or geographical characteristics.

BAS: You do not know of taxa in the boletes that are closely related, showing just minor morphological differences, and that are connected with different trees?

THIERS: I would not say that there are not any, but I do not know of any.

PETERSEN: If we describe or accept fewer genera, then a number of characters become available for subgeneric, sectional or species ranks. Would you care to comment on the value or weight of characters at the species rank versus genus rank?

THIERS: I have a section in my paper on this subject. I think we are very inconsistent in evaluating characters as to whether they belong at the species or generic level. We follow no pattern and I think it is a very personal feeling as to whether a character is worthy of use in generic limitations or not. I personally see little value of returning to large, massive genera. I do not see that such a move would accomplish a great deal, or that it would really release more characters for species distinctions. I question this because it will be necessary to use subgeneric categories before the species level.

PETERSEN: You mentioned the use of the scanning electron microscope (SEM). I think it would be safe to say that so far, SEM-work largely has been limited to spore surfaces. I am wondering what particular structures would you, in your group, find most valuable for SEM study?

THIERS: Of course the spores, because of the difficulty in certain species where some may be very obscurely roughened, Boletus zelleri, as well as several others. I would like to see us focus a great deal of attention, if we can, on the nature of cuticular hyphae to see if we can determine more about the incrustations and other characters; cross walls, etc. Such information would help us, but I hate to see us begin to use Scanning Electron Microscope characters in taxonomic keys.

KUEHNER: Did you look at the microtopography of pigmentation? Some boletes have pigments dissolved in the cell sap, and others have pigments on the cell wall or between the cells.

THIERS: We did not check each species to see where the pigments were located.

SINGER: Dr. Thiers, you used the pronoun "we" with several statements and it was not quite clear to me whether this meant "you" with the pluralis majestatis, or whether it meant your group, or whether it meant "you and Smith", or whether it meant the "boletologists" in general.

THIERS: It either meant my students and I or Alex Smith and I. It did not mean all of the "boletologists", because they might not agree

SINGER: That's OK then, because you said "we are generally not paying much attention to the chemical characters" and "we think that the mycorrhizal relationship is generally neglected". A statement you made was that 99% of Boletes are mycorrhizal. Of course, when you look at the world flora, this statement, probably correct for your state, is exaggerated. When you look at the tropical boletes there is a much lower percentage.

THIERS: But how many species of boletes are there that are tropical?

SINGER: Well, you would be surprised!

BAS: Quite a few of the boletes of SE Asia grow with tropical mycorrhizal trees, like Castanopsis and Lithocarpus.

SINGER: Yes, but if you look at the tropical forests, where there are no ectomycorrhizal trees, you will find nevertheless quite a few genera represented which are non-mycorrhizal, and definitely so; I have proved that for several species of Gyrodon, etc.

BAS: May I say something about terminology? If Dr. A.F.M. Reijnders would have been here, he would certainly have made a remark about the terms "bilateral" and "divergent" as used here, because according to him these are two different things. A divergent trama of the hymenophore is an original situation that has become permanent in some groups, and in the case of a bilateral trama you start with a regular trama that later becomes bilateral. I cannot confirm this, because in the stages of the fruitbodies I usually look at, the trama is already bilateral, but according to Reijnders' observations in Amanita, the trama starts being regular.

BRESINSKY: Some remarks on your paper, and I hope that these remarks will not be understood as criticisms. It seems that at whatever taxonomic rank we are, the use of characters depends on the gaps we find in variation, so that it is the gaps that give us the possibility to use characters for delimitation, and not the characters per se.

SINGER: It is the hiatus that counts.

SMITH: We are talking about variation rather than diversity.

SMITH: In mushrooms we have a problem in dealing with the fruiting bodies from a single mycelium and their differences, and we would assume that that is variation. Differences between fruit bodies from different mycelia may be genetic diversity (and important taxonomically) or simple variation. And between mycelia it may be genetic diversity, the differences that you find.

BRESINSKY: Well, I mean it in the sense of Stebbins who devoted a whole large book on variation.

SMITH: It has been done that way, but in mycology I have found it is necessary, if you are to document your data properly, to distinguish between these two conditions: the fruiting bodies of a single mycelium and their differences, compared to a mycelium of *Lactarius volemus* from Michigan and from Indiana.

BRESINSKY: In this sense I mean diversity. Now, depending where the gaps are, the characters may be very different; there may be morphological ones, there may be chemical ones, macroscopic ones, and microscopic ones. Regarding chemical characters and the techniques used to find such characters, I think that it is very, very difficult to judge from more or less obscure chromatographic patterns, as to whether these patterns are really gaps or not.

THIERS: That is the negative point of view. The positive point is that there are certain spots there.

BRESINSKY: OK, there are certain spots! Now a second problem: these spots come from compounds which are very easily destroyed or changed by oxidation, so I think the rf-values are not reproducible at all if you are running different chromatograms of different collections, at least if different investigators are involved. Minor differences in the methods will give completely different results. So I feel uncomfortable when the methods are not very much standardized to make them really reproducible, and that would include in the boletes, methods to destroy oxidases. The other point is that I feel uncomfortable if the compounds are not identified, because only this would allow us to make judgements.

THIERS: The point is, that in our laboratory we are not equipped to identify compounds. I do not see this as a great deterrent, however, and sooner or later we may make identifications, and then we can put a name in that circle.

ESSER: I think we must make a difference between low molecular weight compounds, and high molecular weight compounds, such as nucleic acids and proteins. As we know during the life cycle of any fungus the enzymes and all protein compounds change. As far as I am aware in your techniques you are dealing with low molecular weight compounds and according to our experience low molecular weight compounds also to a certain extent are formed at a certain metabolic state of the mycelium or the fruiting body, but the probability of error is less in low molecular compounds. As far as I understand Dr. Thiers he just wants to use this technique as a means of description. He has his taxonomic scheme and he wants to see whether this is in agreement with some biochemical criteria, and as far as I followed his paper he did not make major conclusions from this.

BRESINSKY: Distributions of spots on a two-dimensional sheet are difficult to compare. I think it is difficult really to see gaps. And if you know that these spots really move around all over their possible rf-values by very minor influences, then one may feel some doubts.

Herbette Symposium on Species Concept in Hymenomycetes 1976

SPECIES-CONCEPT IN AMANITA SECT. VAGINATAE

C. Bas

Within the Agaricales there are a number of groups in which we are dealing with many apparently closely related taxa on species-level or lower, of which nobody seems to know at what taxonomic level they have to be placed. The swarm of taxa in and around the classical species Amanita vaginata is a good example of such a group.

From the earliest days of mycology up to the present time people have tried to bring some order in the Amanita vaginata complex. The character most frequently used when doing so was the colour of the fruit-body, particularly the colour of the cap.

Fries in his "Systema" (1821) contented himself with the enumeration of six unnamed colour-forms (with the colours albidus, griseo-lividus, caesius, spadiceus, fulvus, and viridus) of Agaricus vaginatus Bull. It is remarkable that he could refer already to more than six specific names for these "colour-forms" in what we call now the pre-startingpoint literature. More of such names existed already that time.

 Secretan (1833) recognized 11 species and 9 varieties,
 Quélet (1888) 5 forms,
 Gillet (1884) 7 varieties,
 Gilbert (1918) 2 subspecies, 9 varieties and 9 forms,
 Vesely (1934) 25 forms,
 Gilbert (1940) 11 varieties,
 Kühner & Romagnesi (1953) 4 species and 2 varieties,
 Moser (1967) 8 species,
 Singer (1975) 7 species.

It is only the rank of subspecies that is poorly represented in this enumeration, but this infraspecific rank is in general rarely used by agaricologists.

It would be incorrect to give the impression here that all the mentioned taxa were distinguished merely on colour. Size of the fruit-body, length of the striation of the cap margin, décoration of the stem and behaviour of the volva played a part here and there; as, from Gillet on, also did the shape and size of the spores.

Apart from Secretan, who was not far from describing and naming individual collections, there is a general trend in the treatment of the A. vaginata-group by the mentioned authors.

In a first period, up to about 1935, more and more colourforms (supported or not by other characters) are described as forms or varieties of A. vaginata. Serious attempts to separate these taxa clearly are hardly made.

In a second period from 1935 up till the present time the number of infraspecific taxa is being drastically reduced, but the number of taxa recognized on species level is slowly growing.

If we look upon the results of the second period, we cannot help to observe that not much really has changed. The most characteristic members of the A. vaginata complex such as A. fulva, A. crocea, A. mairei, etc. are lifted out of the "formpool" and described, but as the rest of the complex remains in the mist the delimitation of even these few taxa is far from satisfactory. Probably every European agaricologist has met with specimens now and then of which he was not quite sure if they had the right red-brown colour for A. fulva or if they were sufficiently orange to be called A. crocea.

When a great number of well-dried collections of members of the A. vaginata complex are brought together in the laboratory and superficially examined, the number of intermediate forms seems to be so great that one is tempted to put them all together in one extremely variable species with a few forma-names for the most conspicuous types.

That this is not the correct way of handling this problem becomes clear when we are in the field and collecting in a region with sufficient ecological variation. It appears then that there is some order in the chaos. Not every form grows everywhere. This can clearly be demonstrated by describing the situation in The Netherlands.

Most of the forests in the Netherlands are planted on poor acid sandy soils. In these forests only one member of the A. vaginata complex can be found and in great quantities; it is <u>A</u>. <u>fulva</u>. Much less common is a grey form with a strong preference for Populus on rich soils (from loamy sand to heavy clay). The third member, usually called <u>A</u>. <u>lividopallescens</u>, is a large to very large fungus with a usually salmon buff to pale buff or greyish buff cap and its base deep in the heavy, slightly calcareous clay of the alluvial woods along the big rivers; the same biotope frequented by the grey-warted <u>A</u>. <u>inaurata</u>. Then there is one locality of a slender salmon buff form that belongs perhaps to the northern <u>A</u>. <u>flavescens</u> Gilbert & Lundell and another of a grey-capped form with a reddish brown stem for which I have no name. This picture did not alter in the last ten years and I am sure that similar situations exist elsewhere. It is evident that the A. vaginata complex comprises a number of taxa that behave like biological entities and therefore have to be described and delimited as clearly as possible.

In my nomenclatural card-index on Amanita there are 160 epithets introduced for taxa in section Vaginatae (Vaginata-, Inaurata-, and Caesarea complexes), many of them used at several levels from forma to species. Nearly 100 of them are created for European taxa and the great majority of these for members of the A. vaginata complex. Therefore it seems pertinent to start to try to disentangle the taxonomy and nomenclature of section Vaginatae in Europe before studying this section for the world. It is very fortunate that American mycologists have been rather reserved in regard to the description of new taxa (no more than 10) in the A. vaginata group, although it is probably not less complex in North America than in Europe. Perhaps the more than frustrating state of affairs in Europe has something to do with this.

Trying to discriminate means trying to find discriminating characters. In this light I want to discuss here some useful and some not so useful characters in section Vaginatae.

C o l o u r

The colour of the <u>pileipellis</u>, although too much stressed in the past, remains a leading character. It seems, however, that we have to

be prepared to admit more colour-variation to the taxa than usually
is done in literature. It seems for instance that there exist white
and very pale forms of several taxa that normally have a coloured
cap. This is definitely the case in a number of grey-capped taxa.

The colour of the stem is mainly determined by the colour and the degree of development of the partial veil and more or less corresponds
to the colour of the edge of the gills which is also covered with
remnants of the partial veil. The latter is therefore not an independent character and of less value than in other genera of the agarics.
Luxuriant specimens of a taxon may have a considerably darker partial
veil and consequently a darker stem and gill edges than normal specimens.

The colour of the volva is usually white or pallid. Rusty spots occur
frequently in several taxa. When the volva has some colour of its
own, this seems a valuable character.

There is little variation in the white to whitish colour of the spore
print. But in many cases information on the spore-print colour is
still missing.

Structure of the pileipellis

The upper layer of the pileipellis is always an ixocutis, but in mature specimens in the field the pileipellis may have a dry aspect in
one taxon and a subviscid appearance in another. Also in herbarium
material this difference is sometimes rather striking. I have tried
to correlate these observations with the thickness of the ixocutis in
radial sections of the pileipellis mounted in a 10 % ammonia-solution. At least in some cases it seems possible to differentiate between rather closely related taxa on account of this character.

The type of arrangement of the hyphae in the cutis underneath the
ixocutis seems useful too. In some taxa the hyphae of the cutis are
strictly interwoven, in others they show a distinct radial arrangement.

The hyphae of the cutis are mostly rather narrow (mostly between 1 and 6 microns wide). When a cap shows an innately, radially fibrillose aspect it is usually because of the presence of radial bundles of coloured hyphae. The pigmentations is always intracellular.

Marginal striation of the cap

The length of the marginal striation (or better: sulcation) is an important character. It is necessary, however, that it is expressed in relation to the radius of the cap (e.g. 0.2 - 0.3 R) and not in millimeters or centimeters. It is certain that there are taxa in which the marginal sulcation generally is longer or shorter than in others. The usefulness of this character is somewhat diminished, however, by the fact that in small fruit-bodies the sulcation is often longer than in large fruit-bodies of the same collection.

The volva

The universal veil or volva in section Vaginatae varies from being fleshy-membranous to completely friable. For the moment it is convenient to distinguish four types: (i) The firm, saccate, fleshy-membranous volva (like in A. crocea). (ii) The soft felted-membranous volva (like in A. fulva). (iii) The breaking-up, submembranous-subfriable volva (as in A. hyperborea). (iv) The floccose-friable volva (as in A. inaurata). But it should be kept in mind that there are intermediates between these types.

One would expect to find important differences in the microscopical structures of these volval types, but thus far the results of my observations are somewhat disappointing. In each of the four types mentioned, large inflated cells (mostly globose, ovoid, ellipsoid or broadly clavate and mostly terminal on hyphae) are present.

The differences in behaviour and appearance of the volva in the four groups seem to be determined mainly by (i) the structure of the outer layer of the volva, (ii) the ratio between inflated cells and hyphae in the tissue inside the volva, and (iii) the thickness of all layers of the volva together.

The difference between the firm saccate volva and the soft saccate volva seems mainly a matter of thickness of the volva. Both these types have an outer layer composed of interwoven hyphae only. In the third type which usually breaks up into patches or warts, the outer layer of hyphae is very thin and inflated cells are visible in a scalp under the microscope even without crushing the mount. In the fourth type the outer layer of hyphae is lacking or so thin that it breaks up in an early stage of the development of the fruit-body.

Although not much material has yet been studied in this respect, I think the first results are indicating that in section Vaginatae differences in structure of the volva are more difficult to put into words and less useful as differentiating characters than in some other sections of Amanita.

Considerable differences exist in section Vaginatae in the degree of attachment of the saccate volva to the base of the stem. In some taxa the lower part of the volva is adnate to a relatively large part of the base of the stem (as in *A. argentea*), in others the volva is attached only to the outmost base of the stem (as in *A. hemibapha*). I belief that this difference is rather fundamental as the latter structure occurs particularly in a type of fruit-body that in my opinion is the most highly developed in Amanita, viz. with a quickly expanding, umbrellalike cap with a long marginal sulcation and a hollow, rapidly elongating stem for which all the material available in the primordial bulb has been used.

Huijsman (1959: 18) pointed out that in the first case the internal limb of the volva is situated more or less at the place where volva and stem meet and that in the latter case this internal limb is situated on the inside of the volval limb. He introduced the term "unitangent" for the first, and "bitangent" for the latter situation.

The character just discussed will certainly play an important part in the final classification of the taxa of the Vaginata complex, but as transitional forms occur, it will rarely help to distinguish two very closely related taxa.

Amanita

S p o r e s

It is true, the size of the spores in section Vaginatae ranges from about 4 - 5 microns in <u>A</u>. <u>aurea</u> to 15 - 16 microns in some taxa in the Vaginata complex, but there is an unpleasant preference for the range between 9 and 13 microns. Add to this the fact that 1-, 2-, 3- and 4-spored basidia may occur on one fruitbody and it is clear that in section Vaginatae spore-size is a tricky character, particularly in the A. vaginata complex. Besides measuring spores very carefully, it is necessary to check per sample the number of sterigmata on quite a few basidia.

The shape of the spores as expressed in the length-breadth ratio seems to be an important character. The differences being fairly small, however, it is necessary to registrate them rather precisely. I measure 10 spores of each sample and determine the length-breadth ratio of each spore measured. Usually I take one sample per collection, but from critical or very variable collections more samples are taken. Per sample the average length-breadth ratio is calculated so that finally per taxon the range of average lenth-breadth ratios can be given.

As far as my experience goes this range is a very great help. Many taxa in the A. vaginata complex have globose to subglobose spores with the average length-breadth ratios ranging between 1.0 and 1.1. For the moment I have them in one group and the species with a range above 1.15 in another group (section Ovigerae with Sing.).

A few words should still be said about characters that up till now were found to be not very helpful in distinguishing taxa on species level and lower in section Vaginatae.

<u>Clamps</u> seem to be present in all species with a ring (Singer's section Caesareae) and absent in species without a ring.

<u>Pigmentation</u> is always intracellular.

<u>Cystidia</u> are lacking. The cells of the marginal tissue of the gills are remnants of the partial veil and very variable; they collaps usually at an early stage.

<u>Mycorrhizal</u> connections seem rarely or not to be confined to one species or genus of woody plants.

<u>Colour-reactions</u> with Phenol and Phenolaniline give red to brown dis-

colourations on the flesh but seem to be not very precise.

After this introduction to the general problems connected with the delimitation of taxa in Amanita section Vaginatae, I want to demonstrate these problems again, but now on the basis of two particular cases.

In the autumn of 1973 Dr. Watling and I had the pleasure to collect agarics with Prof. Dr. P. Kallio from Turku around the Kevo Subarctic Station in Finnish Lapland. I was very anxious to see some of the frequently discussed arctic members of the A. vaginata complex such as A. nivalis Grev. and A. hyperborea (P. Karst.) Fayod. Unfortunately it was not before the last collecting-trip that we found about 10, mostly solitarily growing fruit-bodies of whitish ringless Amanitas on an arctic heath with shrub birch and several species of shrub willows. My first impression was that these fruit-bodies belonged to two different species.

One (No. 6104) had a very thin-fleshed, very pale greyish-buff tinged cap without volval remnants, a nearly smooth stem and a neat saccate volva with a small internal limb.

The other (No 6105) had a somewhat thicker, first white then pale isabellina-buff cap with large volval patches, a fairly floccose stem with some vague floccose girdles, and a usually broken volva with a rather large internal limb.

After measuring 3 spore samples of both collections my first impression was more or less confirmed by the fact that the spores of No. 6104 were globose to subglobose with the average l/b ratio ranging between 1.05 and 1.1 and the spores of No. 6105 subglobose to broadly ellipsoid with the average l/b ratio between 1.15 and 1.2. A study of the volval tissue in both collections did not reveal noticeable differences.

After that I studied the small number of in Leiden available collections of alpine-boreal pale "A. vaginatas" from Swedish Lapland, Scotland, Austrian Alps, Greenland, and Labrador to see if I could recognize the two taxa from Finland in this material. A few appeared to agree with the first type   one from the Alps with the second type. The rest appeared to represent a third type with globose to subglobose spores (aver. l/b 1.1), but the volva breaking up in usually small wart-like patches especially at the centre of the cap. In this third type the outer layer of interwoven hyphae is thin and under the microscope inflated cells are visible right through it. Moreover, the volval limb is slightly coloured and the cap seems to vary in colour from very pale to fairly dark grey-brown.

If I am right in distinguishing these 3 taxa in the alpineboreal group of the A. vaginata complex, what names can be used?

First there is the name A. nivalis Grev. 1822. A type-collection does not exist, but Greville published an excellent plate and a long description. There can be hardly any doubt that the first collection (No. 6104) I described here belongs to this taxon.

It would be nice if the name A. hyperborea (P. Karst.) Fayod could be connected with the second or the third type I mentioned above. I have recently received the type of A. hyperborea on loan and found the spores to be mainly broadly ellipsoid (l/b 1.1 1.35, aver. 1.2 in two samples). This means that only my second type can be considered for this name.

The overall aspect of the type of A. hyperborea is, however, rather different from that of my second type, mainly on account of the many small volval warts on the cap, but this may be a matter of age. A study of more material has to show if intermediates occur. It is difficult to tell if the covering of the stem of the type of A. hyperborea has been floccose and I still have to analyze the structure of the volva.

My third type has been given the provisional name A. groenlandica, because the best collections I have, come from Greenland (collected by P. Milan Petersen).

Perhaps this all looks like scating on thin ice, but somebody has to do this to see how thick the ice really is. After studying more collections I hope to be able to report that it is sufficiently thick to support me with my little theories.

The second case I want to discuss concerns members of the A. caesarea complex.

If one compares pictures of the European and the North American form of A. caesarea, one is struck by the difference in habit of these two. The European form is thick-set with a fleshy, hemispherical, later convex cap rarely with an umbo; and a relatively short stem with a heavy volva. The North American form is slender with a rather thin-fleshed campanulate to plano-convex cap with usually a rather prominent umbo, a slender stem, and a not conspicuously heavy volva. Gilbert (1940/41) introduced the invalid name: "forma specialis americana" for the latter.

From the literature it seems that there is also a difference in spore size between the two forms:

	European form	
Bresadola (1927)	10 - 12 × 6 - 7	microns
Kühner & Romagnesi (1953)	11 - 13.5 × 7.5 - 10.5	
Romagnesi (1956)	9 - 12 × 6 - 7	

	N. American form	
Coker (1917)	7.5 - 8.5 × 5 - 6.5	microns
Pomerleau (1966)	7 - 11.5 × 5 - 8.5	

In the material of the North American form available, I found in three samples taken from three collections the spores to measure (7.5 -) 8 - 10 (- 10.5) × 6.0 - 7.5 microns. These observations seemed to strengthen the opinion that the American form is different from the European and it looks as if there is a name for it in literature.

In 1950 Vassilieva described from the far East of Russia a new species, A. caesareoides, very much like A. caesarea but with the slender form of the American "A. caesarea" and differing from the European A. caesarea by small spores, 8 - 10 × 7 microns. In the two collections Dr. Vassilieva kindly sent me, I found the spores to have nearly exactly the size of the spores of the American form, viz. (7.5) 8 - 9.5 (- 10) × 6.5 - 8 (- 8.5) microns.

The picture seemed clear then. There are two taxa involved, viz. the
thick-set A. caesarea in Europe with spores > 10 microns and the
slender A. caesareoides with spores < 10 microns in Eastern Russia,
Japan (see Imazeki & Hongo, 1957: 44), and N. America.

One problem left was the delimitation of A. caesareoides against A.
hemibapha. The latter has been described by Berkeley & Broome in 1871
from Ceylon. The type plate shows rather slender fruit-bodies with an
umbonate cap with a red centre and a strongly sulcate, yellow margin,
a yellow ring on a white stem and a white volva. Petch (1910: 374),
who collected the species on Ceylon himself, reports the type plate
to be a poor reproduction of the original drawing. According to him
the cap should have no umbo, the gills and the stem should be yellow,
and the stem should show red patches. From this it is clear that A.
hemibapha is very close to A. caesarea and A. caesareoides.

I have not yet studied the type of the Ceylon species, but collected
a very similar fungus in 1963 in northern India. The spores of that
collection measure 8.5 - 10 x 6.5 - 7.5 microns. Bakshi (1974: 10)
gives about the same spore size for A. hemibapha in northern India.

The only characters I have at the moment to distinguish A. hemibapha
from A. caesareoides are the very long sulcation of the cap margin,
the lacking umbo, and the attachment of the volva to only the outmost
base of the stem.

It would be interesting to know the western limit of the area of distribution of A. caesareoides. For that reason I asked Dr. S.P. Wasser
in Kiev to send me A. caesarea from southern Russia, which he kindly
did. Somewhat to my surprise it had spores smaller than 10 microns
and fruit-bodies partly small and rather thick-set and partly larger
and slender, always with a distinct umbo.

I did then what I should have done at the beginning. I measured the
spores of the European collections of A. caesarea at hand. The results were rather disastrous for the ideas developed before:

```
France  (Marchand,  11    XI  1966)  10 - 13.5 x 6.5 - 9    microns
Belgium (Baeq,      26    IX  1949)  8.5 - 11   x   6 - 8
Belgium (Demoulin,  19   VIII 1951)  8.5 - 10.5 x   6 - 7
```

Rumenia (Fl. Rom. Exsicc. 2920)   9 - 10 x 6 - 7.5 microns
Spain   (Marchand, 30   X 1966) 8.5 - 10 x 6 - 7.5

It looks as if there might be a small-spored form of A. caesarea in
Europe also, but a correlation with the habit of the fruit-body does
almost certainly not exist.

Without the support of differences in spore size, I wonder if there
still is sufficient reason to recognize A. caesareoides even on subspecific level.

Regarding the limits against A. hemibapha: Dr. D.T. Jenkins from
Birmingham, Alabama, sent me a colourslide of a very slender A. cae-
sarea-like fruit-body with a long sulcation of the cap and red patches on the stem. Is this A. hemibapha?
Well, this case is certainly not closed yet.

My approach of the subject of this symposium has been a very practical one. I have shown how difficult it is to separate the taxa which
constitute the "form-pool" of the A. vaginata complex on morphological and ecological data. I have also shown what kind of problems
arise when it comes to comparing material from geographically widely
different origin.

In spite of all that, I believe that finding the right solutions is a
matter of studying many well-annotated and illustrated collections
and of more experience with the concerning taxa in the field. I expect that my efforts will lead to the circumscription of a fairly
great number of "small species".

The rank of subspecies I reserve for constant variants within a species, that have a geographical distribution differing from the typical variant; the rank of variety for constant variants within a species occurring mixed with the typical variant and of which the aberrant character is not connected by a series of intermediates with the
corresponding character in the typical variant; the rank of form for
other variants within the species that for some reason need a name to
give it a "handhold".

I believe that is very important for the taxonomy of higher fungi that descriptive mycology precedes any type of experimental taxonomy. The classification built up from all the information gathered by the morphological taxonomist should be the base of the studies of chemists, geneticists, and others who will gradually remodel it into a real biological classification.

It is therefore worrying that so few professional mycologists are working in the field of morphological taxonomy of higher fungi. In regard to the work that still has to be done the progress is very slow. Therefore I think we all should encourage and promote morphological taxonomy of higher fungi at every occasion that it might have a positive effect. If not more work in this field is done than we shall in the near future be in the situation that we have allround biological systems for some relatively small groups of macromycetes but that we are unable to write a well-balanced flora for the higher fungi of even a small country, not to speak of larger parts of the world.

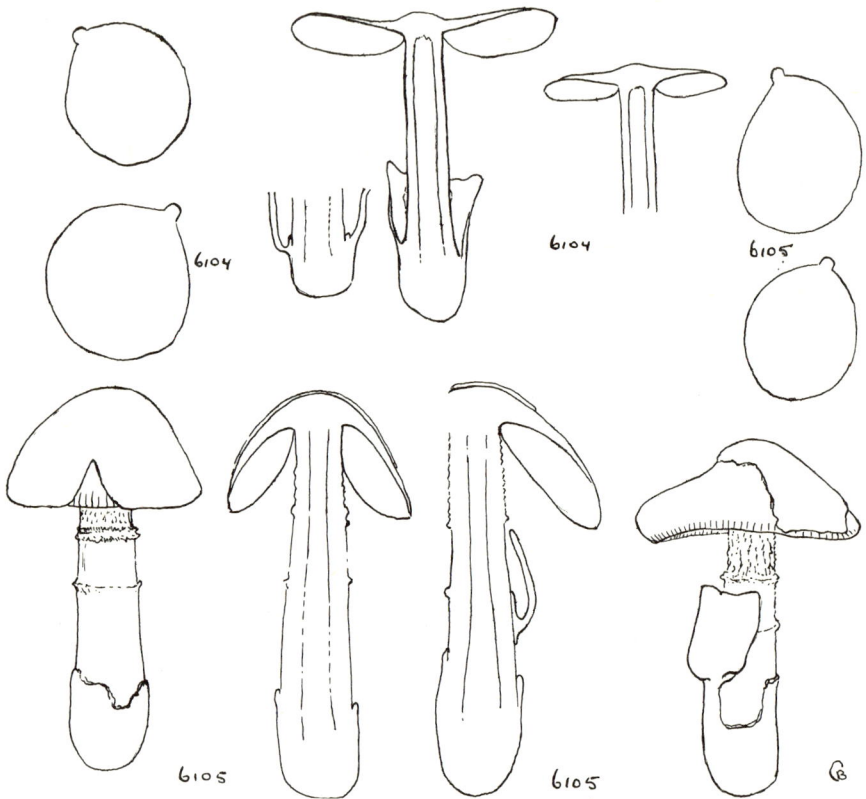

Fig. I. Bas 6104 - Amanita nivalis Greville
" 6105 - A. ? hyperborea (P. karst.) Fayod

DISCUSSION FOLLOWING DR. BAS' PAPER

SMITH: I would like to reinforce what Dr. Bas said of the need for morphological taxonomy.

WATLING: Why are your "small species" not the same sort of concept as my microspecies?

BAS: Eventually they may prove to be the same, but at the moment I have no cultural evidence. Your microspecies are based on cultural evidence, my "small species" are based on observations on morphology and ecology.

WATLING: Yes, but the point about Conocybe and Bolbitius etc. is that when I find some intersterility, and one looks carefully at these collections we can see small differences, albeit perhaps a little smaller than in Amanita. The cultural work is always supporting the morphological work and vice versa. The same sort of thing might be possible in Amanita if one could culture them.

BAS: Unfortunately this is still too difficult. In the grey-brown vaginatas, I think you have to work the classical way and proceed from your own region in outwards circles, getting more and more information on the "forms" you meet, because the available information is a mess. I am far from convinced that there is just one or a few species. Probably many taxa are involved here.

WATLING: The very fact, you see, that you have to introduce your own "small species" indicates that it is necessary to have a concept to cover them, otherwise we would have to call them species.

BAS: Well, I was using the words "small species" in contrast to what they usually call these things; just Amanita vaginata, brown form, white form, etc. That is still the general attitude towards this group.

SINGER: How about the possibility for experimental work?

BAS: My personal possibilities are at the moment practically nihil;

I have to stick to morphological work, and I would be interested in collaborating with someone who is interested in cultural work in this group. I know several people who are cultivating Amanitas, at least from tissue samples. Dr. Petersen, you have quite a few Amanitas in culture. Are there members of the A. vaginata group among them?

PETERSEN: No. We cultured what I am convinced is Amanita hemibapha. The mycelium grows perhaps 0.5 mm from the lamella edge, but it will not grow any farther than that. I wonder whether A. caesarea from Europe has been cultured successfully.

SINGER: We cultured the American one in Virginia and we got exactly the same result as you.

WATLING: Could I suggest that you grow sterile trees and inoculate roots with cultures, because Amanita muscaria shows very interesting parallels between cultures on roots and the morphological diversity that one finds in the world.

PETERSEN: Amanita muscaria grows very well in culture anyway. Another question: it is my understanding from the literature that Amanita caesarea from Europe contains microquantities of amanita-toxins, certainly not in lethal doses. We find the same thing in the "American caesarea" (what I think is hemibapha), and I wonder whether anyone has done analyses in A. caesarioides. A second question: what value would you put on such data, namely amounts and numbers of amanita-toxins?

BAS: I find it very difficult to answer this question. The information on the toxins in the A. phalloides group is growing, but I must say that as a taxonomist I find hardly any help in this information. They go into great detail now, and we hope that we may learn many things about other substances in mushrooms, but I have the feeling that as a taxonomist I have no help at all from this.

PETERSEN: But I think we have returned to precisely where we were yesterday with Harry Thiers' paper. Some years ago it was relatively easy to take a gram of an Amanita fruit-body, squash it and run a chromatogram to get some spots. It may have been quicker to go through that procedure in about two hours than it was to key down the taxon. We have moved past that now, with the identification and fractionation of the spots. The spots have "proliferated". Have we moved beyond the point of practicality for that kind of taxonomic character, and if so, how do we use it now?

SMITH: Well, I think that type of taxonomy has been pretty well vitiated. Some body has found an amanita-toxin in <u>Cantharellus cibarius</u> according to a report I have seen this year. This is the current concept I have, that these toxins occur generallly in fungi in very minor amounts, and in a few of them, like in <u>Conocybe</u> and in <u>Galerina</u> and in others they do occur in rather large amounts. But the taxonomic values in separating any small groups are pretty well vitiated.

PETERSEN: No. Perhaps the presence of amanita-toxins will become of value in <u>Cantharellus</u>. Just because they are called amanita-toxins does not mean that they are found only in <u>Amanita</u>.

SMITH: You should examine the whole spectrum of fleshy fungi. And it is not worth while generalizing before we have done that.

PETERSEN: I could not agree. To say that a character is of no value in one group because it is found in another group is unsound. It may be of value in the second group, in a third, or a fourth, but perhaps <u>not</u> in the first. We have no way of knowing that.

SMITH: Well, it is better to make a judgement on three of four examples than on one, and when you find it in unrelated fungi, then the problem calls for investigation and not for generalization.

PETERSEN: I agree.

BAS: I think it is at the sectional and at the subsectional level where it may be helpful, especially in the <u>Phalloideae</u>, but not when you go into details.

THIERS: I know you have indicated that you have not yet begun to work with North American material, but I would like to ask if in your work with European material you find great variation in the stature and robustness in the basidiocarps of the <u>vaginata</u> group? In other words, on the west coast we have very large, robust forms as well as the small, gray forms. Do you find the same variation over here?

BAS: Indeed. Some taxa have nearly always large fruitbodies, like Amanita lividopallescens, but occasionally you find them also small.

SMITH: What is the name of that black one, Harry, on the coast?

THIERS: This is A. pachycolia, a species provisionally named so by a student. I know this A. pachycolia, which is a very large species. Also associated with this large size, Kees, are orange-colored lamellae, particularly when dried. Do you find the same sort of thing over here?

BAS: Yes, this discolouration of the gills happens quite often in the A. vaginata group. Yet there are some taxa that show it more frequently than others. I think the tendency to discolour can be used as a character. In some taxa the gills sometimes turn pink, in other taxa never. Up to now I have not found the pink colour of dried gills to be a constant character of any taxon. I have looked into this also in Michigan in several taxa of the A. vaginata group. In some collections apparently belonging to the same taxon the gills turned bright pink on drying, in others the gills became a kind of pale pinkish salmon, and in others they stayed just whitish or greyish. Therefore I do not think that the colour change is constant.

THIERS: Do you find Amanita velosa at low altitudes here, or do you find it mostly in higher elevations?

BAS: I think it is restricted to California.

SINGER: The phenol-reaction, a rather deep violet, somewhat like the Russula olivacea reaction, this I frequently confirmed for Amanita crocea, exactly as Melzer had first stated it. And I understood you told (or I saw it in your manuscript) that you saw only brown reactions.

BAS: Oh no, it is a whole range of colours.

SINGER: But did you actually see violet?

BAS: No. But there are two types of phenols in circulation; one without sulfuric acid and one with. I always used it without, but I just found in a paper by Henry that you get better results when you use some sulfuric acid in it. He described a very bright colour reaction in *Amanita crocea*. I tried it in England, and although the colour was fairly bright, there was also quite a bit of variation. Moreover the colour changes so strongly in a relatively short time.

SINGER: Yes. And then, did you ever find any confirmation of what we observed in Virginia once: I found there a annulus in a collection. I called *Amanita fulva*. I could not find any indication on the literature on *A. fulva* with a well developed annulus. But that was very exceptional; there were only two specimens with an annulus. I would be very much interested in your opinion about the taxonomic value of the annulus. In the group as a whole, this character seems to be very differently evaluated.

BAS: A great number of Amanitas have a membranous partial veil; but of course even that consists of tender tissues. Everywhere in the expanding fruitbody there are tensions in all directions, and even a very good membranous ring may break apart. On the other hand you have species that practically always have only a very slight velar covering of the stem in the mature fruitbody; just some fine fibrils derived from the partial veil. There the partial veil is extremely thin, and is nothing more than a separation layer between the gills and the stem.

SINGER: My question was whether you have seen *fulva* with a thick, membranous ring.

BAS: This is one of the few species I really know, and I have never seen it with such a ring.

SINGER: So you would say that this was a new species?

BAS: Yes.

BRESINSKY: I would like to comment on your remark, and possibly you are waiting for this comment, that morphological taxonomy should precede any type of experimental taxonomy. It corresponds to the given development of science, including taxonomy, that descriptive parts precede the experimental parts. But I would like to ask you if I am right that you feel the necessity to know the last morphological speciality of the last _Amanita_ in the last corner of the world before one may start with any experimental work? I do not think that you meant this, but an interpretation of your sentence in this way is possible. I think that everyone should do the work for which he has the facilities, and he should be measured according to the facilities and possibilities he has. It is very essential to do morphological work and to make monographs etc., but I would not see any priority for one or the other type of taxonomy.

BAS: Don't you think that for experimental taxonomy it is of very great advantage when the descriptive part has been done by a competent mycologist?

BRESINSKY: Well, it may be an advantage, but it may be a disadvantage, in the sense that my thoughts and my ideas are due to pre-given concepts which, perhaps, are hindering to my work.

BAS: I would say that when you start experimental work in a group that has not recently been worked up by a morphologist, you would have to restrict yourself to a limited number of collections. The danger is that you may overlook quite a bit of the variation of the taxa involved.

BRESINSKY: Two viewpoints! We are not together, but... OK!

BAS: And one of the reasons for which I made this remark: I really get the impression that the emphasis is going to experimental work at the cost of morphological work, and I think we need both. I believe in remodeling the classification that has been made by morphological taxonomists, with the help of data supplied by experimental work.

SMITH: I think we must realize, and every one does, that systematics is subject to change according to the amount of information available. And I think the same thing applies to experimental work. It has to be done over and over again several times for verification, and both phases must be developed without assignment of priorities.

BAS: Do you think that the classical and the experimental taxonomy of a group should be done by the same person, or do you think that it should be done by different persons? I think the last would be better. Because when I try to decide at what level I shall recognize certain taxa and I know that somebody has been working on this group, and it looks as if this taxon is intersterile with that one, I have a kind of second hand information, and I am influenced by that. Probably I do this work better without that information and by going purely on morphological evidence as far as possible. The mentality of people working experimentally and the mentality of people working morphologically are slightly different. One is not inferior to the other, but they are slightly different, and therefore I think that separate approaches are important.

SMITH: Unfortunately the discipline of one group seldom carries over to the other, and I have seen this in the fungi. The discipline of the physiologists does not carry over, and he is not equally accurate in his principles and operation in the taxonomic area. Taxonomists constitute the infantry in this whole operation. The infantry clears the area and sets up a temporary working system so that the experimentalist has a basis for comparing results. In North America what we need mostly is alpha-taxonomy. What is hindering the work on mycorrhiza is the lack of accurate systematics of Cortinarius.

ESSER: I think I shall come back to the phenol compounds you mentioned and said that it is a very unsufficient criterion. We should know at first, that there are, in most fungi at least, two different phenol oxidases, the tyrosinases and the laccases, and also peroxidases. The amount of phenol oxidases present in each fruit body depends on the age. We cannot be surprised that even in the same species, even in fruit bodies from the same mycelium we have not the same reaction. The change of color is due to a chain of subsequent oxidations until we eventually come to the melanins. This test is a rather valuable test, if it is made under proper conditions. To drop phenolic compounds on the fungus gives no information at all, just that phenoloxidases are present. And the various phenolic compounds go first to a colorless compound and then give different colors, and the extinction coefficients are absolutely different for thoes different compounds.

It is very easy to apply. Just grind your fruit body with a little sand in a mortar and make an extract in a lab centrifuge, and if you have a photometer, a filter photometer, at 436 nm you can do a reading. This is a test very easy, you can even do it in the field. There are small, portable photometers operating with batteries. If this test is properly worked out it could be effective in distinguishing fungi.

CLEMENCON: I have been mildly surprised to see that you rely so heavily on spore size and shape in your species concept. I measured spore dimensions in Hebeloma, Naematoloma and others, and I find that there is correlation between the age of the carpophore and spore size and spore shape much beyond the limits of 1.0-1.1. I am surprised by your attitude because, of course, you know that this happens in Amanita phalloides, it has been published. Why do you use it in this narrow concept?

BAS: Well, you could say by the lack of better. But the length-breadth ratio really works. That's all I can say; I believe it really works.

CLEMENCON: How big is the standard deviation of the length-to-breadth range? It seems to me, that the two groups, below and above $Q = 1.1$ are so close together that they seem to overlap. I did the same thing in Lyophyllum, but here I have from 1-1.2 and from 1.5-2. There is no overlapping. In your case it seems to be so close that I fear a big overlap.

BAS: If you measure spores in the Amanita vaginata group, and you are above 1.2 you get a great range. But if you measure the spores of the group of species with globular spores, very often the range is between 1.0 and 1.05. All I can say: it really works. When you regularly find things that you can recognize in the field, and you see, immediately when you look through the microscope that the great majority of spores are perfectly globose, I would be a fool not to use this character!

HORAK: Where do you take the samples from?

BAS: I always take them from the middle of a gill. I would prefer to take them from a spore print, of course, but as I am working with material from everywhere often without spore prints, I always use the same method. I take them from the gill, and if there is a spore print I check whether there is a difference. I must say that, in general, the spores taken from spore prints are slightly more variable than those taken from the gills. I think that is because, when I look at the gills I automatically pass by a number of spores that I assume to be immature. You would expect a spore print to have only mature spores and that consequently the range of variation is smaller, but actually it is slightly larger than in the spores taken from the gills.

SMITH: The spores in <u>Lactarius</u> from the spore print often measure smaller than the spores taken from the gills. Now, I wonder have you found this in <u>Amanita</u>?

BAS: Not in <u>Amanita</u>, but in other groups, yes. I think it is especially the case in groups with thick-walled spores. I do not know exactly what happens but it seems that after the spores have been dropped by the fungus, something happens to these spores.

SMITH: It shrinks!

BAS: They shrink a little, yes. Maybe some of the material is used to build the thick wall, I do not know. I just had a colleague a few days ago coming to me with the same story. He found the same phenomenon in <u>Rhodophyllus</u>.

CLEMENCON: I have strong indications from electron microscopy that a freshly shed spore is not mature, and that the building up of the wall is still going on. So when you take a fresh spore print you very probably have different stages of maturing spores.

BAS: When you measure spores from a spore print you seldom have a spore print that is fresh; it is always a few hours old.

CLEMENCON: When you take a spore print on paper or on glass, I think it dries out relatively quickly. That stops all physiological activity.

BAS: Oh, you mean that the process of maturing is not going on then, because they dry out.

CLEMENCON: Yes, it is more or less stopped or slowed down. When the spores are liberated in nature, then they may go on maturing for a day or two or so, but when you collect them on paper, and then you take the paper and scrape them off, the spores are dried out rapidly.

SMITH: The argument then would be, the spores in the dried tissue in Lactarius, which measure larger, are the proper sample to take, rather than the spores from a deposit on a paper.

BAS: Another experience I had is that if a fruitbody is too young, you get a much larger variation in spore size than in a mature one.

PETERSEN: I find myself reluctantly obliged to defend the physiologists. I have heard comments about the mindset of physiologists and of taxonomists not being inferior to each other but both being different from one another. When I find physiologists sitting around this table they are also taxonomists. Taxonomists include the morphologists, the micromorphologists, the physiologists, the geneticists, when they are describing characters, and when they are sorting taxa. I do my kind of taxonomy, which is practical, classical; they do their kind of taxonomy, perhaps more recently evolved, but no less valid than the kind of taxonomy that I am doing. I am describing one kind of character, they are describing different kinds of characters. I must be open to great criticism for occasionally describing one collection as a new taxon. And I therefore cannot attack the physiologist who describes one set of measurements of a chromatogram as being a character. I find it difficult to say to the physiologists you must do it over, and over, and over again, and corroborate, corroborate, and corroborate when I use one collection as a new taxon. I have not done it over and over and over again, either.

SMITH: You expect that somebody will do it over and over again in order to finally establish the value.

KUEHNER: Permettez moi deux observations en langue française: la première relative au groupe A. vaginata, la seconde aux conclusions générales.

Concernant le groupe A. vaginata, il est possible que sur la planche que j'ai publiée sur les Amanites de zone alpine, se trouvent figurées sous le même nom A. hyperborea, à la fois A. hyperborea et A. nivalis. La plupart des figures de cette planche se rapportent sans doute au A. nivalis tel que le comprend Mr. Bas; la figure de droite, en bas reproduit un croquis de Mlle. D. Lamoure; il n'est pas impossible qu'elle se rapporte à une espèce différente; c'est surtout le fait que de dernier champignon ait été récolté parmi Salix herbacea, où nous trouvons souvent A. nivalis qui nous avait fait supporter qu'il s'agissait de la même espèce.

Dans ses conclusions d'ordre général, Mr. Bas dit qu'il pense que les recherches morphologiques doivent précéder les recherches expérimentales. Sur ce point je ne suis pas d'accord, sauf cas particuliers comme par exemple celui du groupe A. vaginata; en effet, dans l'impossibilité où nous nous trouvons actuellement de cultiver des formes de ce groupe, et où, par conséquent, aucune recherche expérimentale n'est encore possible, il est bien évident que les recherches morphologiques vont nécessairement précéder les recherches expérimentales à venir.

Mais pour les champignons dont nous savons actuellement obtenir des cultures monospermes, cet ordre ne me parait pas souhaitable. Si vous remettez à plus tard les recherches expérimentales, ne risquez vous pas, au moment où vous vous déciderez à les entreprendre, de ne pouvoir le faire pour certaines espèces très rares dont vous aurez étudié la morphologie, mais que vous ne retrouverez peutêtre plus jamais?

Pour débrouiller certains groupes complexes d'Omphalina, Mlle D. Lamoure a conduit avec succès simultanément les recherches morphologiques et les recherches expérimentales faisant appel aux tests d'interfertilité - interstérilité. Ces tests lui ont été fort utiles dans nombre de cas pour délimiter des espèces si proches que les caractères morphologiques qui les distinguent ne sautent pas toujours aux yeux immédiatement; souvent c'est en constatant l'interstérilité entre deux souches qu'elle a été incitée à rechercher et qu'elle a généralement trouvé des caractères morphologiques permettant de distinguer les deux espèces auxquelles ces deux souches appartiennent.

Ayant suivi de près les recherches de Mlle. D. Lamoure, je suis convaincu que chaque fois que c'est possible, les recherches morphologiques et expérimentales doivent être conduites simultanément, et si possible par la même personne qui supervisera attentivement le travail expérimental nécessairement confié, au moins partiellement, à une technicienne.

DIFFERENTIATION OF SPECIES IN CLITOCYBE

Howard E. Bigelow

The problems involved in the interpretation of species of <u>Clitocybe</u> do not seem particularly unusual or unexpected, and they are familiar to many. However, due to the interest of several investigators there may be more controversy about <u>Clitocybe</u> than some other agaric genera. This provides an interesting and healthy state though, and has led to more information and knowledge about the taxa and their relationships to one another.

Despite over 20 years familiarity with <u>Clitocybe</u> and countless basidiocarps, I found it was a formidable task to decide how best to discuss differentiation in this large genus of perhaps 300 species. The diagnostic characters are critical of course, but any value judgments upon these really must involve some speculation upon evolutionary pathways and the possible genetic basis of some characters of the basidiocarp.

As a point of departure, the general features of the clitocyboid basidiocarp are relevant for consideration. Since these are present in a majority of species, this basidiocarp may be called "typical". With this basidiocarp in mind, the characters which reveal the development and existence of various taxa may be discussed more conveniently. It is coincidental that the following list seems uncontroversial when considering the generic limits of <u>Clitocybe</u>.

<u>Pileus</u>: depressed when expanded, nonhygrophanous and nonviscid, glabrous, colored; context thin, firm, whitish
<u>Lamellae</u>: decurrent, close, narrow, occasionally forked, whitish
<u>Stipe</u>: central, solid, cortex fibrous, surface with appressed longitudinal fibrils
<u>Spores</u>: elliptic, small (circa 6 x 3 microns), smooth, inamyloid, acyanophilic, white in deposit
<u>Basidia</u>: 4-spored, clavate, lacking siderophilous granules
<u>Hyphae</u>: with clamp connections; cylindric and interwoven in the pileocutis, cylindric to slightly inflated in the context and hymenophoral trama, generally thin-walled but repository

hyphae not infrequent; arranged in an undulate-subparallel fashion in the hymenophoral trama
<u>Cystidia</u>: absent
<u>Macrochemical reactions</u>: negative
<u>Basidiocarp</u>: gymnocarpic in development

Terrestrial, in woods

There is an obvious and deliberate omission of any particular color in the pileus and stipe of this "typical" basidiocarp for there is none in a majority of species. At the least though, it is possible to state that the typical color is certainly not yellow, orange, red, purple, blue or green!

The previous list consists of characters which are very frequent throughout populations of many species of <u>Clitocybe</u> and this implies some stability in nature to whatever genetic mechanisms are responsible for the presence of each character. Thus, when the developmental stages of a character are known, and the influence of environmental factors have been determined, a character which differs from the list has the probability of indicating a taxonomic unit. Not a surprising conclusion to be sure, but what are these units in <u>Clitocybe</u>?

As it exists today, <u>Clitocybe</u> appears to have several major developmental lines within it. Whether or not these represent a polyphyletic origin, or merely divergences from a common ancestor is not of present concern. The fact is that several groups of species exist - as viewed by even the most conservative investigators, past or present. The precise number of groups, which represent infrageneric taxa, is a subject for debate, but whatever the opinion, the groups (subgenera and sections) are defined by major diagnostic characters. These include colors of the basidiocarp, or lack of them, the degree of fleshiness to pileus and stipe, and certain spore types. There are also three sections which are primarily distinguished by specialized structures.

It is rather difficult to comment upon diagnostic characters in a sequential order, as, more frequently than not, every level of taxon involves more than one character. Nevertheless, the order in the discussion attempts to emphasize a degree of relative importance for the

Clitocybe

largest number of species.

Color and Pigment

Color in the basidiocarp would seem to be the result of the action of a number of genes and thus must receive major emphasis in the delimitation of taxa. As far as I know, in the hymenomycetes a color results from a mixture of pigments, and each separate pigment of this mixture has been derived from a biochemical sequence in which a series of enzymes and genes was involved. While it is true that observations on color in Clitocybe are not supported by much analytical data on pigments, the assumptions seem reasonable in view of research done by several workers on the colors and pigments of other agarics and higher fungi.

What appear to be general types of color in the pileus are critical in identifying subgenera, while particular colors and their distribution in the basidiocarp are vitally involved in the diagnosis of sections and species. Eventually when more is known about the formation of pigments in the colors of Clitocybe, it will probably prove necessary to alter the position of some taxa, as certain pigments may be more complicated to form than others.

Of course there are also white species in Clitocybe, and others where tints of color appear to be due to physical rather than chemical properties. The differentiation of these species in the genus is often more difficult to interpret than those which have distinct pigments. Since albinism can result from the rather simple occurrence of a pair of recessive genes, the property of whiteness in common between taxa may be less significant than the possession of a color in common. Indeed, it appears that certain white species may be less closely related to one another than to some colored species. However, until such relationships can be deciphered or proven, most of the white and lightly tinted species need to be treated together in a group in order to allow recognition from one another. (The section Candicantes encompasses the majority of these).

Below the subgeneric level, species apparently with a color in common, may not be found in the same section if there is additional specialization. For example, the development of spore ornamentation separates the species of section Eulepistae from those of section Infundibuliformes.

Species within a color group may be distinguished by particular colors (e.g. pinkish cinnamon vs. reddish brown). Analysis of pileal pigments should reveal some in common with proportions or additional pigments providing the difference between species. The presence of color in the stipe and/or lamellae represents an additional development for a species. Color also may vary between species by its microscopic manifestations - whether the pigments are encrusted, intercellular, intracellular, or in smooth and thickened walls. Thus far, colored repository hyphae have not been found to play a significant role in Clitocybe, although they probably would be viewed as diagnostic if found in some undescribed species.

Hygrophanous pilei are notable in the species of several sections, (e.g. Candicantes, Pseudolyophyllum), and this quality is appropriate for note here as color is involved. Harmaja (1969) has discussed the nature of hygrophany and emphasized its diagnostic role. This property in pilei is particularly critical for distinguishing a number of species. It is a consistent attribute, although not necessarily easy or convenient to observe on all material. Two types seem to have evolved - one in which the light tints of pinkish and yellowish appear to be produced physically by water films between hyphae; the other, which is responsible for dark gray to gray brown colors at first, more likely is due to pigments. Both types of course are changed to paler states by the loss of moisture.

There is much to be done on the biochemistry of pigments, but it would seem that they provide an index for relationships that can be determined with precision. All problems will not be solved of course. As the present information on betalain distribution attests, more problems can be created.

Flesh thickness

In conjunction with color, the thickness of flesh in the pileus and stipe is a major character of species and higher taxa. The extremes are quite obvious in such species as Clitocybe geotropa and C. pyxidata, but there are often difficulties in judging this feature for a particular species. Not only do diminutive or gigantic forms occur occasionally, or etiolated or distorted ones, but some species seem to vary in fleshiness from collection to collection. Probably environmental influences are often the causes of this variability, for collections of a species from the same site in different years can show the same range of variation as found in collections from different sites. However, there is certainly a possibility that there are populations of a taxon which have some genetic difference as a cause for variability between the populations. Some species of agarics in western North America do not look quite identical to their counterparts in eastern North America, and it seems reasonable to wonder if some specializations haven't developed on both sides of the Great Plains barrier.

In Clitocybe, there are two places where it may be particularly difficult to decide about the proper position of a few species on the basis of fleshiness alone. Between Section Clitocybe ss. meo and Section Candicantes there is nearly a continuous series to be found in North America, and the same situation exists between Sections Pseudolyophyllum and Pusillae nom. ad int. While the presence or absence of hygrophany, or encrusted pigments, seem to provide the proper clues to the position of most species, there is no question that there are several which really require a new parameter for correct placement.

Spore Characters

Since the small, smooth, inamyloid, elliptic spore is the most prevalent in Clitocybe, spores which have other properties appear to distinguish species or even taxa of higher rank.

Of major diagnostic importance, i.e. for distinguishing groups of species, are ornamented spores. These are found in what I now recognize as sections Verruculosae and Eulepistae. Ornamented spores may also be present in section Subvelosae nom. ad. int., but this needs verification under the SEM. Clitocybe leucophylla (Fayodia striatula ss. Singer) has ornamented spores as well, but I am undecided about its proper position at present.

The inclusion of species with ornamented spores in <u>Clitocybe</u> of
course has engendered controversy. The issues do not question the basic role of spore ornamentation in the differentiation of species,
but rather the limits of certain genera in the Tricholomataceae. At
present there seems to be a stalemate between the opinions (Singer,
1975 and Harmaja, 1974 vs. Bigelow and Smith, 1969), although the additional proponents for each are not equally distributed. While it
might be interesting, perhaps exciting, to review the data emphasized, it does not seem that much could be added about species now.
There is more to be determined about the species described of course,
and the paper of Drs. Bresinsky and Schneider (1975) on the distribution of nitrate reductase was most welcome. (To deviate briefly, I do
not think that we have yet found all the clitocyboid species with ornamented spores either. For example, I know of specimens from North
America which appear to be additional variants of <u>C</u>. <u>tarda</u>, and a population of apparently <u>C</u>. <u>nuda</u> which is quite puzzling in some characters.)

My view of <u>Clitocybe</u> requires some comment upon another property of
some spores - the amyloid wall. Although the point will again be
controversial, I think that species with amyloid spores have developed in <u>Clitocybe</u> along with those having inamyloid spores - a situation accepted by all for <u>Amanita</u> and <u>Mycena</u> at the least. It would
appear that only the presence of one enzyme, possibly two, is responsible for the accumulation of the polymer which gives the bluing reaction in iodine (Bailey and Whelan, 1961). While a single gene difference can be considered sufficient to separate species, it would
seem that the differentiation of genera should require more. The characters of the basidiocarp in common between Clitocybes with inamyloid spores and those with amyloid spores far outweigh the presence
or absence of starch synthase in my opinion. Thus, I view the amyloidity of smooth spores as a species or subsectional character. <u>Clitocybe</u> <u>cyathiformis</u> is then placed in subgenus Pseudolyophyllum, <u>C</u>.
<u>candida</u> is placed near <u>C</u>. <u>robusta</u>, and <u>C</u>. <u>felleoides</u> goes into the
Candicantes, to name a few examples.

This judgment on amyloidity is not intended to be extended to other
genera of the Tricholomataceae in which the ornamentation appears
amyloid (e.g. <u>Leucopaxillus</u> ss. str., <u>Melanoleuca</u>). There are features of the spore structure and basidiocarp which contradict any

possibility that the species are closely related to those of Clitocybe (Lepista ss. Singer, Harmaja) which have inamyloid ornamentation.

The color of spore deposit has been used as a diagnostic character
for recognizing species by myself and others. I now have strong reservations that this may lead to error if utilized as the sole distinction between species. Apparently, there is some genetic mechanism
or environmental effect which can lead to a species depositing spores
which are not typical. I now have collections of several Clitocybes
(e.g. gibba, squamulosa, hydrogramma, sudorifica, truncicola, phyllo-
phila, variabilis, sinopica, americana, subalutacea) which rarely
have produced the "wrong" deposit. That is, a white-spored species
produced a cream colored or buff deposits or one with tinted spores
yielded a white print. This phenomenon is in terms of one or two collections out of twenty or more of the total studied. Of course the
thickness of deposit and effect of aging have been considered, and
while it is possible these factors may be involved in the data of a
couple of species, the occurrence of divergent spore deposits has
been too often to discount. All other observable characters of the
basidiocarp agreed in the species noted, and on present data it would
seem that the atypical spore deposit cannot indicate another species.
I still consider the color of deposit one useful parameter of taxa,
but my faith in its reliability and diagnostic importance has been
shaken. It is desirable to check up on the deposits yielded by all
the basidiocarps of large fruitings, and to continually monitor "common" species and not make assumptions about their deposit, and to
take more than one spore deposit from a cap when possible.

There is a tedious nature to examining the typical small elliptic
spore of Clitocybe under the microscope, but there is the advantage
in that such spores cannot vary much without an obvious change in
size and shape. With few exceptions, the survey of innumerable spores
of all species has revealed that spore size and shape are consistent
for a species and a reliable character. The range of size is usually
small (e.g. 1-3 microns in length, 1-2 microns in width), and thus it
is easy to recognize a divergence. Problems do occur, and I have
found what appear to be small-spored forms for a couple of species.
There are examples of species with polymorphic spores which can be
confusing, and occasionally some species have 1-, 2-, or 3-spored

basidia as well as 4-spored ones, and the former can distort the normal size range of spores. The occurrence of 2-spored races of species has not been a problem to contend with in *Clitocybe*, fortunately.

Cyanophilic and acyanophilic spores have been found in *Clitocybe* (Singer, 1972), but more species really need to be examined in order to determine if there is much significance to the occurrence of cyanophily. I would certainly agree with Dr. Singer (1972) that cyanophily "should (therefore) not by itself automatically and consistently be considered decisive in taxonomical problems ...".

Substrate and Habitat

The utilization of a particular substrate is important in defining a number of species, at least as they produce basidiocarps in nature. This seems reasonable as there are many enzymes in an organism which are involved in the digestion and utilization of compounds. A few species seem to be able to use several kinds of substrata, but generally humicolous species differ from lignicolous ones, those growing in grassy open areas compose a third group, and probably arenicolous or muscicolous ones constitute other types. Species which utilize conifer debris usually differ from those which occur on hardwood debris, although a few appear to be capable of the breakdown of either type. No *Clitocybe* is coprophilous that I am aware of. It is generally assumed that Clitocybes are nonmycorrhizal. While generally true, there are a few species whose growth pattern may indicate a relationship with tree roots. However, even if true, this will not affect the recognition of a species for it would have been noted in the particular habitat in the first place.

There are a few cases where the organism seems to be capable of growth on several materials. For example, in North America, *C*. *cyathiformis* is usually found upon wood, but occasionally it is terricolous. *Clitocybe epichysium* typically grows on coniferous logs and stumps in western North America, but prefers hardwoods in eastern North America. Usually, *C*. *clavipes* grows under conifers, but an occasional collection is found under hardwoods. There are other Clitocybes whose requirements do not seem specific, but the majority seem to require a particular substrate for fruiting. The growth of mycelium in the natural habitat may well be much more general. In culturing some two dozen Clitocybes on twenty carbohydrates, I found little selectivity. Of course there were differences in growth rate of a species on different carbohydrates, but apparently each had a broader spectrum of enzymes available than I had anticipated. Both agar and liquid media were used in testing.

Miscellaneous diagnostic characters

The absence of clamp connections on hyphae of the basidiocarp is a reliable index of species distinction. Clampless species are in the minority, but appear in several infrageneric taxa. At present, there are no species known which vary on the occurrence of clamps.

Cystidia are rare in Clitocybe. Indeed, the presence of elaborate types would bring a serious reservation about whether a species was a Clitocybe. The simple type of filamentous cystidia found in a few species of the Candicantes ss. meo and in section Aberrantissimae Singer does not appear to dilute the concept of Clitocybe, and aids in recognizing a few species.

Another valuable attribute of some Clitocybes is the presence of a particular odor. Those recalling anise, herbs and spices, fishinooo, or that known as farinaceous, are not uncommon. Sweet or bitter or rancid tastes can be found. These qualities can be difficult to communicate, but represent synthesized compounds which are significant in the genesis of a species. When more is known about these compounds it may be that emphasis above the species level will be warranted.

Other features of the basidiocarp may indicate a specialization. A few species have a pileus cutis consisting of a turf of cystidioid end cells, and the unusual structure of Clitocybe hydrogramma has been recognized at a species to a generic level. Rhizoids at the base of the stipe are known to characterize another small group. Of use, but often relative and variable, are such features as pileus shape, stipe texture, the spacing or width of lamellae in conjunction with degree of attachment to stipe, the proportions of pileus and stipe, striations on the moist pileus, surface texture of pileus and stipe.

of little or no use in species distinction are: forking of lamellae and venosity between, arrangement of hyphae in the hymenophoral trama, macrochemical reactions, details of mycelium or tomentum at the stipe base, hyphal size in general, basidial size, patterns of discharged spores.

Although they may lack the glamour of some genera, species of Clitocybe have no lack of characters when the specimens are found in prime conditions.

The preceding diagnostic characters, usually in combination, do allow the separation of Clitocybe species. The practical problems which can occur in working with specimens are not due to any intrinsic occult nature of clitocybes, but are usually from unsuitable material and/or inadequat observations. However, there are problems which cannot be minimized.

As with most of the large fungi, when a large number of collections of a particular Clitocybe have been examined, an awareness begins to dawn that certain collections have variations which do not seem to be present in most populations. The distinctions may be rather subtle as found between C. clavipes in western North America and eastern North America. On present evidence, the same would seem true for C. gibba, C. epichysium - there are likely others but I am not positive since I have not seen fresh specimens of the species concerned in both regions. A judgment must be made on one of these in relation to the "typical" species - is one a geographical race, a form, a variety, a subspecies, of the other? In other words, what parameter(s) determine a particular subspecific taxon? There are certainly definitive differences between populations which lend themselves more to description, for example, the presence of a fishy odor to C. inornata in Europe, while the North American representatives of the species appear to lack these features. What level of taxon best expresses such a distinction? Should the North American taxon be viewed as a variety of the European? Or, would it be more fitting to describe the North American one as a "new species", since the parent species is apparently absent and thus a variety could not be segregated from it?

Then there are the observable intermediate specimens between two "species" which lead to speculation about hybridization. There are puzzling specimens in North America which seem to combine characters of C. geotropa, C. maxima and possibly also C. subsalmonea Lamoure. In a particular region, one, or maybe two, of these species seem quite definable. However, on examining specimens from Washington, Utah, Michigan, Massachusetts, Montana in the United States as well as England, France, Sweden, and Greece, there are problems in identifying particular collections. Sometimes such dilemmas are eliminated by seeing more material, but this complex of species has become more difficult with more specimens and information.

I am sure that every specialist can cite problems parallel to these mentioned, but all do not face quite the number of previously described taxa. There are over 1,300 names for Clitocybe, and these present obvious difficulties when something "new" is encountered. The genus is by no means completely known for the world, although we are certainly closer to this than thirty years ago. There are more taxa to be discovered and/or described, I am certain, and considerable experimental research is necessary on those we do know.

"Differentiation" of species can imply more than delimiting the means for recognizing species from one another, and it would seem appropriate to speculate upon that from which Clitocybes differentiated. It seems to me that they probably came from Hygrophorus ss. lato for in this group are all major clitocyboid characters. There is no other genus or group which has so much in common with Clitocybe. It is tempting of course to attempt to connect certain extant species in both genera, but the view should be general and consider only diagnostic features of the basidiocarp. The clitocyboid fruiting structure would seem to be a successful modification of the hygrophoroid type by which a larger spore producing surface was attained. The addition of lamellae to the basidiocarp of course required support and varied structural modifications in the pileus and stipe took place to accomplish this. Whether the pigments of Clitocybe evolved during the period of alteration of anatomy, or whether they were sustained from existing hygrophoroid pigments, cannot be postulated until there are more analytical studies of the pigments in both groups. Probably both situations will be evident.

Whatever the genetic processes were, Clitocybe as it exists today has several major groups of species. Students of the genus view these as subgenera and/or sections. In comparison with many other genera, Clitocybe might seem to be quite heterogeneous. I view this as a reflection of its antiquity. At the same time, the heterogeneous nature of Clitocybe would appear to provide a gene pool from which it is not difficult to trace connections to other genera in the Tricholomataceae - Tricholoma, Collybia, Lyophyllum, Mycena. There are others possible, and even a sequence to the Rhodophyllaceae via Rhodocybe.

Over the years my idea about the limits of Clitocybe and its infrageneric taxa have changed. Probably they will be modified again as new material of Clitocybe and relatives is found, and more is determined about the biology of the species. The genus continues to be interesting - if it were not so, perhaps my efforts might have been completed several years ago.

Literature cited.

Bailey, J.M. & W.J. Whelan, 1961: Physical properties of starch I. Relationship between iodine stain and chain length. J. Biol. Chem. 236: 969-973.
Bigelow, H.E. & A.H. Smith, 1969: The status of Lepista - A new section of Clitocybe. Brittonia 21: 144-177.
Bresinsky, A. & G. Schneider, 1975: Nitratreduction durch Pilze und die Verwertbarkeit des Merkmals für die Systematik. Biochem. Syst. & Ecology 3: 129-135.
Harmaja, H. 1969: On hygrophany of basidiocarp in the genus Clitocybe Kummer. Karstenia 9: 51-53.
Harmaja, H. 1974: A revision of the generic limit between Clitocybe and Lepista. Karstenia 14: 82-92.
Singer, R. 1972: Cyanophilous spore walls in the Agaricales and agaricoid Basidiomycetes. Mycologia 64: 822-829.
Singer, R. 1975: The Agaricales in modern taxonomy, ed. 3. J. Cramer, Weinheim. 912 p.

DISCUSSION FOLLOWING DR. BIGELOW'S PAPER

BLAICH: I doubt whether one may find two fruit bodies which are genetically completely identical. Within a mycelium one can isolate quite a lot of nuclei carrying mutations which normally are repressed due to the heterocaryotic state. So you might say that each mycelium contains, let's say, a hundred potential species. So a monogenic difference is certainly not sufficient to make a new species.
It is about the same as with hair color. You have dark hair and my hair is light. There are some genetic differences, though we are the same species.

ESSER: I think it is about time to go into detail with the species concept. What geneticists are concerned, we must be aware that we do not deal in fungi with individuals, like we do in the higher plants. Because in the higher plants there are only certain cells which propagate sexually, but in the fungal mycelium each nucleus may be able, under certain conditions, to come into a fruit body, to undergo karyogamy, subsequently meiosis. According to our experimental experience we consider a mycelium as a population of nuclei which may be genetically different. We very often have obtained from one fruit body in basidiomycetes as many as ten or fifteen genes which we could show to originate from different nuclei. They just happen to be together in a population, as Dr. Blaich mentioned, which complements their efficiency. And they are held together in most mycelia by the incompatibility factors A and B. When these incompatibility factors are not any more effective in monokaryons these genes express themselves in absolutely different ways which we should keep in mind. About some 20 years ago we were able to introduce the spore color in ascomycetes as marker genes, Podospora and Sordaria. Colored spores occur very often in nature. But we should not be surprised that in coprophilous ascomycetes they should not persist, because usually the spores of Sordaria pass through a herbivore(e.g. a horse) and the colored spores have no melanin pigment and are digested. So using, colored spores in ascomycetes which are single gene mutations, as a criterium in taxonomy, I would say these are just simple mutations. I am no mycologist, but I would a little bit hesitant to use color of the spores as a criterion for species delimitation.

BLAICH: There might be different reasons why the amyloidity you mentioned disappears, and it is not sure that amyloidity of spores is identical with the starch reaction in other plants.

BIGELOW: My information indicates that the starch-like compounds which give the bluing of fungal spore walls have a lower molecular weight than the starch of higher plants, i.e. the chain is shorter, with fewer glucose units. The bluing can start with about 40 units whereas in higher plants the reaction is due to 500 to 600 units. After this there comes a point where the chains are too long and they fall apart. One enzyme is responsible for the synthesis of the chain though, regardless of length.

BLAICH: The lack of one enzyme is certainly enough to make amyloidity disappear. But to form a new species, other mutations must be added, and the new mutations must be isolated etc. It is not sufficient to consider just one mutation. As you say, you have other differences in this case.

CLEMENCON: If my memory is correct, in Schizosaccharomyces, a yeast, the wild strain has pseudoamyloid spore walls getting dark brown in Melzer's, and there is a mutation with inamyloid spores. This is case where just one mutation causes an positive iodine reaction to disappear.
It has been mentioned several times here, that the substance giving a blue reaction is something like starch. I did some experiments with Chroogomphus and Rozites caperata. There are wall incrustations which give a dark blue color in Melzer's, but they are not degraded by amylases. So they are certainly not starch. When we degrade starch and come to shorter chains, the blue reactions does not just disappear, but changes to brown-red etc.

BIGELOW: According to my reference the red-brown, dextrinoid or pseudoamyloid reaction, is much more specific. No one, I think, claims that the bluing substance of fungi is starch. It is in the same progression, but has a much shorter molecule.

CLEMENCON: I do not think so. Because a short molecule of the same progression should be degraded by amylase. But this bluing incrustations were not attacked at all, so they cannot have the same structure, they cannot have the same chemical linkage we find in the starch-progression.

BIGELOW: Was your solution buffered?

CLEMENCON: Yes, and the controls were very convincing, the amylase solution degraded starch in a control experiment within 20 minutes, but the same solution, under the same conditions, did not attack the incrusting substance at all. It remained strongly bluing in Melzer's after 24 hours. They just are different polymers.

SINGER: There is no doubt, to me, that what we call amyloidity of the spore may include chemically different things. Amyloidity is mostly but certainly not always very constant. Dr. Kühner may remember what I later called Pseudohiatula dorotheae. When I had studied the spores fresh, and this is the only case I know, I thought they were amyloid, and I said so. Then Dr. Kühner wrote back (and I could confirm that) that the spores were inamyloid in the dried material. So I took new material from the greenhouse, and the spores were amyloid again. So here it seems there is a certain degradation taking place in the herbarium. Yours may be such a case. On the other hand, I have studied material from the Persoon herbarium, which is 150 years old, and which gives beautiful blue reactions.

KEMP: My contribution is not only concerned with Coprinus but is, I hope, more widely applicable. It is not surprising that we find variability in the characters of the fruit body especially in outbreeding species. This variability is needed for evolution and adaptation but these are not the same as speciation. Why should we expect constancy in characters which have nothing to do with speciation? Speciation, in most cases, probably involves the interactions of the hyphal tips which are embedded in the substrate. The shape of the fruit body and the spores and all other fruit body characters are the result of speciation and not the cause of it. As species must be able to evolve they must vary both in fruit body characters and in "speciation characters" such as nuclear migration and lethal hyphal interactions.

PETERSEN: I think, Roger, you may just have approached a difference in the use of the words "species" and "speciation". Unless I am mistaken, I think we have been talking about species as taxonomic units,

as separable from one another on rather easily defined characters. I
think you are talking about species as a different type of entity, a
biological species.

KEMP: I am trying to consider species in relation to the mechanisms
of speciation.

BLAICH: There seems to be some danger, because the species concept
of mycological taxonomists seems to differ from other species concepts in higher plants or in animals, if it is kept too narrow. In
higher plants, genetic work was not possible if individuals would not
differ in a lot of genes, thus allowing genetic analyses. I think the
fungal taxonomist should make an allowance for this kind of variation, and not try to separate the slightest units. Mycologists talk
of species in quite a different sense than other taxonomists.

SMITH: In reply to Dr. Blaich: I think that, if you take the higher
plants, like the Ranunculaceae, or the mosses, or the lichens, you
find that the pattern of speciation is markedly different in each
group. So why should the fleshy fungi have to conform to some other
group? They are a group unto themselves.... Now, talking about single
character differences, when we are talking about taxonomic characters
and the taxonomists, at least the taxonomists that I know, have no
idea whatever whether this is the only difference between the two
taxa. The other differences are for the geneticists and physologists
to find out. We are merely trying to properly map what occurs in nature, so that other people can take it up, work with it intelligently. If we lump everything in the genus _Agaricus_, it is going to be
very difficult to make people think that _Agaricus_ _campestris_ is the
same as _Cortinarius_ _coeruleus_.

BRESINSKY: I think that the question of the species is not open as
to how many genes participate in differences between species. The
main thing is that we have a hiatus, and that this hiatus is not
overbridged by any kind of continuous variation, finally that we
have any isolating mechanism. Mostly, in fungi, the isolating mechanisms are depending on the disability to interbreed. So the differences may be very, very minor, also in terms of how many genes participate, but we are not able to make any statement on how many genes
are making differences between species.

SINGER: May I make the proposal that we postpone this discussion because I have the impression that we have some communications coming which show that mycologists actually should have, other and different species concepts. Even in the higher plants there is not one type of species in all genera, but you get different types of species. For mycology the situation is such that we are quite justified to think that it is possible that the species concepts may be different in different groups of fungi.

SPECIATION IN LACTARIUS

Alexander H. Smith

One of the crying needs in the systematics of the Agaricales is the need for types (as collections) to establish concepts for those species described early in the history of mycology long before the type system became established as part of the International Code of Botanical Nomenclature. I refer particularly to the species described during the Friesian era. As it is now, one name is often being applied to two or more species, or a name is being applied to a species in contradiction to its characters as given in the original description. Consider <u>Agaricus helvus</u> Fries. In the Systema he described it as having, among other characters: "... Odor debilis, sapor acris, lac album immutabilis...". In Europe, at present an "American" species (i.e. described from North America), <u>L. aquifluus</u> Peck, has a pronounced odor, mild to bitterish taste, and water-like latex. We cannot intelligently discuss the evolution of species, and a basic pattern of species concepts for a genus, until the above type of situation is cleared up. It is with this in mind that I make the following suggestions relative as to what is the best procedure for designating types for these "old" species. The need for type collections in agarics is particularly acute. Descriptions (especially the older ones) and illustrations are not sufficient to serve this purpose properly, and it is unreasonable to insist that a single specimen, let us say, sent by Fries to one of his correspondents, and which turns up a hundred years later in some herbarium, is going to be, actually, the species Fries thought it was at the time he sent it. Species concepts at this period in agaric systematics simply were not accurate enough for modern purposes. This is no reflection on the early authors -- they were pioneers, and, I feel sure, did not assume that their word would be final for any species. We should, however, build on the foundations they laid, and we should be certain that any lectotypes for their species have the important positive characters which they included in their descriptions. This is what the International Code of Botanical Nomenclature is all about. My suggestions for the establishment of type collections for the species in question are as follows.

1) Only species named in accordance with the binomial system of

nomenclature and for which a description is given by the author should be typified. Just what constitutes a description will need some discussion.

2) An ample collection of fresh material fitting this description and obtained in the country of origin of the original material, should be made.

3) This collection should be described as if it were to be the type of a "new" species. All the data, macroscopic, microscopic and chemical characters should be recorded <u>from this one collection</u>. This will ensure that all the characters noted were included in a single genome. Photographs and/or paintings should be taken or made, and microscopic characters illustrated.

4) The collection must show the positive features as given in the original description. For instance, <u>Agaricus helvus</u> must be an agaric with a weak odor, an acrid taste and white unchanging latex. A collection with a pronounced odor, mild to slightly bitterish taste, and a water-like latex unchanging when exposed, simply would not be eligible for consideration as the type for <u>Agaricus helvus</u>.

5) The collection should be dried by modern methods (dry heat at relatively low temperatures or over activated silica gel for small fruit bodies). Freeze drying causes many specimens to become very fragile.

6) The International Botanical Congress should be asked to establish a commission to review these "types" for the purpose of rejecting or accepting them.

7) When a collection is accepted as a type by the commission, an official description of it in Latin should be published in a widely distributed journal and I would favor a translation being published along with the Latin in a second language. The specimen(s) should be deposited in an herbarium that is known to be actively taking good care of its materials at present.

8) The mycological society, or club or association in each country should be asked to undertake this work for their region since the project would of necessity be a long-range one (more or less 25 years). If no such group exists in a country, then investigators from neighboring countries should do it by whatever arrangements seem feasible.

9) Species published prior to 1821 but in accordance with the binomial system of nomenclature, should be typified according to the proposed program.

10) The types, once established, should not be sent through the mails or by package services. One wishing to consult them should go to the herbarium where they are deposited. The published account should be sufficient for almost all purposes. At present, many agaric types have been literally destroyed by loaning them and exposing them to the hazards of handling by the transportation services.

11) The designation of these types, officially, by the International Botanical Congress should in no way be regarded as a step toward conserving species names. The International Code of Botanical Nomenclature should apply to them the same as for any author-designated type.

Comment: The European mycologists should be the ones to push this program since the work of necessity falls in their collective lap. I should like to see this proposal placed on the agenda of the next botanical congress.

Turning now to Lactarius in North America. I should first like to introduce the classification Hesler and I are proposing for the North American species. We recognize six subgenera, 18 sections, 200 species and 60 varieties.
1. Subgenus Lactarius: type, L. deliciosus
    No sections recognized
2. Subgenus Plinthogalus: type, L. lignyotus
    Sect. Plinthogalus
    Sect. Fumosi: type, L. fumosus
3. Subgenus Lactifluus: type, L. volemus
    Sect. Lactifluus
    Sect. Allardii: type, L. allardii
    Sect. Piperati: type, L. piperatus
    Sect. Albati: type, L. vellereus
4. Subgenus Piperites: type, L. torminosus
    Sect. Atroviridi: type, L. atroviridis
    Sect. Aspideini: type, L. aspideus
    Sect. Piperites

5. Subgenus _Tristes_: type, *L. argillaceifolius*
          Sect. _Violaceo-Maculati_: type, *L. subpalustris*
          Sect. _Pseudomyxacium_: type, *L. kauffmanii*
          Sect. _Tristes_
          Sect. _Colorati_: type, *L. glyciosmus*
6. Subgenus _Russularia_: type, *L. subdulcis*
          Sect. _Triviales_: type, *L. affinis*
          Sect. _Pseudo-Aurantiaci_: type, *L. substriatus*
          Sect. _Thejogali_: type, *L. thejogalus*
          Sect. _Russularia_
          Sect. _Subsquamulosi_: type, *L. alpinus*

The species concepts of Hesler & Smith as proposed in our forthcoming work are much broader than in our earlier contributions. The problem of how much diversity should be included in the concept of a species will undoubtedly be the subject of debate for years to come. We make no claim to having proposed a final solution. When we started our work, we subscribed to the idea that we would not recognize infraspecies taxa in this genus. The manuscript thus produced, however, did not satisfy us as more populations were sampled. We were confronted, figuratively speaking, with the situation that diversity in nature cannot be packaged into units of equal size and weight. In taxonomic work, diversity is represented by "characters" and since these can be counted and dealt with statistically, this approach has also been tried by way of formulating species concepts. Such an approach is not successful unless some means of evaluating the significance of characters is introduced into the study, and in the end we are right where we started: relying on the judgment of an investigator who has studied the group critically, and who has a "feel" for the intangibles in that group.

In _Lactarius_ the problem we encountered involved sorting out genetic diversity between populations from variation within a single plant (mycelium). It must be kept clearly in mind that in all taxonomic work on Agaricales we are often dealing with the fruits of an individual plant even though we have a hundred or more basidiocarps. The differences between these fruit bodies we shall refer to hereafter simply as _variation_ -- as contrasted with genetic diversity. In agarics both rest on an assumption: first, for variation, that it has indeed been ascertained from a single mycelium, and that separate

occurrences of this character or characters, on basidiocarps widely
distributed geographically, are indeed the same kind of variation as
observed from an individual mycelium. An example of such variation in
Lactarius may be cited for L. volemus. Frequently, in large fruitings, some basidiocarps have yellow pilei, the majority being the
typical orange-brown. In our southern states, however, we have populations featuring yellow pilei. Our assumption here is that in the
southern "variant" the character is fixed genetically, and in the
northern variant it is not. All this is based on field observations.
Since we cannot test our assumptions experimentally, we accept them
as long as they remain supported by numerous and continued observations. If it is found that in the South we also have populations in
which the basidiocarps in a fruiting contain some yellow specimens,
but mostly the normal orange-brown ones, the question arises, is this
simple variation in our sense, or the result of gene exchange with
the characteristically yellow southern variant? We do not know the
answer. Should the southern yellow variant be recognized as a taxon?
If so, at what level? In a system in which infraspecific taxa are not
recognized, the genetically based yellow variant would be lumped with
one (by contrast) apparently environmentally or physiologically based
(in the latter case, such a variant might be among the last basidiocarps to develop in a large fruiting). The difference between the two
yellow variants we are discussing, however, could ultimately be important to those doing experimental work if it turned out that in the
genetically based variant other hidden differences were present. For
instance, all yellow pigments are not the same chemically, and there
might be a difference in the effects on the mycorrhizal host. The
systematist should not conceal the existence of such diversity nor
overemphasize it to such a degree that it will not be recognized by
others for what it is. Infraspecific taxa, in our estimation, serve
this purpose well in a system of necessity erected on assumptions.
Consequently we recognize L. volemus var. volemus as a taxon in which
color variants occur in individual fruitings -- possibly from various
causes. This is covered in the description. In L. volemus var. flavus
we place fruitings with consistently yellow pilei. In short, in .
Lactarius in particular, we are convinced that infraspecies taxa
serve a very important function. We have recognized this by redoing
our entire manuscript to conform to this conclusion. This represents
over two years of additional work.

It is trite to say that in <u>Lactarius</u> the situation in any one species is complicated -- they all are. In <u>L</u>. <u>volemus</u> we have found in some fruitings (from a single mycelium?) individual basidiocarps with the gills forked repeatedly -- reminding one of <u>L</u>. <u>piperatus</u>. This has been observed in a number of localities miles apart. This variant, if found by itself, might be described as a "new" species by some since it is readily distinct, and it is extremely doubtful if the difference can be attributed to environmental conditions in the habitat or quantitative differences in some compound such as a pigment. Our forked-gilled variant may actually be a distinct species --˙for a while we carried it as such in our manuscript. However, since it has so far been found only <u>with</u> <u>some</u> fruitings of var. <u>volemus</u>, we finally decided to simply discuss it as a variant or "sport" from the mycelium of the type variety.

In the course of our study of the structure of the stipe of <u>L</u>. <u>volemus</u> it was found that numerous caulocystidia were present. This is nothing new. One interesting feature, however, is that in our material very frequently we found that these cystidia became secondarily septate near the base and that the cells thus formed finally enlarged somewhat. The endpoint in this development has been found in a number of basidiocarps in which a dense covering of caulocystidia occurs. The basal cells of the cystidia had inflated greatly and by mutual pressure were packed together to form a "cellular" layer. The apical cell thus became the "cystidia". But in <u>L</u>. <u>volemus</u>, as that species occurs in southeastern Michigan, all variations from a naked stipe to one velvety from caulocystidia, to some which show varying degrees of the formation of the cellular caulocuticle have been found. We have observed the same pattern in <u>L</u>. <u>thejogalus</u>. In the material of <u>L</u>. <u>volemus</u> studied to date, we have refrained from naming taxa (or a single taxon) on the basis of the caulocystidia and caulocuticle, though we seriously considered doing so on the basis of the formation of secondary septa.

All of you, I feel sure, are aware that the cellular cuticle of the pileus in many species of subg. <u>Plinthogalus</u> is also of trichodermal origin (see <u>L</u>. <u>lignyotus</u>). Other types of pileus cuticle, such as the rounding of the inflated cells of the cuticle in many species of section <u>Thejogali</u> indicates basic trichodermal origin but with evolution having progressed to the point of more or less obscuring that arrangement. The important point here is that developing diversity is evident in this character state.

Variation in *L. volemus* is also found in the latex. In most basidiocarps it stains the gills brown but in some does not become brown itself. In others it changes slowly or rapidly to brown and in a few collections the stains are near "avellaneous" to "wood brown" or slightly more vinaceous and the latex dries more or less the same color. In the literature there are some records of it staining tissues yellow, and records of it being whey-like to watery. In var. flavus it does not change on exposure as far as known at present. It is obvious, that if we found a "*Lactarius volemus*" with a latex changing to avellaneous or staining the gills this color, gills forked as in *L. piperatus*, caulocuticle a cellular layer, a consistently yellow pileus, and watery latex, it would certainly be recognized as a new species. To date, however, I am not aware that this combination of characters exists in a single population of *L. volemus* s.s. lato. By treating the expression of these characters as we have, the diversity involved is presented to the interested public, but the burden of new names is not increased. It is worthwhile to note that all the characters mentioned are in the world gene-pool of *L. volemus*. It is not unreasonable to visualize such a process as described here as indicating how species have evolved in this genus: Characters may arise by mutations thus adding new characters to the gene pool and these become combined or recombined in other populations, and finally distinct species, as we know them, would be born. Also, as apparently happens, characters may simply drop out. We have, for instance, found a few basidiocarps of *L. kauffmanii* in which the acrid taste was absent, and we have what we believe is a white variant of *L. mucidus* (an albino?). This pattern, I believe, represents the dynamics of speciation in *Lactarius* as postulated solely on the basis of observation.

Let us now consider some taxonomic characters from the standpoint of the genus as a whole. Spore features have been regarded as of primary importance since the microscope came into general use, and will be considered first.

Spore size. It is a standard character in the order Agaricales. In *Lactarius* we have found, for practically all the species for which we have large numbers of collections, that a small-spored variant sooner or later is encountered. In many collections we have found that the spores from deposits are smaller on the average than those from crushed bits of gill tissue. Spore deposits, from young pilei, usually give smaller more subglobose spores than those from mature caps. Deposits from old caps may give an erratic picture and tissue mounts often show a small to more or less significant number of exceptionally large spores. In these features our observations substantiate reports in the literature on spores of the Agaricales and on Lactarius in particular.

Spore ornamentation. In the last ten years a great change has come about in our evaluation of this character in Lactarius taxonomy, courtesy of the SEM, and by more field work on which to base our assumptions on the meaning of the characters that can now be studied.

First, we have found that patterns of ornamentation are often evident in subgenera such as Plinthogalus, where other characters than spore features are used to define the taxon -- such as chemical features, basidiocarp anatomy and "field characters". In Plinthogalus many of the species have peculiar and conspicuous spore ornamentation, with L. pterosporus of Dr. Romagnesi leading the list. In many others the spores are simply coarsely reticulate with the height of the elements 1-2.5 microns high. It is worth mentioning here that the SEM photomicrographs show more ornamentation than is evident in Melzer's sol. under the light microscope, but the basic pattern is usually evident, especially on coarsely ornamented spores. Although in Plinthogalus a large number of species have heavy coarse ornamentation, certain species such as L. pseudogerardii have large spores (10 x 8 microns more or less) with very fine ornamentation (0.2 micron high) of warts connected only by faint lines (in Melzer's). In other words, in Lactarius, spore ornamentation as we have found it is just another character and should not be overweighted in speculations on phylogeny. We find it most useful in Lactarius at the level of subsection or stirps (the latter an unofficial but very useful category). In our classification we have used fewer patterns than in the system proposed by Singer. We recognize three major categories: reticulate, warty, and zebroid. We also recognize intermediate patterns such as an incomplete reticulum (with some meshes complete) and a broken reticulum (in which branched ridges do not quite connect to form complete meshes. Short ridges and warts together form the ornamentation on many species, but here one must be careful because immature spores of other types may show this pattern to some extent. If in the reticulate pattern, no additional isolated particles are present the pattern is classified as "clean". If warts and minor particles are present in many of the meshes of the reticulum, we describe the ornamentation as cluttered in addition to being reticulate. In both classifications (Singer's and ours), there is a great deal of overlap between types. In Lactarius, spore ornamentation is of greatest value at the level of species.

Slime-layers of Pileus and Stipe. The problem of ascertaining the degree of gelatinization of the cuticular layers of the pileus and stipe in Lactarius is complicated by the fact that in most species the hyphae of the tissues, as revived in weak (2.5 per cent) KOH have more translucent walls than is generally true for species in the Agaricales. This might be indicative of a chemical difference in the hyphal wall in this genus, as, for instance, compared with members of the Tricholomataceae. Aside from this, and as limited to cuticular layers, the situation for the genus is routine except for those on the stipe. The excretion of slime and the gelatinization of hyphal walls on the basis of our observations, are two independent characters which apparently may or may not be linked since both processes may occur on a single species. Slimy to merely viscid layers occur on some species in all subgenera. They are a feature of subg. Tristes, common in subg. Lactarius, Piperites and Russularia, and in Plinthogalus and Lactifluus their presence, in wet weather is very limited to doubtfully present (see L. acris in Plinthogalus). If considered on the anatomy of the cuticle, we find a slimy and dry type for each of most categories, i.e. a cutis and an ixocutis, a trichoderm and an ixotrichoderm, a lattice and an ixolattice. In defining infrageneric categories, the presence or absence of slime in the pileus cuticle has been used at various levels. We found it impractical to use the feature of a viscid stipe as a major character as arbitrarily as in Cortinarius or Mycena, for instance. The reason for this is that in Lactarius the stipe is often scrobiculate (with large shiny spots). The cuticle in these areas is usually a slime cuticle of some sort, and the routine "touch" test for a slimy stipe can be misleading. Also, we have some evidence that an ixocuticle of one type or another can occur in patches not visible as shiny spots.

Turning again to the latex, we find considerable diversity here even
if the changing and staining reactions are disregarded. Taste of the
latex is a difficult feature to accurately ascertain because in most
species it is difficult to collect the latex free of influence of the
tissues of the basidiocarp. Where we tried to work with this feature,
we found that in some species with yellowish latex, the latter was
mild when in the white stage, but acrid after the change to yellow
had taken place. We still have this feature down for further study as
material becomes available.

Physical features of the latex offer characters of limited value to
the taxonomist. We have species with a clear water-like latex, see L.
aquifluus and L. subserifluus. We have many in which the latex is
milk-white in young specimens but whey-like to (finally) watery in
old ones (see section Thejogali). In most, the latex is simply milk-
like in appearance and finally dries as droplets of hardened latex.
The extreme situation relative to viscosity is found in L. subpaludosus in which the latex exudes as viscous droplets which harden in a
short time, usually less than 10 minutes.

Change in color of latex on exposure to air has long been used as a
standard character in Lactarius taxonomy. We have continued our study
of species showing such changes, particularly the change to yellow.
Because of the nature of the early observations -- using the eyeball
test on fresh basidiocarps, the character was first used in distinguishing those species in which it was readily and obviously evident
when latex was exuded. It was then found that in some species the
change was slower than in others, and it was also found that the
change in some was very slow. In others it was not only slow but the
tint was rather faint, as in those in which the latex was slightly
cream-color at first. We also observed that in some basidiocarps the
latex exuded from injured gills might remain milk-white whereas that
from the apex of the stipe changed to pale dull yellow. In some species the yellow change occurred more rapidly when the latex was in
contact with white paper, and this started us on checking the latex
of all species in the genus to see how widely this change occurred.
Final results are as yet not available, but the evidence is clear
that in many species in which the latex does not yellow on exposure,
it does stain white paper yellow. The reaction was particularly important in L. thejogalus. So far we have found it mostly in Piperites

and Russularia. Although incomplete, our study clearly indicates that the yellowing of the latex or staining tissues yellow is not, as a single character worthy of emphasis at the level of subgenus or section if a natural arrangement of species is one's goal.

The staining reaction to violet, lavender, pink or reddish cinnamon in Plinthogalus has been a most helpful one because here it is correlated with the velvety to unpolished pileus and the dark (for the most part) pigmentation. But it also appears in the Piperites and Lactifluus as well as in Tristes (section Violaceo-Maculati). In other words the reaction is not confined to one related group of species.

The Heteromerous Trama. One of our "discoveries" in working on Lactarius is that NO character is free from deviations. The family Russulaceae is based on the heteromerous context of the pileus and stipe, and the amyloid spore ornamentation. To date it is only in the Astrogastraceae that we find species with no or almost no amyloid material on the spore ornamentation. As to the heteromerous trama, in a few species of Lactarius we find it greatly reduced to, for all practical purposes, absent. In L. subserifluus, for instance, the nests of sphaerocysts are absent from the tissue of the stipe and pileus. This may account for the hard consistency of the stipe. Parenthetically, may I add that the term rosette for the groups of sphaerocysts in the stipe of Lactarius is a misnomer. Since the time of de Bary it has been known that these cells in the stipe are often present in long columns which only in cross section give the appearance of rosettes. In other species (of more fragile consistency), the sphaerocysts may be so numerous as to be the main component of the pileus trama. The majority of the species fall between these two extremes.

Color of pileus in Lactarius has long been a major feature in species recognition. It runs the gamut of no color (white) to green, violet, yellow, red and various mixtures of all of these too numerous to mention to (finally) brownish black at the opposite extreme from white. We have found it to be a relatively constant feature for each species, or if changeable, the pattern of change is constant. The color changes as the pileus fades, as one would expect, but many of the changes are very characteristic. In L. atrobadius which is blackish red becoming (slowly) mahogany red, the color on fading changes to pale drab. Generally speaking however, color of the pileus is as constant as the pattern of spore ornamentation. I wish to be very emphatic on one point: It is sheer nonsense to maintain that macroscopic characters (such as color) are less important in the systematics of higher fungi than are microscopic characters. Taxonomic characters should be evaluated on a less arbitrary basis.

For a long time to come "what is a species" in Lactarius will be determined, as it has in the past, by the opinion of an experienced investigator. It will be years before we can accumulate data on all the species of say Lactarius or Cortinarius or Psathyrella in sufficient detail that we can advantageously use modern sophisticated methods of data analysis to give an unbiased answer as to where a genus came from and where it is going. But this approach, if followed, will materially change our pattern of procedure in field work since one will be more interested in the large fruitings than in single isolated occurrences, and "complete" data for each collection must be recorded in order to make the collection valuable to the collector or others at a later date, for one will not find all the material he needs in any given season. At this point the function of a well-cared-for herbarium in an institution is vital.

In Agaric studies in the future some of the points to be emphasized are:
1) Designation of type specimens from well described fresh material from the country of origin, if no type was designated by the original author.
2) Intensive field work is needed in areas as yet very incompletely explored for mushrooms. I know of no areas, yet, which are "completely" sampled for any large genus of gilled fungi.
3) Continue work on the chemistry of the basidiocarp generally of the sort that some of you here today are doing.
4) Data recording: Building up data banks on all the fungi collected. This is a tedious task, but herbarium specimens, without the data on the fresh state of the fruit bodies, renders them 50 per cent useless. We found this lack of data a great handicap in our study of Lactarius.

Lactarius   135

KODACHROMES FOR **LACTARIUS** IN NORTH AMERICA

A l e x a n d e r H . S M I T H

Subgenus Russularia
1. L. camphoratus
2. L. imperceptus
3. L. oculatus
4. L. rimosellus
5. L. thejogalus
6. L. thejogalus
7. L. subserifluus
8. L. subserifluus
9. L. obnubilis
10. L. frustratus
11. L. subviscidus
12. L. alpinus var. mitis
13. L. aquifluus
14. L. aquifluus
15. L. rufus
16. L. rufus
16a. L. mutabilis

Subgenus Piperites
17. L. controversus
18. L. atroviridis
19. L. psammicola
20. L. alnicola
21. L. scrobiculatus
22. L. payettensis
23. L. torminosus
24. L. pubescens
25. L. subpaludosus

Subgenus Plinthogalus
26. L. gerardii var. rubescens
27. L. fallax
28. L. fuliginellus

Subgenus Lactifluus
29. L. volemus
30. L. pervelutinus
31. L. corrugis
32. L. hygrophoroides
33. L. tomentoso-marginatus
34. L. subvellereus
35. L. allardii
35a. L. allardii

Subgenus Tristes
36. L. pallescens
37. L. pseudomucidus
38. L. subpaludusus
39. L. uvidus
40. L. hysginus var.
41. L. argillaceifolius

Subgenus Lactarius
42. L. rubrilacteus
43. L. deliciosus
44. L. subpurpureus
45. L. thyinos
46. L. paradoxus

DISCUSSION FOLLOWING DR. SMITH'S PAPER

SMITH: What did we learn from this investigation? For one thing we learned that characters must be standardized. For spore deposit color one must obtain a heavy deposit, and the latter must be viewed for color as the pileus is removed from the set-up, and read again after the print has dried for about 15 minutes. We found that for some species the deposit which is white at first will yellow slightly in that time. This is pertinent relative to notations by collectors in general who, of course, record what they observe. On a number of occasions we had a fungus we recognized, but the collector's data on the color of the spore deposit did not "fit". We finally came to disregarding such color notations if they were within the A^*B color range (white to cream-color).

We also found a problem with spore size and shape. Make a spore deposit from a young cap and let us say the spores are nearly globose. On a length-width ratio one comes out with a value of 1.1-1.2. Set up the same cap again and get another deposit. From this deposit one finds a value of more or less 1.2-1.75 approximately. What does this indicate statistically? If the same cap was used for both, the only meaning of the difference that seems acceptable is that it indicates the stage of maturity of the pileus used. This was verified by taking a continuous spore deposit over a period of time - the way in which Lactarius spore-deposits are often taken. The spores up to a point, become more ellipsoid as the set-up ages. When the cap passes its prime, the spores produced often have odd shapes and sizes and should not be considered in the normal range for taxonomic purposes. I think Dr. Bas had these considerations in mind this morning when he suggested taking material for the spore sample to be measured from the middle of the gill and more or less half way between the stipe and edge of the pileus. This establishes a reasonable set of standardized conditions.

We also found that our spore deposits generally showed smaller spores than those from macerated tissue of the same fruit body after the latter was dried. We have a problem here and one solution which is suggestive is that in macerated-tissue mounts it is common to find more large spores than in deposits, and many are of odd shapes, as already mentioned. Possibly, in <u>Lactarius</u>, the discharge mechanism does not function properly and the spores merely continue to grow.

the frequency of such spores on a single pileus appears highly variable.

In nearly all species of Lactarius where we had large numbers of collections, we found a "normal" size range and also a few pilei showing a smaller size range as well as sometimes a few showing a larger size-range. In fact we have found this to apply to common species throughout the Agaricales. The magnitude of the differences is about of the following order: normal, 7-9 x 5.5-6.5 microns; small, 6-7.5 x 4-5.5 microns, and large, 8-10.5 x 6-7 microns. The problem arises when the small size was discovered first and the species described on that basis. If the large spored form were discovered next it would, by most of us, be regarded as a distinct species and it may be years before enough collections are accumulated to allow attention to be focused on the true situation.

In our study we also found that the height of the spore ornamentation needs some standardization to be used effectively as a taxonomic character. It varies with the degree of maturity of the spore. Again, following Bas' lead, use spores of the size near the peak of the range. The most peculiar abnormality we found ralative to spore ornamentation is that every once in a while in a given species, in mounts made in Melzer's, big globules of amyloid material, roughly 4 x 5 microns, adhere to the spores or were free in the mount or both. Also, one finds spores that are smooth and nonamyloid. Different stages in the process of the removal of the amyloid ornamentation can be readily found.

As to the height of the normal ornamentation, the maximum heights are most distinctive taxonomically, but do not overemphasize the warts or ridges near the apex of the spore which are often higher than those elsewhere. Patterns of ornamentation are distinctive if grouped in such major types as reticulate, of isolated warts, zebroid, or with partial to broken reticulum.

Relative to macrocystidia, we found that in Lactarius they have the amazing habit of collapsing and disappearing by maturity in some groups. To study macrocystidia use gills just approaching maturity. At this stage both mature and immature cystidia are found and the changes in shape from young to old cystidia can be observed.

In the pileus cuticle and adjacent area beneath it, in our subgenus Tristes, we found the dark granules mentioned by Pearson and by others. They were an aid in defining the subgenus. These granules are soluble in KOH and are readily missed in revived mounts. If sections are mounted in slightly acidified tap water they are more readily demontrated. They are slowly soluble in Melzer's reagent.

To us, Lactarius appears to be a relatively young genus. The indications for this are the amount of diversity in a given region. In most common species of the genus in North America we find more diversity and intergradation than in other genera. Psathyrella, by contrast, presents rather well defined species - even though the characters on which they are based may seem trivial to some. Returning to Lactarius, it appears to us to be in a vigorous stage of biological activity and active speciation.

BLAICH: How about the frequency of your abnormal fruit bodies?

SMITH: We really do not have any data we could subject to statistical analyses. They come in occasionally. And we had no opportunity to survey a section of woodland to see how many are there. In Lactarius volemus, if that is the one you are thinking of, we find one or two in a big fruiting. It is difficult to draw conclusions from that. It can be argued either way - the variant with repeatedly forked gills represents a poor fruiting of a true species living in the same habitat, or a mutation from typical L. volemus on the other.

BLAICH: That would be the same frequency as in laboratory. If you isolate spores and look for mutants you find about 1%.

SMITH: I think we have assumed, by discussing this as a variant, that this is a mutant.

BLAICH: If the spores of this mutant are distributed over the area this character will disappear because of hybridization with other forms.

SMITH: Yes, that is what happens with Psilocybe mutants where. You have a white spored mutant and a dark spored type - and you still have only dark spored populations. The white spored variant disappears.

HORAK: Did you ever count the nuclei in the abnormal big spores you have.

SMITH: Well, no! You see, we found all of those things after the collections had been dried, after years. That is the next thing to do, when we find these things fresh and can get enough of these big spores, we can do some staining right there on the spot, and determine the number of nuclei. We have not done it yet, but it would be interesting to do. But my guess would be that they probably would have 3,4,5 nuclei.

HORAK: You said that you consider _Lactarius_ a rather young genus in the sense of evolution.

SMITH: A very expanding genus.

HORAK: When you think of _Nothofagus_, which is considered phylogenetically a very old genus, so you would not find any _Lactarius_ in the _Nothofagus_-forests in South America. This would support your theory. But in New Zealand there are about a dozen or more different _Lactarius_ species. And the same applies, more or less, to the occurrence of Boletaceae. So far there are only about two species of _Boletus_ and related genera from South America, in the meantime there are five, and again in New Zealand there are at least twenty different species. So, I cannot judge.

SMITH: It is very difficult. My feeling is that the genus is still an aqressor relativo to "Lebensraum", but in the present state of our knowledge this cannot be verified one way or the other. It might be argued that the species lacking truly heteromerous tissue are primitive, but I for one do not believe they are. Species duch as _L._ _subserifluus_ are obviously related to members of subg. _Russularia_, not to other known agarics.

WATLING: Could I just make a comment on that particular aspect? If you grow _Lactarius_ _rufus_ in culture and put them onto roots of various conifers they sometimes form primordia or fruiting bodies. The vegetative mycelium is just like that of any other agaricoid Basidiomycete. The primordium is also made up of filaments and there are no spherocysts there at all at this stage. It would seem to be quite possible to develop Lactarius species without spherocysts.

SMITH: We have three like this, Roy. Three is enough to pose the problem for serious consideration. If there is another agaric genus that seems to have Lactarius character, it would be an easy thing to propose some relationship to this group and hence possibly a point of origin for the genus. But I doubt if the problem of origin of the genus could be solved that easily.

SINGER: I can name it!

SMITH: Bondarzewia??

SINGER: Yes.

SMITH: Oh no! That's...

SINGER: It is a polypore.

SMITH: I do not think it has anything to do with Lactarius!

SINGER: Dr. Horak will remember that there was in Patagonia a Gasteromycete which you could not put anywhere but in these "laticiferous Hydnangiaceae". It is an enormous, big thing. We called it Hybogaster. And it is so absolutely identical with Bondarzewia in every characteristic excepting its gastromycetous spores and development, and sometimes, when you look from far, you cannot distinguish it from the South American Bondarzewia guatecasensis. It has a very strong and a very visible white latex. So that, with this gasteromycetous "bridge", you cannot help but connect this Bondarzewia with the astrogastraceous series, only that in this case the hymenophore comes out polyporoid and there are no spherocyst nests. Now, since you mentioned that you have some species without spherocysts, this would somehow confirm the affinity of the Bonderzewiaceae with the Agaricales via Russulaceae.

SMITH: Well, something to keep in mind!

SINGER: There is another such thing. One of the basic characteristics of the Russulaceae is the lack of clamp connections. There is one tropical species in Lactarius where there are clamp connections, but in the lower part of the stipe - not the mycelium - all the hyphae have clamp connections. That is Lactarius quercuum.

PETERSEN: This is a very haunting conversation. Some years ago, I talked with Donk, conversing about a genus of clavarioid fungi called Amylaria, which has amyloid, reticulate ridges on the spores. Donk was commenting that if one starts with Amylaria, and looks for that kind of spore and those kinds of hyphae, then surely one must arrive at the Bondarzewiaceae. "And if you begin there, where do you end", he said, "Lactarius? Russula?" So here is the germ of the idea again.

WATLING: I would like to comment on Egon's (Horak) observations. In Australia the distribution of Nothofagus is very odd; it is on the tops of mountains, from northern Queensland down to the eastern range. Sometimes there are different species on the different mountains. When you go through a tropical rain forest and come across just a few isolated Nothofagus trees there are Lactarius, Russula and Boleti. The vegetation seems to be so specialized on these mountain tops that one could introduce a plant geographical argument and suggest these fungi probably also might be very ancient.

SMITH: Well, it seems to me, that if you assume that a genus has recently evolved, and it is spreading out, it takes advantage of any niche it can find. It is really inconsequential to whether it finds that niche in an old plant association or one that has been destroyed and has come back to the current higher plants. That is, if the genus is fairly new, and is looking for survival (in a sense, that is pretty teleological) the age of the association is relatively inconsequential. It depends on the circumstances to whether the fungus establishes itself. Did I make myself clear?

WATLING: Yes, it could be true, but phytogeographically I do not know of any example that would support this. I think we have to understand the relationships of the species on the tops of isolated mountains.

SMITH: That is a very difficult question to work out. You can find a relationship, but to predict the sequence that was involved is quite different.

SINGER: Dr. Horak, about the New Zealand question, could you name some older and now perhaps rare mycorrhizal fungi on Nothofagus that might have been replaced by Lactarius?

HORAK: No. But another question: Did you ever come across a secotiaceous _Lactarius_ in the States?

SMITH: Oh yes! _Arcangeliella_.

CLEMENCON: Let us go back to the question of what a species is. Alex, it is exactly 10 years now that I asked you a question. I do not know whether you remember it, and because it is 10 years now, I would like to repeat it: What, in your mind, is a species of higher fungi?

SMITH: Well, in the first place, it has to meet certain requirements. It has to be collectable in nature and characters resonably constant. Our whole fund of information is based on that one principle.

CLEMENCON: So we have an ecological notion in it!

SMITH: Yes, it will have a certain ecological character. I think one thing I have learned in the last ten years, Heinz, is, I think, ecology is a lot more important than I thought it was originally. We cannot predict, on the bases of number of genes or anything like that, when we are working entirely from circumstantial evidence. Our purpose is the classification of the species on the basis of the characters available, and it is to be regarded as a tentative classification. I must recognize species on their features as they occur in nature.

CLEMENCON: Do you look at nature as a giant and almost perfect physiological laboratory?

SMITH: Oh, heavens no! It is a catch-as-can brawl.

BLAICH: You said "reasonably constant". Who proves that this character is "reasonably constant"? You collect and look at your characters, and you do not prove whether it changes with cultural differences.

SMITH: Come out of the test tube! Come out of the test tube and go into the woods!

BLAICH: Yes, but there you cannot prove that I do not need a test tube. You describe two species, one growing on wood, the other growing on humus. If you could change these substrates, then, perhaps, you would see that the two species are one species, because these character depends on the substrate. There is no test tube necessary.

SMITH: There is another way approaching this, however. If there are two species in the same area, one growing on wood, the other growing on humus, you can pretty well judge that one is on humus because it could not survive on wood, and vice versa. And so that this selection in nature has worked out to separate them.

BLAICH: How did you prove that this would not survive on humus?

SMITH: Well, as I have insisted from the beginning, systematics is a tentative science, and it does not necessarily claim to offer proof. We build up our information on observations of facts. Now, when you have a hundred observations of Mycena hemisphaerica on oak logs, and you have a hundred and fifty observations of Mycena stannea on the ground, on what basis could you put any other interpretation on the ecology of these species other that one grows on wood and the other grows on the ground? If they can interchange you would find some in both habitats.

BLAICH: No, they change their characters if you change the substrate!

SMITH: There is no evidence for that.

BRESINSKY: Not necessarily.

BLAICH: Not necessarily, if it is possible, I would not say that every case....

SMITH: I grew Mycena leaina, back in the thirties, and had a lot of fun with it, and I grew it on a lot of different media, and it was a very routine and boring subject, it did not change much from one medium to the other. You starve it and it does not grow, you overfeed it and it does not fruit, just keeps on growing. These things are organisms with some built-in ability to ajust to a particular circumstance, and they do it! And the one that has become adjusted to it, is not going to suddenly jump over and grow rapidly on humus, unless there is a major mutation in its physiological genome.

BLAICH: Well, you cultured your fungi, then!

SMITH: Oh yes, I cultured fungi long before it was considered the thing to do in taxonomy.

ESSER: There was one sentence I would like you to repeat. I did not get it the way you wanted it. "Taxonomy or systematics does not need to be proved".

SMITH: No, it is all based on assumptions. It needs verification from as many approaches as possible.

ESSER: That is not science, it is religion!

BRESINSKY: It is a proposal for an order which is existing in nature. It is a proposal which is subjected to controls and studies, and we are able to change our concepts of this order.

ESSER: I want to come to the concept of the species, or to the title of this symposium, the species concept. And another question did you consider the genetic species concept?

SMITH: I do not think we know enough yet of the genetics of fungi to generlize to that extent.

ESSER: Ah, that depends.

SMITH: You are basing your whole genetics on the detailed study of two field species. There are thousands of species that have not been studied at all.

ESSER: No, by any means, there are many that have been studied. So I hope that I get your remarks after my talk. Because I think we must start with few species or even genera to be integrated in a concept. A species does not exist, it is a category!

SMITH: A species exists in nature!

ESSER: No, the individuals exist, the species does not exist! It is just a classification, a description bringing together all the individuals.

SMITH: You should go out in the woods and look at fungi! I have been teaching amateurs, and you can teach species, you can teach them to recognize species! If you have a practical concept of that type, there must be something to it! The problem is, that a species is a labile thing. You see, there is nothing in nature, in living organisms, that says that any species is static, but is constantly changing. Hundreds of years from now, <u>Cantharellus cibarius</u> will not be the same as it is to-day.

ESSER: All right. So we see single individuals, and they have certain similarities, and now we put them together into taxa, and the problem is, what are these characters, what are they in scientific terms, what are the parameters which we agree on to put them together in one drawer?

SMITH: This is why the species themselves are continuosly changing as we have more information. In the early days the species were first described and recognized on one or two characters, now we recognize them on many more characters. In the future, when we understand genetics and physiologies we may lump a lot of species back together into relatively few species. Homo sapiens is a good example, we lump all sorts of things under this one species.

BLAICH: Of course we need species for scientific communication, but we are talking now two days and no one has ellucidated this concept. There were given some characters but no description of a concept.

SMITH: Now as a geneticist or physiologist, how are you going to begin an investigation of an organism if you do not start on the assumption that you have a species of one kind or another?

ESSER: When they are able to exchange genetic material!

SMITH: You just cannot go out and study any isolate of a fungus, study it genetically and pin a label on it from what you have learned. No one else can duplicate what you have done by way of verifying your work.

ESSER: Certainly, everybody can do it.

SMITH: No, they cannot go out and get the same thing.

ESSER: OK! Look, you want to have a concept, and we are talking about a species concept. A concept means that you start with something. The DNA was first discovered, and then it was found that DNA was in all organisms. And the fine structure of chloroplasts was first described from one organism, and them people found that there are similarities. I do not want to have a dogma from the other side, too. I think there should be no dogmas that cannot be proved.

SMITH: Are you not confusing the origin and the structure of life with the expression of species in populations?

ESSER: No, I am not! look, look ...

SMITH: I know you are not!

ESSER: I think we talk in different languages. Somebody who is not familiar with taxonomy may think really that the fungi have been punished by the taxonomists by having given so and so many names. So you come to Flammulina, Collybia and others, and they are getting more and more names. For a normal biologist this is absolutely not understandable.

SMITH: The term "species" itself is a contradiction of generalities, is it not? Because a species means something fairly definite, and in talking general terms I do not see how you are going to bring the two together. A species is the result, perhaps, of distinct populations in nature as they are recognized by a competent investigator. And that is probably the closest we can come to a definition.

PETERSEN: (to Dr. Esser, about science vs. religion): Personally I agree with you, that largely I participate in my profession as a religion, because certainly I make faith statements constantly. I have faith in the integrity of this character and the constancy of that character. I have little way of proving this, and I must accept it on faith.

SINGER: Say confidence.

PETERSEN: Yes! With some predictability perhaps. So it become a secondary religion. Unless I was mistaken, I think I heard agains, two different uses of the work "species". One of them, by Alex's parameterization ("these are some of the things I think a species must have"). What he was saying, I think, describes a taxonomic unit. I can sort one from another, from another, from another, on some type of character. I think Karl was talking about biological species using different, or additional criteria for dealing with the word "species". Perhaps the answer will be that we need a new term for one of these two concepts. This is what some of the higher plant people have done.

SMITH: I do not know about the idea of separating the concepts, the biological species and the so called taxonomic species. I think it is better, if possible, to continue the emphasis on one kind of species, study its physiology and its genetics and everything about it, and then you focus upon a name at the end.

PETERSEN: Ruy and I were talking about the advantages and disadvantages of being the first and the last speaker. Roy said he thought there was a disadvantage in being the first speaker, but later he decided that, perhaps, it was a great advantage. I am beginning to think that it would be very advantageous to be the first speaker, too and I still hold to that, except that I must speak on the last day. My conclusion is that while the taxonomists use the taxonomic unit as a species concept, perhaps there is another, more insightful, deeper concept of a species in the biological sense. Unless I am mistaken, the people who are dealing with genetics, physiology, and biochemistry recognize to some extent what we taxonomists are talking about. I think we have dealt with it now for three hundred years or more, and I think we will continue to deal with it, whether or not this symposium comes to a different idea for the species. We will still collect specimens and will sort them on the bases of fruitbody characters. But I think the taxonomist must recoqnize that there are other stages besides fruitbodies; there are other things on the tree besides apples. We must describe and exploit new characters and new terminology where the character we have already fails.

OBERWINKLER: The species concept in my sense is that, probably, it is not possible to give a definition, because, a species is a thing that has to be defined by very closely related organisms, and it is a question whether you can give a theoretical concept which covers all species. I do not think that this exists at all. But you can give exact morphological data which can differntiate species in several groups. And probably there can exist species in nature, probably, in other groups, they cannot exist, because the speciation has not gone so far, that you can differentiate.

ESSER: I like to propose my species concept to-morrow. But on the other hand, what I wanted to find out, are kind of "traffic laws", which people can memorize. And that is what we need, from my, point of view as being not a taxonomist, for teaching botany. When asked by students about taxonomy, I think we should agree on something like "traffic laws".

WATLING: I am a morphologist because I think by talking in morphological terms one can transmit information from one person to another more easily. So, as a morphologist, I will give you my definition of a species. I would say that in the Agaricales two species which are intersterile are undoubtedly two different species. But as a morphologist I put together all those groups which are intersterile under one heading (microspecies) if I cannot see any morphological differences. If I, however, can see morphological differences, which I can transmit to my colleagues, I would call these macrospecies, and I feel that these latter should be given a latin name, description and a type specimen designated. How I deal with the lower microspecies, is to apply the deme structure.
The deme structure was introduced by J.S.L. Gilmour and J.W. Gregor to try to put into some order the array of biological information - physiology, chemotaxonomy, ecology etc. found in populations of flowering plants. As this information was not easily transmittable in the normal way they considered it necessary to suggest a new terminology to cover the correlation between ecology and a whole array of other factors. They created a base word, deme, and added to it va-

rious suffixes. (Nature London 144, 33 1939)

DISCUSSION FOLLOWING DR. SMITH'S PROPOSAL

SMITH: You have all read this, haven't you? So we can start talking about it. One of the first things that I wanted to re-emphasize is that my proposal is not an attempt to have species names conserved. Any species which is typified by this plan and the type accepted by the commission of the Botanical Congress is subject to the rules of the International Code of Botanical Nomenclature. It can be placed in synonymy with another species or not as the case requires. When species typified according to my plan are compared, the oldest name would have the priority without regard to any starting date. Starting dates as presently designated would apply only for species not yet typified. Any species described and to which a binomial was given would be eligible for typification. For instance, some of Secretan's species would be eligible for typification and some would not.

PETERSON: Alex, I think I find two different possibilities in what you are saying. First: Smith approaches a species in 1976, described by Fries for the first time in 1821, and there is no type for that name. Smith now takes it upon himself (under his own proposal) to describe a type.

SMITH: No! Smith does not! Smith is not going to do this work. The European mycologists in the different countries are going to do it!

WATLING: Some of this already has been done!

SMITH: Have you got a type collection?

WATLING: Lundell and Nannfeldt! The specific purpose behind Lundell and Nannfeldt' exsiccata was to collect around Fries' collecting area.

SMITH: OK, validate it! Right now they are not validated.

PETERSEN: I still have two interpretations, and I still need to clear them up. We have a name, described in 1821 A European mycologist assumes the responsibility of proposing a type.

Proposal

SMITH: He sets up a type and a commission of the congress aproves it.

PETERSEN: Within the proposals that you have presented here, I understand that. And that <u>name</u> will have as its starting date, the date that the "commission" approves the association between the name and the type specimen.

SMITH: No! No, no, no! The starting date is the publication in which it appeared.

PETERSEN: Unless I am mistaken, you might want to use the expression "representative specimen".

SMITH: Well, that would be OK. I would accept that. But "representative specimen" as designated as a type by the international congress.

PETERSEN: We would have to figure out some other sort of type. It is not a holotype, not a lectotype, not a neotype.

SMITH: We would have to have another category... well, some designation that would not conflict with any of the current definitions.

PETERSEN: Nevertheless, regardless of when the type was designated, the starting date of that taxon...

SMITH: The starting date is 1821 for Fries, in volume I of Systema Mycologicum; when a name was published in Elenchus Fungorum 1828 would be the date for that species.

PETERSEN: And when it was published by Linnaeus in 1753, as <u>Agaricus</u>?

SMITH: If the species was described in 1753 as <u>Agaricus</u> with a single species epithet, that would be the starting date for that species in matters involving priority of names.

PETERSEN: So we wind up with 1753 as the starting point?

SMITH: No! Because only the species that were typified would be excluded by the starting point. You are not going to typify all the species in two or three years. This is going to be a long-range project.

PETERSEN: Let me try another example. A species is described by Bulliard, the name is validated by Fries (in 1821), but there is no type. Until a type is given, the starting date of that taxonomic name is 1821. But once a type is selected, the starting date is Bulliard's publication.

SMITH: Right. And then we can determine synonymies and priorities on really very accurate bases.

PETERSEN: What are we allowed to use as types; Bulliard's plates?

SMITH: No, no, no! You must get a collection of specimens fitting Bulliard's concept, describe it fully and publish the description - then it would, when accepted by the commission, be the "type" for that species. French mycologists collecting in or near Bulliard's collecting areas would decide on the collection - whether or not it really represented Buillard's species. The Commission would then confirm or reject it.

WATLING: This often disagrees with Fries in 1821. There would not only be different starting points, but also (perhaps) different species!

SMITH: You do not need one starting date for all of these retypified old species; it may well be different for most. Each must be recognized by the new protologue. If you have a good case for a particular species Linneaus described, 1753 might well be the date for it.

BAS: I think the greatest problem that you are creating now is that you get two types of names. We have already names with holotypes, and lectotypes; for these names you maintain the starting point. For the type of names you want to introduce, there would be no starting point. I think you should stick to the original proposal; there have to be neotypes for names without a holotype or lectotype, and you may work out a way for these neotypes to be fixed, but you should not also introduce this change in the starting point, because I think it would bring a lot of trouble.

SMITH: Well, I have been against changing the starting point, too, but I have been thinking about finding a way to increase our accuracy. If you go back to the early descriptions you run into all sorts of problems, and we cannot really solve them until someone says "this is what it really means". And mycologists of the respective countries should be the ones to nail down that designation for a species described from their country.

BAS: Imagine that this is happening. There is a commitee that has to judge whether this designation is correct. Now, here comes a proposal for <u>Rhodophyllus</u>. Next day there comes a proposal for <u>Psathyrella</u>. These people who have to judge whether the designations are correct or not, can they do that without knowing anything about <u>Rhodophyllus</u> or <u>Psathyrella</u>?

SMITH: All they have to have before them is the original description to see whether the indicated characters in that description agree with the positive characters describing the newly proposed type.

SINGER: I fully sympathize with what Alex wants to accomplish, but it sounds to me like the dream of paradise of a monographer. I had some of these dreams myself which, unfortunately, I am afraid, will never be accepted by an international congress and will also not become reality because - who will be able and willing to sacrifice enough time and effort at the expense of his real scientific work in order to help find the correct and practically useful neotypes which you wish to be acceptable and binding as some sort of lectotypes; and for how many groups of fungi will you find such specialists capable of doing this work in a country like Sweden?

SMITH: Actually, it would not be too irksome, because it is not going to take a person very long to read the original latin description and to check whether there is anything wrong with the specimen.

SINGER: It may not take long, but it might take long to find the person to do that work.

WATLING: I see a very great advantage. If we do select types based on new material, then with the capabilities of cultural techniques and liquid nitrogen, we will be able to have a growing type, with which we will then be really able to fix the species.

SMITH: I think there is some merit to this idea, because, what happens to mycologists in North America and some other continents, is that we encounter a lot of taxa very close to European taxa as judged by their descriptions, which is all we have for comparison. The result is that in many cases we confuse an American taxon by using a European name for it. The specimens preserved under a single name at random in herbaria almost invariably represent a number of different taxa; or, we find that current concepts of a species in Europe do not conform to the original author's concept of his species. There is the case of <u>Armillaria luteovirens</u>, for instance.
Under these circumstances it is better if investigators on other continents simply describe as new the species they find. This will make the synonymy cumbersome eventually, but one major difficulty would be avoided.

HORAK: I think we would achieve some kind of progress if we would agree that a description of a new species should never be accepted without drawings.

SMITH: I would support your idea of laying down more astringent requirements for the publication of new species.

PETERSEN: If <u>I</u> were to happen to find something valuable in Europe <u>I</u> would have no trouble at all in <u>my</u> conscience describing a neotype for a European name.

SMITH: The minute you do that, somebody else finds something else that <u>he</u> thinks is the right thing, and the neotype does not hold.

PETERSEN: I really want to continue my first point. If I want to find a neotype for <u>Clavaria flava</u> Schaeffer, I cannot go to Sweden. I must go to the vicinity of Regensburg, Germany. When I want to find a neotype for something that Bresadola described I don't go to Paris. I must go where Bresadola collected. I must collect as those authors tell me, e.g., under beech. And I must find something which agrees with their description. But if no European has done that, and if no European has neotypified something which I consider important because it is in my group (whatever "my group" means), then I am going to do it, because the only way I can know the North American species is to neotypify the European species names.

SMITH: Well, that is the way you are proceeding. Your specimens would meet my requirements. But I want the European investigators to do this work, not a foreigner like myself. Neither of us can spend the next 25 years in Europe tracing down all these old species. The neotypes I object to are specimens pulled out of an herbarium and so designated but on which there is no factual data as to the actual characters of that particular specimen.

PETERSEN: My second point is that I think, with rather small differences, we already have a mechanism for what you are proposing. We could call the neotype a "representative specimen" or a "fixotype". The jury that judges right or wrong is the mycological public, which ought to do the job anyway. If the "fixotype" violates the rules as arbitrary, or it does not agree with the original description, under the International Code it ought to be thrown out by someone who has a specimen which is not arbitrary and does fulfill the original description. These are clearly set down in the Guide for the Determination of Types and Article 7. Types can be thrown out if they violate the following principle: It ought to agree with the original description, and it ought to be well documented. I think you are saying exactly the same thing. The International Code says the same thing, but it says it under neotypification, and it uses a jury of the mycological community.

SMITH: The present system is not working. I submit this proposal to you by way of getting around problems that are coming up relative to starting dates, and those with some current trends in neotype typification. My proposal if it is to accomplish its purpose, must be carried out by European investigators.

A PROPOS DE LA DELIMITATION DES ESPECES DANS LES
HYGROPHORUS FRIES DU SOUS-GENRE HYGROCYBE FRIES.
DEUX CARACTERISTIQUES PEU OU NON UTILISEES.

R . K ü h n e r

Dans le présent mémoire ont été mentionnés divers résultats de recherches d'ordre anatomique et cytologique effectuées par l'auteur
sur une quarantaine d'Hygrocybe.
On en retrouvera facilement les noms dans la liste suivante où elles
sont classées dans l'ordre alphabétique; dans cette liste, nous indiquons, comme il se doit, les noms de leurs auteurs; ils n'ont pas été
mentionnés dans le cours du texte, afin de ne pas alourdir celui ci.
Pour d'autres détails sur des espèces de cette liste et sur quelques
autres espèces d'Hygrocybe, nous renvoyons le Lecteur à notre mémoire
intitulé: "Vers un système phylogénétique des Camarophyllus et Hygrocybe".

Hygrophorus acutoconicus (Clements) Sm.
H. acutopuniceus (Haller)
 -(=Hygrocybe acuta Møller)
H. aurantiolutescens Orton
H. aurantiosplendens (Haller)
H. brevisporus (Møller) Orton
H. calyptraeformis Berk. et Br.
H. cantharellus (Schw.) Fr.
H. ceraceus (Wulf. ex Fr.) Fr.
H. chlorophanus (Fr.) Fr.
H. citrinovirens (J. Lange) Konr. et Maubl.
H. coccineocrenatus Orton
H. coccineus (Schaeff. ex Fr.) Fr.
H. conicus (Scop. ex Fr.) Fr.
H. flavescens (Kauffm.) Sm. et Hesl.
H. fornicatus Fr.
H. glutinipes (J. Lange) Orton
H. helobius Arnolds
H. ingratus Jens.-Moell.
H. insipidus (J. Lange) Lundell
H. intermedius Pass.

H. laetus (Pers. ex Fr.) Fr.
H. langei (Kühner) Pearson
H. marchii Bres.
H. miniatus (Fr.) Fr.
H. miniatus, f. longipes Sm. et Hesl.
H. mollis (Berk. et Br.) Kauffm.
H. ovinus (Bull. ex Fr.) Fr.
H. parvulus Peck
H. psittacinus (Schaeff. ex Fr.) Fr.
H. puniceus (Fr.) Fr.
H. quietus Kühner
H. reai R. Maire
H. salicis-herbaceae Kühner, s.p. nov.
H. splendidissimus Orton
H. squamulosus Ell. et Ev.
H. strangulatus Orton
H. subceraceus Murrill
H. subglobisporus Orton
H. subminutulus (Murr.) Murr.
H. substrangulatus Orton
H. turundus (Fr. ex Fr.) Fr.
H. turundus, var. sphagnophilus (Peck) Hesl. et Sm.
H. vitellinus Fr.

Les résultats mentionnés ci après ont été obtenus, tant sur nos récoltes personnelles, que sur des exsiccata aimablement communiqués par plusieurs Mycologues.

L'expérience nous ayant convaincu de la difficulté de détermination de nombreuses espèces d'Hygrocybe, nous nous sommes presque uniquement adressé à des spécialistes ayant publié des descriptions d'Hygrocybe. Parmi ceux à qui l'on doit des travaux d'ensemble sur ce groupe, citons L.R. Hesler et A.H. Smith qui, dans leur belle monographie des Hygrophores d'Amérique du nord, donnent de précieux renseignements sur la structure du revêtement piléique. Pour l'Europe, citons d'abord E. Arnolds, qui a traité tout récemment des espèces des Pays Bas dans un mémoire descriptif inédit, qui tranche sur la plupart des travaux modernes concernant les Hygrocybe par l'utilisation de tous les caractères anatomiques, y compris ceux de la trame des lames, un peu négligés dans la monographie de Hesler et Smith, et par un intéressant essai de classification naturelle, qui n'a malheureusement pas été tenté par les auteurs américains; pour l'Europe, citons encore P.D. Orton, à qui l'on doit une très précieuse clé analytique des espèces britanniques et plusieurs excellentes descriptions d'espèces nouvelles ou critiques, et qui nous a aimablement communiqué, outre ses exciccata, de nombreuses notes manuscrites accompagnées d'aquarelles, qui nous ont été fort utiles pour mieux comprendre le sens dans lequel il a pris de nombreuses espèces figurant ou non dans sa clé.

Que ces personnalités veuillent bien trouver ici, pour leur aide inestimable, l'expression de notre reconnaissance.

Nos vifs remerciements vont également à R. Haller, M. Moser et D. Reid, qui nous ont transmis des exsiccata de quelques espèces nouvelles ou critiques, dont ils ont publié des descriptions, enfin à nos amis M. Josserand et H. Romagnesi, pour quelques collections et descriptions inédites, en particulier du groupe confus de H. miniatus.

Nous intéressant particulièrement, dans le cadre d'une révision des Agaricales de la zone alpine, aux espèces qui croissent au dessus de la zone des forêts ou au nord de celle ci, nous avons amplement utilisé le beau travail descriptif de F.H. Møller sur les champignons des Faeroës et étudié les exsiccata de plusieurs de ses récoltes, dans la mesure où elles correspondaient à des espèces non rencontrées par nous ou critiques; les exsiccata de cet auteur étant très fortement moisis, nous n'avons malheureusement pu obtenir de leur étude tous les résultats escomptés.

Nous remercions enfin les Conservateurs des collections de divers établissements, qui nous ont aimablement communiqué un abondant matériel d'étude: Rijksherbarium de Leiden (Pays-Bas); Royal Botanic Gardens d'Edinburgh et de Kew (Grande Bretagne); Universitetets Botanisk Museum de Copenhague (Danemark); Herbarium of the University d'Ann Arbor (U.S.A.).

Les deux caractéristiques auxquelles il est fait allusion dans le titre de ce mémoire sont d'ordres fort différents, puisque l'une relève de l'anatomie, l'autre de la cytologie.

I. UNE CARACTERISTIQUE ANATOMIQUE TROP RAREMENT UTILISEE

Il s'agit de la structure de la trame les lames. On sait que, dans le genre <u>Hygrocybe</u>, V. Fayod a défini une section <u>Conicae</u>, en utilisant notamment l'allure de la trame des lames: "très régulière à hyphés très allongés", l'opposant à deux autres sections, caractérisées par la brièveté des articles de cette trame.
Caractérisée en partie par la structure de la trame des lames, la section <u>Conicae</u> a été admise par Kühner et Romagnesi ("Flore analytique", 1953), et, tout récemment (1974) par E. Arnolds, qui a fait très justement remarquer que les très longs articles de la trame des lames des <u>Conicae</u> ne sont cylindriques qu'en apparence, leurs extrémités étant plus ou moins longuement atténuées.
Dans ses excellentes descriptions d'espèces nouvelles ou critiques d'Hygrocybe, P.D. Orton n'a malheureusement donné aucune indication sur la structure de la trame des lames, et il n'a pu, de ce fait, tenir compte de celle-ci dans l'établissement de la précieuse clé qu'il a donnée des <u>Hygrocybe</u> britanniques. Dans cette clé, les espèces vivement colorées sont réparties en deux groupes, suivant que leurs lames sont décurrentes à largement adnées ou au contraire étroitement adnées à libres.
La limite entre ces deux groupes manque évidemment de netteté puisque l'auteur de la clé a rappelé H. <u>puniceus</u>, H. <u>quietus</u> et H. <u>strangulatus</u> dans chacun des deux groupes.
L'étude de la trame des lames permet de distinguer deux ensembles beaucoup plus nettement tranchés. En effet, seul le second groupe de Orton renferme des espèces dont les articles fondamentaux de la trame sont très longs et atténués aux extrémités (les <u>Conicae</u> de Fayod). Chez H. <u>puniceus</u>, H. <u>quietus</u> et H. <u>strangulatus</u> ces articles sont courts.
Certaines espèces, placées par Orton uniquement dans son second groupe, ressemblent aux trois précédentes par la brièveté des articles de la trame des lames: il s'agit de H. <u>psittacinus</u>, qu'il est difficile d'éloigner de H. <u>sciophanus</u>, placé par Orton dans son

Hygrocybe 161

premier groupe, de H. acutopuniceus Haller, que Mø̈ller, qui l'appelait H. acutus, avait d'abord considéré comme variété de H. puniceus, de H. splendidissimus que Orton, son auteur, comparait à H. coccineus du premier groupe et à H. puniceus, enfin de H. aurantiosplendens, qui a été comparé à H. quietus, et qu'il ne faut pas confondre avec H. aurantiolutescens, espèce publiée par Orton après la parution de sa clé et qui s'éloigne de toutes celles nommées ci-dessus par la grande longueur et la forme des articles de la trame des lames, qui en font indiscutablement un Conicae.

Pas plus que P.D. Orton, L.R. Hesler et A.H. Smith, dans leur belle monographie des Hygrophores d'Amérique du nord, n'ont utilisé la structure de la trame des lames pour délimiter leurs subdivisions d'Hygrocybe, et leur série Conici est uniquement définie par la forme du chapeau: conique ou a mamelon conique.

H. flavescens n'est évidemment pas placé par eux dans les Conici, pour la raison qu'ils décrivent son chapeau comme largement convexe, puis étalé ou légèrement déprimé au disque; ils le placent dans une autre série, à côté de H. marchii, dont ils le disent très voisin, et qu'ils ne différencient dans leur clé que par la couleur du chapeau: celui-ci, orangé à jaune chez H. flavescens, serait initialement rouge-sang ou écarlate chez H. marchii, mais pâlissant vite à rouge-orange et finalement à jaune. En réalité ces deux espèces sont profondément différentes l'une de l'autre par la structure de la trame des lames, même au sens de A.H. Smith, de qui nous avons étudié des exsiccata: chez H. marchii les hyphes de la trame des lames se présentent comme des chaines de saucisses, aux articles de longueur faible ou moyenne, 38 - 105 x 10 - 20 microns, alors que, chez H. flavescens, la structure est typique des Conicae par ses articles si longs, par exemple 750 microns, que sur des dissociations, on en repère difficilement les deux extrémités, qui sont souvent longuement atténuées. La grande longueur des articles des lames avait déjà été remarquée, sur des récoltes européennes étiquetées flavescens par J. Favre comme par E. Arnolds, ce dernier auteur ayant très judicieusement placé flavescens dans la section Conicae (qu'il appelle Hygrocybe).

II. UNE CARACTERISTIQUE CYTOLOGIQUE NEGLIGEE: LE NOMBRE DE NOYAUX DE LA SPORE

Travaillant sur des coupes de matériel inclus à la paraffine après fixation, coupes colorées à l'hématoxyline ferrique, nous avons montré, il y a plus de trente ans, que les spores des Hygrophoracées sont uni- ou binucléées suivant les espèces.

A. Origine du stock nucléaire de la spore. Variations intracarpiques et variations d'une espèce à une autre.
   1. La sporogénèse dans les conditions normales de vie.

Utilisant les techniques évoquées à l'instant, un chercheur de notre laboratoire, N. Durand, a étudié le comportement nucléaire de la baside à la spore dans une vingtaine d'espèces d'Hygrophoracées, représentées par des carpophores tétrasporiques, non parthénogénétiques. Nous résumons ci-dessous les résultats de ses recherches restées inédites.

Jusqu'à la poussée des stérigmates le comportement nucléaire est le même pour toutes les espèces; c'est le comportement classique de toutes les Agaricales: le noyau de fusion de la baside se divise en deux et chacun des noyaux fils se divise à son tour une fois; c'est seulement quand la baside renferme 4 noyaux issus de ces deux séries de divisions que poussent les stérigmates.

Pour 8 espèces d'Hygrophoracées N. Durand a dénombré les noyaux des spores sur frottis de sporées, colorés au Giemsa; ses dénombrements montrent que le nombre de noyaux des spores est rarement constant dans toutes les spores d'une même sporée, mais que la prédominance d'un des types de spore (uni- ou binucléé) est généralement écrasante puisqu'elle correspond à au moins 94 % des spores; c'est ainsi que 99 % des spores étaient uninucléées chez H. calyptraeformis et qu'il n'a été vu que des spores binucléées chez H. ovinus.

Ces dénombrements sur sporées à l'aide de la coloration de Giemsa ont uniquement porté sur des Hygrocybe. Dans les mêmes conditions nous avons dénombré les noyaux sur 2 espèces d'Hygrocybe non étudiées par

N. Durand et nous arrivés à des résultats comparables concernant
l'importance de la prédominance d'un des types de spores. L'étude de
deux récoltes de H. marchii peut donner une première idée de l'amplitude des variations susceptibles de se manifester d'un lot à un autre
de la même espèce: les proportions de spores à 2 et à 1 noyau étaient
respectivement de 94 % et 6 % pour l'une des récoltes et de 88 % et
12 % pour l'autre.

A en juger par les images observées sur des coupes de matériel inclus
dans la paraffine, les spores uninucléées sont prédominantes chez les
2 Camarophyllus et chez les 3 Limacium étudiés par nous, de même que
chez les 8 espèces de Limacium (différentes des nôtres) étudiées par
N. Durand.

L'étude de telles coupes, réalisées sur les Hygrocybe à spores binucléées prédominantes, révèle toujours la présence d'une division
nucléaire dans certaines spores, division longitudinale ou fortement
oblique; l'état binucléé noté sur les spores chues peut donc résulter
du fait que le noyau cédé par la baside à la spore se divise dans
celle-ci. Mais il n'est pas exclus que les divisions nucléaires de
cette troisième série se produisent parfois dans la baside car, dans
3 espèces, ont été vues, sur des coupes de matériel inclus dans la
paraffine, 8, 7 ou 6 noyaux dans certaines basides stérigmatées.

Chez les Hygrophoracées ou prédominent les spores à un seul noyau, il
y a aussi 3 séries de divisions nucléaires aboutissant à la formation
de 8 noyaux à partir du noyau de fusion de la baside, mais les 4
spores ne se partagent que 4 d'entre eux; les 4 autres sont destinés
à la baside et y dégénéreront lorsqu'elle aura projeté ses spores; ce
sont les noyaux dits résiduels.

Ces noyaux résiduels ont été repérés par N. Durand chez 10 espèces
d'Hygrophoracées à spores uninucléées étudiées par elle. Sur les
coupes de matériel inclus dans la paraffine de 11 Hygrophoracées à
spores uninucléées prédominantes ont été repérées quelques spores
montrent un noyau en division; il est probable qu'un seul des noyaux
fils reste dans la spore, l'autre retournant dans la baside, ce qui
est facilité par la disposition longitudinale ou très oblique du fuseau de division. Sur les coupes de 2 espèces de Limacium ou ont été
repérées des spores renfermant un noyau en division, ont été également repérées quelques basides stérigmatées renfermant plusieurs
noyaux en division, 4 par exemple; dans une autre espèce de Limacium
a été repérée une baside renfermant 8 noyaux résultant de telles divisions.

Il est donc certain que, tant chez les Hygrophoracées ou prédominent les spores uninucléées que chez celles ou prédominent les spores binucléées, les divisions de la troisième série, celles qui se produisent après la poussée des stérigmates, peuvent avoir lieu, soit dans la spore, soit dans le corps de la baside, soit encore dans les stérigmates.

L'expérience nous a appris que lorsqu'on ne possède pas de sporée, mais que l'on dispose seulement d'exsiccata de carpophores, on obtient, pour les spores, le même nombre dominant de noyaux qu'avec une sporée, en colorant le frottis par la méthode de Giemsa. Le frottis peut être obtenu en agitant, dans une goutte d'eau déposée sur une lame porteobjet, un fragment de feuillet de manière à en disperser les spores mûres; on peut aussi, à l'aide d'une aiguille lancéolée, broyer le fragment de feuillet dans la goutte d'eau et colorer le broyat. L'avantage de cette dernière méthode est que des touffes de basides, portées par des articles du sous-hyménium, accompagnent les spores dans le frottis, ce qui permet de reconnaître, au contenu des basidioles et des articles du sous-hyménium, si un carpophore non bouclé est ou non parthénogénétique; dans le cas des Hygrophoracées un tel contrôle est généralement superflu, car tous les représentants de cette famille se sont révélés bouclés, au moins au pied des basides et aux cloisons du soushyménium, autrement dit l'absence totale de boucles suffit ici à reconnaître la parthénogénèse, au moins chez les espèces européennes que nous avons pu examiner. Un désavantage certain de la méthode consistant à travailler sur des broyats de feuillets est qu'à côté des spores mûres, comparables à celles des sporées, se trouvent des spores immatures, détachées des stérigmates par la manipulation, et dont le nombre de noyaux peut être différent de celui des spores mûres; malgré cet inconvénient, nous avons habituellement utilisé la dernière méthode, l'expérience nous ayant appris qu'elle permet presque toujours de déterminer avec certitude le nombre dominant de noyaux, tel qu'on le déterminerait sur une sporée; le nombre caractéristique est en général représenté dans au moins 70 % des spores présentes dans la préparation issue du broyage d'un fragment de feuillet.

Dans les carpophores non parthénogénétiques le nombre de noyaux des spores dépend dans une certaine mesure du nombre des stérigmates; c'est ainsi que, dans les carpophores tétrasporiques de _H. coccineus_, dominent fortement les spores uninucléées (94 % dans une sporée étudiée par N. Durand), alors que, dans les carpophores de la même espèce où les basides tri-, bi- ou monosporiques sont particulièrement nombreuses, le pourcentage de spores binucléées, déterminé sur un broyat de feuillet, peut être du même ordre de grandeur que celui des spores à un seul noyau.

Par contre, bien que les carpophores parthénogénétiques des Hygrophoracées se distinguent des carpophores normaux de la même espèce par la présence de 2 stérigmates au lieu de 4, le nombre dominant de noyaux est le même que dans les carpophores normaux, ce qui s'explique par le fait que, dans les carpophores parthénogénétiques, le noyau de la baside ne donne que 4 noyaux au lieu de 8 dans les carpophores normaux; il n'y a qu'une division avant la poussée des stérigmates.

2. Incidence du mode de préparation des exsiccata sur les caractères des spores.

Plus la dessiccation des carpophores est rapide, plus faible est naturellement le risque que des spores continuent à se former ou à se développer au cours de la déshydratation, c'est à dire dans des conditions plus ou moins anormales.

Une dessiccation rapide exigeant souvent un appareillage relativement encombrant, dont on ne dispose pas toujours en voyage, nous avons très souvent laissé les carpophores se dessécher à l'air libre à la température ordinaire; même en les réduisant en petits fragments, leur déshydratation peut demander plusieurs jours et il n'est pas exclus que, durant cette période, se produisent des sporogénèses anormales, éventuellement liées à une réduction du nombre des stérigmates formés par les basides, ou des modifications des spores tombées, sur les feuillets encore relativement riches en eau.

Dans cet ordre d'idées, signalons que, dans quelques préparations de feuillets ou de sommet de stipe de certaines espèces d'Hygrocybe de la section type, nous avons trouvé, à côté d'une majorité de spores à contour régulièrement arrondi et absolument lisse, comme il est de règle chez les Hygrophoracées, des spores plus ou moins rares dont la coupe optique était franchement épineuse. Nous avons noté le fait particulièrement dans la série Squamulosi, mais aussi dans la série Conici.

Sur nombre de spores les épines ne peuvent échapper, car elles sont à la fois très élevées et très déliées; elles se présentent sous forme de baguettes de 1 - 1,7 x 0,2 - 0,3 microns, subcylindriques ou cylindro-coniques; sur certains lots on observe des termes de passage entre les spores lisses et les spores épineuses, par exemple des spores noduleuses ou des spores simplement bosselées ou subanguleuses.

Les spores ainsi ornées ont, sans les ornements, des dimensions très proches de celles des spores lisses qu'elles accompagnent, de sorte que nous pensions tout d'abord à des spores ayant subi une évolution anormale après leur chute, jusqu'au moment où, chez H. intermedius, nous avons repéré des spores ornées en place sur les stérigmates. Le nombre des ornements par spore est fort variable; il nous est arrivé, dans un carpophore du groupe miniatus, de découvrir une spore n'ayant que deux nodules, et même une spore ne présentant qu'une seule épine - un véritable aiguillon d'ailleurs - localisée au sommet. Quoiqu'il en soit, les spores ornées sont si rares par rapport aux spores lisses qu'elles ne sauraient correspondre à des stades obligatoires du développement aboutissant à celles-ci.

Soulignons que les spores épineuses dont il vient d'être question ont été découvertes sur exsiccata et que le temps nous a manqué pour les rechercher sur matériel vivant (carpophores ou sporées). Il est possible qu'elles aient développé leurs ornements au cours de la lente déshydratation à la température ordinaire du matériel ayant servi à la préparation des exsiccata.

Nous avons simplement voulu attirer l'attention sur une particularité qui semble avoir échappé jusqu'ici et dont la signification ne pourra être bien comprise que grâce à des observations sur matériel vivant.

B. Preuves de la constance intraspécifique du stock nucléaire de la spore

1. Récoltes personnelles.

La prédominance des spores à 1 noyau a été vérifiée dans toutes nos récoltes des espèces suivantes:

       H. miniatus (11 lots)
       H. coccineus (8 lots)
       H. quietus (7 lots)
       H. psittacinus (5 lots)
       H. salicis herbaceae (5 lots)
       H. puniceus (4 lots)
       H. reai (2 lots)
       H. fornicatus (2 lots)

La prédominance des spores à 2 noyaux a été vérifiée dans toutes
nos récoltes des espèces suivantes:
- H. conicus (22 lots)
- H. langei (11 lots)
- H. marchii (8 lots)
- H. laetus (7 lots)
- H. cf. coccineocrenatus (5 lots)
- H. flavescens (4 lots)
- H. cantharellus (3 lots)
- H. ingratus (2 lots)

2. Récoltes de divers Mycologues.

a. Récoltes désignées sous le même nom spécifique que certaines des nôtres.

Les spores à 1 noyau prédominent dans toutes les récoltes précisées ci-dessous et désignées sous les noms spécifiques qui suivent:

H. ceraceus (6 lots: Arnolds, 7 et 22-10-74; Hadley; Henderson, 1436; Orton 1190; Thiers, 3348; Watling, 7026; il faut ajouter à ces lots un lot personnel).

H. glutinipes (2 lots: Orton, 483; Watling, 857C; il faut ajouter à ces lots un lot personnel).

Les spores à 2 noyaux prédominent dans toutes les récoltes précisées ci-dessous et désignées sous les noms spécifiques qui suivent:

H. cantharellus (5 lots: Moser, 69.62; Orton, 471 et 472; Romagnesi, 72-129; Smith, 21370 et 63554; concordance avec nos 3 lots personnels);

H. coccineocrenatus (2 lots: Arnolds; Orton 783, type; concordance avec nos 5 lots personnels);

H. flavescens (2 lots: Bigelow, 3296; Smith, 63717; concordance avec nos 4 lots personnels);

H. chlorophanus (2 lots: Smith, 9464 et 10147; concordance avec une récolte personnelle).

b. Champignons désignés sous des noms spécifiques non encore utilisés par nous.

Prédominance des spores à 1 noyau.

H. insipidus (4 lots: Arnolds; Orton, 821, 890 et 1549);

H. miniatus f. longipes (3 lots: Smith, 42560, 42561,

42562, type et paratypes)
H. parvulus (3 lots: Moser, 66.225 et 66.303; Smith, 9663)
H. subminutulus (3 lots: Orton, 811, 1221 et 4-11-50)
H. vitellinus (3 lots: Arnolds, 204 - environ 30 % de spo-
res binucléées -; Orton, 2366; Henderson, 6853)
H. splendidissimus (2 lots: Orton 1215; Reid)
H. subceraceus (Smith, 20118)

Prédominance des spores à 2 noyaux.
H. brevisporus (3 lots: Møller, 14-8-38, Strömö, type;
Orton, 810 et 1548)
H. citrinovirens (3 lots: Orton, 3595; Watling, 831C et
868C)
H. subglobisporus (3 lots: Orton, 1558, 3295, et 3296)
H. squamulosus (2 lots: Smith, 1045 et août 1942)

C. Connait on des espèces chez lesquelles le nombre dominant de
noyaux de la spore peut varier
d'une récolte à une autre?
1. Les faits.
Il est certain que le même nom spécifique a parfois été attribué
à des champignons différant les uns des autres par le nombre dominant
de noyaux des spores. Le tableau I en donne des exemples.

2. Discussion et conclusion.
Dans nombre de cas au moins le fait qu'un même nom spécifique ait
été attribué à des champignons différant les uns des autres par le
nombre dominant de noyaux des spores, loin d'illustrer une variabi-
lité intraspécifique de ce caractère, traduit simplement les confu-
sions faites entre espèces voisines.

Pour suivre notre discussion, il est important de se souvenir du re-
groupement des espèces du tableau précédent, tel qu'il a été proposé
par Arnolds, sur des bases en grande partie anatomiques. H. acutoconicus est un représentant typique de la section Conicae Fayod (que
Arnolds appelle Hygrocybe), à la fois par la forme pointue de son
chapeau et par la structure de la trame de ses lames. Les autres
espèces du tableau qui précède peuvent être regroupées d'après la
structure de leur revêtement piléique. Arnolds range H. miniatus,

H. mollis et H. turundus dans une subsection Squamulosi, caractérisée par le fait que les articles terminaux de la surface du chapeau sont larges, x 4.5-15 microns selon lui. Il range H. strangulatus et H. substrangulatus dans sa subsection Coccinei, qui s'oppose à la précédente par la présence d'hyphes grêles à la surface du chapeau. Orton, auteur de ces deux espèces, a bien noté, dans les descriptions qu'il en a données que la "cuticule" du chapeau y est formée de grosses hyphes x 6-16 microns, surmontées d'hyphes grêles (peu nombreuses chez H. substrangulatus). Arnolds n'a pu situer H. marchii, ne l'ayant jamais récolté.

H . a c u t o c o n i c u s
Hesler et Smith synonymisent H. langei Kühner (= H. constans Lange, non Murr.) à acutoconicus.

Dans les 11 lots européens de H. langei examinés par nous, qu'ils soient normaux ou parthénogénétiques, prédominent toujours de façon fort nette les spores à 2 noyaux, alors que parmi les trois lots nord américains de acutoconicus transmis par Smith, deux différaient de H. langei par la nette prédominance des spores uninucléées; ils semblent avoir le chapeau sensiblement plus pointu que la plupart de nos langei, et représentent sans doute des vrais acutoconicus; le troisième lot, aux spores binucléées prédominantes, correspond peut être à notre H. langei.

Nous n'avons jamais remarqué, chez notre H. langei, que la base du stipe noircisse avec l'âge comme ce serait habituellement le cas chez H. acutoconicus; chez notre H. langei, aucun noircissement n'a été observé sur le frais, même sur carpophores un peu passés; c'est seulement en herbier que le stipe grisonne, brunit ou bistre, tantôt à peine, tantôt nettement, au moins inférieurement.

H . t u r u n d u s
Il est clair que les lots de turundus à spores uninucléées sont spécifiquement différents de ceux à spores binucléées.

Ceci ressort déjà du fait que, chez les types à spores uninucléées, les spores sont plus petites que chez les types à spores binucléées.

Chez le turundus à spores uninucléées de Moser, les spores mesurent

7-9 (10.5) x -4-5 microns selon son auteur, 6.5- 7.5 x 4-5 microns
d'après nos mesures sur l'exsiccatum qu'il nous a transmis; chez le
turundus de Arnolds, dont les spores sont également uninucléées, cel-
ces-ci mesurent 7-8 x 4.5-5 microns selon Arnolds, 7-8 x 4-5.5 microns d'après l'étude de l'exsiccatum communiqué.

Chez le turundus de Hesler et Smith, dont nous avons reconnu que les
spores sont binucléées, celles-ci mesurent 9-14 x 5-8 microns d'après
ces auteurs, 8.7-10.5 à 9-11.5 (12.5) x 5-5.7 à 5.5-6.5 microns d'après nos mesures sur les trois lots transmis. Dans la var. sphagnophilus, que Hesler et Smith rattachent à H. turundus, les spores sont au
moins aussi grandes, (9)-10-14 x 5-9 (10) microns d'après ces auteurs, 8.5-10.7 x 6.5-7.2 microns d'après notre étude de l'unique lot
communiqué, et également binucléées.

Il est difficile de comprendre pourquoi Hesler et Smith ont écrit que
H. coccineocrenatus est très proche de cette variété, sinon identique, puisque Orton, auteur de coccineocrenatus, a dit que son chapeau
présente des écailles brunâtres, qui deviennent d'un brun plus foncé
ou noircissent, alors que var. sphagnophilus se distingue précisément
de var. turundus par le fait que les squamules piléiques s'assombris-
sent seulement quelque peu, ce qui avait fait que Peck, auteur de
var. sphagnophilus, l'avait rattachée à H. miniatus. Orton puis
Arnolds ont néanmoins admis la synonymie suggérée par Hesler et
Smith. Il nous parait encore prématuré d'affirmer que le sphagnophilus de Hesler et Smith est le même que le sphagnophilus de Peck.
Quoiqu'il en soit il est certain que turundus et sa var. sphagnophilus au sens de Hesler et Smith sont proches de H. coccineocrenatus,
dont les spores mesurent, d'après Orton, son auteur, 10-13 (14) x 6-8
microns, et sont binucléées d'après nos observations sur le type de
l'espèce.

Les articles de la trame des lames nous ont paru remarquablement
courts et larges, elliptiques, 40-90 x 16-32 microns, chez coccineocrenatus. L'un des turundus de Smith (73337) nous a donné des articles également de fort calibre et souvent courts ou très courts, 32-
150 x 13-25 microns, mais dans le turundus du Mt Rainier (29586),
dont Hesler et Smith écrivent que la teinte des écailles piléiques
n'est certainement pas due à une décoloration, ces articles, qui sont
également larges, sont souvent bien plus longs, 72-225 x 10-28 mic-

rons et, dans le turundus var. turundus 50893 les articles fondamentaux des lames sont plus étroits que dans les duex précédents, en même temps que peu longs, 38-80 x 8-12 microns; ceux de var. sphagnophilus sont comparables, 52-100 x 7-10 microns, et donc nettement plus étroits que ceux de coccineocrenatus. Nous n'indiquons ces dimensions qu'à titre documentaire, car il est possible qu'elles varient passablement avec l'âge, et, pour la longueur, avec la largeur atteinte finalement par les lames.

Les spores de H. coccineocrenatus sont ellipsoïdes selon Orton; celles des turundus de Smith (var. sphagnophilus comprise) sont, d'après nos observations, régulièrement elliptiques à subcylindriques. Il n'en est pas de même pour les spores du turundus de Moser, dont la forme particulière permet, selon cet auteur, de distinguer cette espèce des espèces voisines: il précise que les spores sont légèrement étranglées vers le milieu de leur longueur et, sur la vue de face, élargies dans la région basale, ce qui donne une forme en poire. D'après nos observations une telle forme, en vue de face, est fréquente chez H. reai, champignon d'identification facile, au reste fort éloigné de turundus par sa viscosité et par l'absence d'écailles.

Dans le turundus de Arnolds, à spores petites et uninucléées, comme le turundus de Moser, de très nombreuses spores sont elliptiques; seules certaines sont ovoïdes-subtrigones de face. Nous pensons néanmoins que le turundus de Arnolds et celui de Moser appartiennent à la même espèce car, sur les exsiccata de ces deux auteurs, nous avons vu se glisser, entre les gros articles terminaux du revêtement piléique, 25-50 x 5-12 microns, de petits poils grêles, x 2-3,5 microns, régulièrement cylindriques, venus de la profondeur, qui nous ont paru manquer dans le turundus de Hesler et Smith, comme dans le coccineocrenatus de Orton, deux espèces ou nous n'avons pas vu d'hyphes grêles au dessus des articles terminaux du revêtement piléique, dont la largeur était respectivement de 7-15 microns et 8-11 microns. Comme Moser, Arnolds pense que son turundus est le turundus de Orton. Moser a fait remarquer que si Orton donne des spores un peu plus grandes, il précise que certaines peuvent apparaître étranglées, ce que l'on voit sur la fig. 286 qu'il en donne, et qui a été prise sur le lot 473. Ayant étudié ce même lot 473 du turundus de Orton, nous sommes en mesure d'affirmer que ce champignon est bien différent des turundus de Moser et de Arnolds par ses spores binucléées, plus grandes, 8-11 (12) x 4.5-5.5 microns selon Orton, 9-11.5 x 5-5.7 microns d'après nos mesures; dans ce lot 473 les spores sont subcylindriques, non sensiblement étranglées, et nous n'en avons vu aucune qui soit élargie transversalement à la base. Par les dimensions des spores, 9-12 x 5.5-6 microns, le turundus de Møller, que Arnolds considère comme étant le même que le sien, paraît beaucoup plus proche du turundus de Orton; en raison du très mauvais état de conservation du matériel de Møller, nous n'avons pu malheureusement déterminer le nombre de noyaux de ses spores.

Par ses spores relativement grandes et binucléées, le turundus de Orton est évidemment très proche du turundus de Hesler et Smith; il serait surtout remarquable par la marge du chapeau lacérée-dentée et par l'absence de franches couleurs rouges sur le stipe et sur le chapeau, qui seraient jaune-orange, mais selon Hesler et Smith, le turundus nord-américain aurait le chapeau écarlate à orange et jaune.

H . s t r a n g u l a t u s
Telle que définie par Orton, son auteur, cette espèce se distingue microscopiquement de H. miniatus par le fait que des spores étranglées n'y sont pas rares et qu'au dessus des grosses hyphes du revêtement piléique se trouve une mince couche d'hyphes grêles x 1-2 microns.
Nous avons effectivement repéré ce type de revêtement dans les 2 lots déterminés par Orton que nous avons pu examiner. Lot 3619: articles terminaux 20-40 x 7-10 microns; des poils grêles x 2,5-2,7 microns, flexueux, venus de la profondeur. Lot 4162: articles terminaux 18-63 x 7-13 microns; on y voit çà et là des poils grêles x 2-2,5 microns, longuement filiformes, venus de la profondeur.
Ces deux lots diffèrent brutalement l'un de l'autre par le nombre dominant de noyaux des spores et la différence ne peut être attribuée à une différence dans le nombre des stérigmates, le lot binucléé 3619 étant purement tétrasporique comme le lot uninucléé 4162.
Nous pensons que ce dernier lot est le plus typique, non seulement parceque les spores y sont uninucléées comme dans les exsiccata déterminés strangulatus par d'autres auteurs, mais aussi et surtout parce que ce lot est le plus remarquable par la forme de certaines de ses spores: si de nombreuses spores sont subcylindriques, à peine étirées, certaines sont franchement étranglées de face comme de

profil; de profil elles mesurent 6,5-9 x 4-5 microns; il n'est pas rare de rencontrer des spores nettement élargies transversalement à la base x 4,5-6,5 microns; ce sont celles qui se montrent le plus nettement étranglées de face, l'étranglement n'ayant que 3,5-5 microns de large; ce sont des spores du type "en poire", ou, si l'on préfère, de type "reai".

Dans le lot 3619, aux spores binucléées, la plupart des spores sont de forme banalement cylindrique ou cylindrique-elliptique, ou tout au plus très faiblement étirées-étranglées de profil comme de face; s'il n'est pas rare d'en rencontrer qui soient un peu plus larges en bas (ovoïdes), il est tout à fait exceptionnel d'en trouver qui soient "en poire". Leurs dimensions, 6,5-7,7 x (3) -3,7-5 microns, sont très proches de celles du lot 4162, uninucléé.

La partie supérieure du stipe n'a pu être étudiée que chez 4162; on y trouve pas mal de poils grêles x 2 microns, mais pas très longs, x 30-100 microns; dressés perpendiculairement à la surface, flexueux, non septés ou à une cloison.

Dans les deux lots, les hyphes de la trame des lames sont en chaînes de saucisses, mais les articles, de 70-150 x 8-13 microns chez 4162; sont franchement courts, 20-40 x 6-10 microns chez 3619.

Le H. strangulatus uninucléé de Arnolds est remarquable par l'abondance d'hyphes filiformes-grêles, x 2-3 microns à la surface du chapeau, dont la structure n'évoque pas celle d'un Squamulosi; dans la partie supérieure du stipe on trouve seulement quelques poils x 3 microns, peu longs, 45 microns ou moins, qui sont de simples extrémités d'hyphes plus ou moins redressées. Spores subcylindriques; si celles qui sont encore en place sur les stérigmates sont, souvent étirées tibiiformes, les autres ne sont généralement pas étranglées et nous n'en avons pas vu qui soient élargies transversalement à la base; dimensions: 7-8 x 3,7-4,5 microns selon nous, 7-10 (11,5) x 3,5-5,5 microns selon Arnolds.

A l'opposé se trouve le H. strangulatus de l'herbier Watling (récolte Kay Edwards), qui figure dans l'herbier d'Edinburgh sur la même feuille que les exsiccata de Orton. Bien que son revêtement piléique soit parfaitement conservé, nous n'avons pu y reconnaître qu'une structure typique de Squamulosi: articles terminaux 27-72 x 7-14 microns; aucune hyphe grêle n'a pu être détectée au dessus d'eux. Spores 8-8,7 (-10,5) x 4,5-5,7 microns, cylindriques-ellitiques, parfois un peu étranglées de profil comme de face; quelques unes montrent un élargissement transversal de la base (tendance modérée vers le type en poire).

Dans sa description de H. strangulatus, Reid ne fait pas allusion à la présence d'hyphes grêles au dessus des hyphes x 10 microns, qui constituent les écailles du chapeau. Il reconnaît l'espèce à ses spores 6,5-9x 4,2-6 microns, souvent étranglées de profil comme de face, les vues de face variant d'ovales à subtriangulaires.

H. s u b s t r a n g u l a t u s

Le lot 1560, étiqueté substrangulatus par Orton, diffère des deux autres, non seulement par le fait que les spores uninucléées y sont prédominantes, mais aussi par le fait que ses spores ne mesurent que 7,2-9 x 4,7-5,5 microns, dimensions bien plus faibles que celles attribuées à substrangulatus par l'auteur de l'espèce, soit 9-12 x 5-7 microns.

Par les dimensions de ses spores ce lot 1560 est donc bien plus proche de H. strangulatus; le nombre important de spores binucléées, 30 %, ne s'explique pas par la présence de basides à nombre de stérigmates réduit, car nous n'avons vu que des basides tétrasporiques. Les spores sont subcylindriques, quelquefois un peu étranglées, mais souvent non; la vue de face de certaines spores est du type en poire. Le revêtement piléique montre, au dessus d'articles x 7-8 microns, de rares poils grêles x 1,5-2 microns, cylindriques, de 20-55 microns de long, venus de la profondeur; il s'agit donc bien d'un champignon de la stirpe Strangulatus. Les articles de la trame des lames sont peu longs 25-60 x 6-8 microns et se rapprochent donc par leurs dimensions de ceux du strangulatus où prédominent les spores binucléées (lot 3619).

Dans les trois autres lots examinés sous l'étiquette substrangulatus, les dimensions des spores étaient par contre très proches de celles indiquées par Orton pour cette espèce, 8,5-11 à 9,5-11,5 x 5-6,2 à 5-6,7 microns; dans ces trois lots dominent nettement les spores à 2 noyaux. Les spores sont subcylindriques à cylindriques-elliptiques; non étranglées dans le lot 2365, elles le sont rarement un peu dans le lot 931c; elles sont étranglées ou non dans le lot 477, avec la base élargie ou non, mais même dans ce lot, les spores en poire sont rares.

Le revêtement piléique des lots 477 et 2365 de Orton est de type Squamulosi, avec des articles terminaux de 68-90 x 9-12 microns; seul le lot 2365 nous a montré, au dessus des gros articles quelques hyphes grêles x 2,5-4 microns, flexueuses, septées et ramifiées, à

Hygrocybe 175

bouts libres piliformes.
Le revêtement de la partie supérieure du stipe ne montre pas de longues hyphes-poils; il ne présente que de rares poils courts, 40 à 60 microns par exemple.
Les articles de la trame des lames mesurent 35-80 x 7-16 microns (lot 2365 de Orton) ou 45-102 x 7-12 microns (lot 477 de Orton); ils sont beaucoup plus longs, (45) -90-200 (-240) x 8-18 microns dans le lot 931c de Watling.

H . m i n i a t u s
Sous cette étiquette ont été conservés des exsiccata différant brutalement les uns des autres par le nombre dominant des noyaux sporiques.
Chez les types renfermant deux noyaux par spore, celle-ci est de forme banale; elle n'est ni étranglée, ni en poire lorsque vue de face.
Cet ensemble n'est certainement pas homogène et, par exemple le miniatus 22.46 de Josserand est remarquable par la structure de la trame de ses lames, qui tend vers celle des Conicae, avec ses très longs articles, 155-325 x 9-13 microns, fusiformes, atténués ou même très atténués au bout; cette structure le distingue bien du miniatus 9.30 du même Mycologue, dont la trame ne comporte que des articles bien plus courts, (22) -36-95 (190) x 10-15 microns, cylindracés ou un peu en tibia. Le miniatus 834c de Watling est certainement conspécifique du miniatus 22.46 de Josserand par ses spores binucléées et par les articles de la trames dont plusieurs sont remarquablement allongés, (65)-140-425 microns x 8-14 microns, cylindracés sur une grande longueur, mais à extrémités longuement et progressivement atténuées; par les longs articles atténués de leurs lames ces deux derniers lots font penser à H. helobius Arnolds, dont les articles seraient encore plus longs et souvent bien plus larges, (67) -98-670 (-710) x (8,5) 9,5-34 (37) microns. Spores 8,5-9,7 x 5-6 microns (Josserand 22.46); 7,5-9,5 x 5-5,5 microns (Watling 2834c).
Du miniatus 9.30 de Josserand se rapproche le miniatus 63108 de Smith, également à spores binucléées et également à articles des lames courts, 40-110 x 8-16 microns, mais le miniatus 9.30 de Josserand a les spores assez grandes, 8,5-10,5 x 5,5-6,5 (-7) microns, alors que le miniatus 63108 de Smith a des spores petites, 6,2-8-5 x 3,7-5 microns, dimensions qui cadrent avec celles que donnent Hesler et Smith pour leur miniatus, soit 6-8 (-10) x 4-5 (-6) microns.
Le miniatus 17546 de Smith ressemble au précédent par les articles

des lames de longueur seulement moyenne, 40-145 x 10-20 microns, mais
il en diffère par ses spores uninucléées, de 7,7-10 x 4,7-5,5 microns
de profil, parfois étranglées, nombre d'entre elles subtrigones de
face, ou elles se montrent élargies à la base parfois subtronquée
x 5,2-6 microns (type en poire); c'est sans doute à des spores de ce
type que Hesler et Smith font allusion dans leur description de miniatus, qui précise: "a few apparently abnormal spores shaped more or
less like corn kernels are sometimes found".

Entre ces deux lots de Smith nous n'avons pas pu repérer d'autres
différences que celles indiquées ci-dessus. Articles terminaux du revêtement piléique 40-90 x 6-12 microns, sans hyphes grêles par dessus,
ni petits poils chez 63108; 39-53 x 6-10 microns chez 17546. Dans la
partie supérieure du stipe des poils dressés ou obliques, très nombreux, grêles, x 2-3 microns, mais très courts, 30-90 microns chez
63108; des poils peu longs, 45-60 microns, qui ne sont que des bouts
d'hyphes tendant à se redresser plus ou moins chez 17546.

Du champignon sphagnicole appelé H. miniatus f. longipes Smith et
Hesler, nous avons étudié 3 lots, dont le type; dans les trois lots
les spores sont uninucléées et leur profil subcylindrique ou cylindrique-elliptique, mesure 6,5-8 à 7,7 -9 x 4-5 à 5-5,7 microns; sur les
trois lots de très nombreuses spores ont une base fortement élargie
sur la vue de face x 5-5,5 à 5,5-6,7 microns (type en poire). Ce
champignon est au moins très proche du miniatus 17546. Articles fondamentaux des lames 37-117 x 5-11 à 8-13 microns. Revêtement piléique
à articles terminaux de 20-45 à 42-80 x 7-9 à 8-13 microns, sans hyphes grêles par dessus. En haut du stipe poils de calibres variés
2,5-7 microns, toujours cours 17-70 microns, qui se réduisent parfois
à des bouts d'hyphes un peu redressés.

6 lots européens rapportes à miniatus par divers auteurs montrent
aussi une nette prédominance de spores uninucléées; dans tous les articles de la trame des lames sont peu allongés.

Le miniatus var. miniatus 948, transmis par Arnolds se rapproche des
miniatus également uninucléés de Smith par ses spores 7,5-8,5 x 4,7-
5,2 (-6,5) microns, dont pas mal sont élargies transversalement à la
base (type en poire); d'autres sont seulement plus ou moins étranglées à mi hauteur. Chez le miniatus var. mollis 774 du même Mycologue, spores 7,5-9 x 5-5,5 microns; beaucoup sont de forme banale,
mais quelques unes vues de face se montrent subétranglées ou obscuré-
ment trigones, à base subtronquée. Ces deux champignons diffèrent
toutefois des miniatus américains par le fait que les poils filiformes x 2,5-3-4 microns de la partie supérieure du stipe sont beau-

coup plus longs, 220-400 microns pour var. miniatus, 300-320 microns
pour var. mollis.
Dans aucun des deux lots étiquetés miniatus par Romagnesi nous n'a-
vons trouvé de spores étranglées ou du type en poire. Spores 7,5-9,5
x 5-5,7 (-6,5) microns, elliptiques.

Les deux lots étiquetés miniatus par Orton ont en commun la prédominance des spores a un noyau, mais ils diffèrent sensiblement l'un de
l'autre par la forme des spores: Orton 816 se rapproche du miniatus
de Arnolds par le pourcentage important de spores de type en poire;
profil 7,5-10 x 4-5,2 microns, cylindracé, phaséoliforme ou légèrement étranglé; face de 4,2-5,7 microns de large à la base. Orton 3608
se rapproche du miniatus de Romagnesi par ses spores 8-9,5 x 5,2-6,7
microns, elliptiques-pruniformes, ne montrant aucune tendance à la
strangulation ou à un élargissement transversal de la base. La différence d'allure entre les spores des deux collections de Orton est
telle que dès le premier coup d'oeil nous avons pensé qu'elles correspondent à deux espèces différentes.

H . m o l l i s

Il est clair que le H. mollis aux spores uninucléées (celui de
Arnolds) est spécifiquement différent du H. mollis de Moser, dont les
spores sont binucléées.
Selon Arnolds le vrai mollis ne serait qu'une variété de H. miniatus,
différant de ce dernier, au chapeau rouge-écarlate ou rouge-orange
dans la jeunesse, par son chapeau d'emblée jaune-orangé ou jaune,
alors que le mollis de Moser serait H. helobius Arnolds, espèce bien
distincte de H. miniatus, notamment par les grandes dimensions de
plusieurs articles fondamentaux de la trame des lames.
Selon Arnolds, les articles fondamentaux des lames mesurent (37)
43-124 (-192) x (6,5) -8,5-19 (-22) microns chez var. miniatus, (28)
52-146 (-201) x 6-17 (21) microns chez var. mollis, (67) -98-670
710 x(8,5) -9,5-34 (-37) microns chez H. helobius.
En fait l'exsiccatum de mollis transmis par Moser nous a frappé par
les dimensions considérables des articles des lames, (90) -160-325 x
14-34 microns, donnant une structure tendant vers celle des Conicae.
Les exsiccata aux spores bunucléées étiquetés avec doute mollis par
Orton nous ont également frappé par le fort calibre et la très grande
longueur de nombreux articles des lames, 70-545 x (8) -13-35 microns,
les plus grands fusiformes ou atténués au bout; il s'agit sans doute

du _mollis_ de Moser, d'est à dire de H. helobius.

H . m a r c h i i

Ce nom recouvre au moins deux espèces, qui ont été souvent confondues, parfois par un même auteur, ce qui suppose qu'elles doivent se ressembler beaucoup.

Dans le champignon dont Reid donne une image coloriée, toutes les spores sont binucléées d'après nos observations. Les coupes transversales du chapeau montrent un revêtement d'articles de calibres variés, de 2,5 à 11 microns, non gélifié, les sections rondes de ses hyphes étant rigoureusement en contact, ne ménageant que des méats liés à leur forme; par places se décollent de ce revêtement des hyphes x 2,5-5 microns, septées. Les 4 récoltes à spores binucléées, rapportées par Orton à H. marchii correspondent vraisemblablement au H. marchii tel que conçu par Reid. En scalp, le revêtement piléique est formé d'articles cylindriques allongés, x 3-4,5 à 3,5-6 microns, les articles terminaux mesurant 37-50 à 65-100 x 4-8 microns.

Celle des récoltes de Orton qui diffère des autres par la prédominance des spores uninucléées, en diffère également par la gracilité des hyphes du revêtement piléique, qui sont filiformes, x 1,7-4 microns, les très longs articles terminaux n'étant pas plus larges. La coupe transversale du chapeau montre que,sur une épaisseur de 20-30 microns, le revêtement piléique est sensiblement gélifié, bien que les couches mitoyennes soient peu gonflées, 1,5-2 microns. De même calibre, x 6-12 microns, dans le type à spores binucléées que dans celui à spores uninucléées, les articles fondamentaux de la trame des lames sont peut-être un peu plus courts, 25-80 microns dans les récoltes binucléées que dans les lots uninucléés, 50-145 microns.

Par ses spores uninucléées le H. marchii de Haller se rapproche du lot aberrant de Orton; il s'en rapproche aussi par le fait que la coupe transversale du chapeau montre un revêtement de 25-30 microns d'épaisseur, formé, dans ses parties externes par des hyphes filiformes-grêles, x 2-2,2 microns, fortement gélifiées (couche mitoyenne 3 microns et davantage); dans la partie profonde, les hyphes sont moins grêles, x 5 microns et beaucoup moins écartées par gélification, le revêtement passant ainsi progressivement à la chair; articles de la trame des lames 38-105 x 5-13 microns.

Deux récoltes à spores uninucléées transmises sous l'étiquette
H. marchii par Smith, rappellent le marchii de Haller par le revêtement piléique formé d'hyphes filiformes-grêles, x 1,5-3 microns, à bouts libres non renflés ou à peine très progressivement clavulés, formant un revêtement peu épais, 20 microns, mais franchement gélifié, voire même très gélifié. Les articles de la trame des lames semblent toutefois plus larges, 38-105 x 10-20 microns (Hesler et Smith donnent x 7-14 microns). Il nous paraît hautement improbable que ce marchii nord américain soit le marchii de Bresadola, dont il diffère par le chapeau rouge-sang ou écarlate dans la jeunesse et les lames adnées à adnexes.

Comme Orton, Møller a du confondre, sous l'étiquette marchii, deux espèces différant par le nombre dominant de noyaux de la spore, et peut-être par les dimensions de celles-ci, qui semblent plus élevées dans le type uninucléé.

Conclusions

En résumé, dans l'ensemble Hygrocybe, le dénombrement des noyaux des spores est un moyen d'approche de la notion d'espèce que l'on aurait tort de continuer à négliger.

C'est tout particulièrement dans l'ensemble des Squamulosi et espèces affines que la connaissance du nombre dominant de noyaux des spores semble devoir rendre le maximum de services; conjointement à l'étude anatomique précise, en particulier des revêtements, non seulement de celui du chapeau, mais de celui, encore trop négligé, du stipe, elle aidera certainement à débrouiller l'ensemble encore confus des champignons étiquetés miniatus, strangulatus et substrangulatus.

D. Lamoure a montré que, dans un autre ensemble d'Agaricales, Omphalina Q., on trouve des espèces affines différant par le nombre dominant de noyaux des spores (1ou 2); c'est le cas notamment dans le groupe des petites espèces lichénisantes dont la plus commune est O. ericetorum (Fr. ex Fr.) M. Lange = O. umbellifera (L. ex Fr.) Q., sensu J.E. Lange, et dans celui des petites espèces gris-brun noirâtres par incrustation pigmentaire de la paroi des hyphes. Dans ce dernier groupe, dont les espèces sont morphologiquement très semblables, la connaissance du nombre de noyaux des spores fournit un élément de diagnostic non négligeable; par exemple O. velutipes Orton à spores uninucléées, ne peut être confondu avec O. obscurata Kühner, à spores binucléées, la valeur spécifique de ces deux taxons étant attéstée par leur interstérilité.

Ici, comme dans les Hygrocybe, c'est le nombre dominant de noyaux qui compte. D'après les dénombrements effectués sur sporées par D. Lamoure, le pourcentage des spores dominantes oscille, entre 95 et 60 % suivant les lots, éventuellement pour une même espèce suivant les sporées.

Bibliographie

Arnolds, E. 1974: Taxonomie en Floristiek van Hygrophorus, subgenus
Hygrotrama, Cuphophyllus en Hygrocybe in Nederland.
Rijksherbarium Leiden. Thèse, ined.
Durand, N. 1960: Recherches caryologiques sur l'hyménium des Agaricales: Bolétacées, Cantharellacées, Hygrophoracées, Pleurotacées. Dipl. Et. Sup. Lyon, ined. 98 p.
Haller, R. 1954: Beitrag zur Kenntnis der schweizerischen Hygrophoraceae. Schweiz. Zeitschr. Pilzk. 32: 81-91.
Haller, R. 1955: Contribution à l'étude du genre Hygrocybe. Schweiz. Zeitschr. Pilzk. 33: 169-172.
Haller, R. 1956: Beitrag zur Kenntnis der schweizerischen Hygrophoraceae. Schweiz. Zeitschr. Pilzk. 34: 177-180.
Hesler, L.R. & Smith, A.H. 1963: North american species of Hygrophorus. 416 p. Univ. of Tennessee Press, Knoxville.
Kühner, R. 1977: Vers un système phylogénétique des Camarophyllus et Hygrocybe. Revue de Mycologie 41: 73-90.
Lamoure, D. 1975: Agaricales de la zone alpine. Genre Omphalina, 2. partie. Trav. scient. Parc nat. Vanoise 6: 153-166.
Møller, F.H. 1945: Fungi of the Faeröes, I: 1-295. Copenhague.
Moser, M. 1967: Beitrag zur Kenntnis verschiodener Ilygrophoren. Zeitschr. Pilzk. 33: 1-23.
Orton, P.D. 1960: in Dennis R.W.G., Orton P.D. et Hora F.B. New check list of british Agarics and Boleti. III, Notes on genera and species in the list. Trans. Brit.Myc.Soc. 43: 159-439.
Reid, D.A. 1968: Coloured Icones of rare and interesting Fungi, part 3. Suppl. to Nova Hedwigia 15: 5-8 et pl. 18.

Tableau I : Nombre dominant de noyaux par spore

	1 noyau	2 noyaux
H. acutoconicus	Smith, 62992. Thiers, 979.	Thiers, 7-13-50.
H. marchii	Haller, 166. Møller, Nolsö, 27-8-39. Orton, 1551. Smith, 56213, 63713.	Møller, Nolsö, 25-8-38. -- , Strömö, 30-7-38. -- , Syderö, 28-8-38. Orton, 813, 814, 1550, 2199 Reid,
H. miniatus	Orton, 816 et 3608. Romagnesi, 68.64 et 72-33. Smith, 17546.	Josserand, IX-30 et XXII-46 Smith, 63108. Watling, 2834C.
H. mollis	Arnolds, 7742.	Moser, 66/64. Orton, 1969 et 2632. Watling, 836C.
H. strangulatus	Arnolds, 544. Orton, 4162. Reid, Watling, 5271.	Orton, 3619.
H. substrangulatus	Orton, 1560. (33 % à 2 noyaux)	Orton, 477 et 2365. Watling, 931C.
H. turundus	Arnolds (pas mal de spores à 2 noyaux) Moser, 48/569.	Orton, 473 Smith, 29586 (Mt Rainier) -- , 50893 et 73337.

DISCUSSION FOLLOWING DR. KUEHNER'S PAPER

BAS : Une question de technique. Avez-vous essayé de colorer les noyaux pour le microscope à fluorescence ?

KUEHNER : Je n'ai pas essayé parce que la coloration selon Giemsa donne bien plus facilement de très beaux résultats.
Pour faire face aux difficultés que j'ai rencontrées, j'ai imaginé certains détails techniques qui ne figurent pas dans tous les manuels de technique microscopique.
Concernant les spores et les basides des Hygrophoracées, et plus généralement de tous les champignons dont les spores et les basides sont particulièrement riches en lipides; il est indispensable de dégraisser les frottis à l'éther après l'hydrolyse chlorhydrique qui prédède, comme l'on sait, la coloration de Giemsa; la forte réfringence des gouttes d'huile qui subsisteraient nuirait beaucoup à la visibilité des noyaux.
Il est préférable de rechercher les noyaux sur du matériel (exsiccata ou sporées) recueilli récemment. Plus le matériel est ancien, plus les noyaux sont difficilement mis en évidence, parce que le cytoplasme se colore trop fortement, mais il est parfaitement possible d'obtenir de bons résultats avec des sporées ou exsiccata âgés de 10, 20 ou même 30 ans, à condition d'utiliser la technique simple que voici : la préparation surcolorée est inondée d'alcool absolu; l'alcool ne doit agir que pendant un temps très court, souvent de l'ordre d'une fraction de seconde, sous peine de décolorer entièrement le matériel; le mieux est d'effectuer ce traitement par l'alcool près du jet d'eau d'un robinet qui permettra d'éliminer rapidement et complètement l'alcool.
J'observe toujours les frottis colorés entre lame et lamelle dans une goutte d'ammoniaque; ce réactif différencie un peu la préparation et surtout il permet, lorsqu'il s'agit d'un frottis de fragments d'hyménium, d'en réaliser plus facilement la dissociation par percussion, ce qui peut être fort utile si l'on désire dénombrer les noyaux dans les jeunes basides et dans les articles du sous-hyménium, afin de reconnaitre si le carpophore étudié était ou non parthénogénétique; on sait que, dans ce dernier cas, ces articles ne renferment qu'un seul noyau.

PETERSEN: Did you see any nuclear behavior other than chiastic? Were all nuclear divisions chiastic?

KUEHNER: Oh yes indeed, always chiastic in Hygrophoraceae.

PETERSEN: In a more general sense: What importance would you put on chiastic versus stichic behavior in all of the higher fungi?

KUEHNER: Je pense que la distinction a un intérêt systématique certain puisque toutes les Agaricales étudiées à ce point de vue se sont montrées chiastobasidiées. A condition de ne pas prendre chiastobasidié dans un sens trop strictement géométrique, c'est à dire comme correspondant à une orientation exactement transversale des fuseaux de première et de seconde division, avec parallélisme rigoureux des fuseaux de seconde division. Il n'est pas rare que les fuseaux de seconde division ne soient pas exactement parallèles; la caractéristique essentielle est qu'ils soient situés au même niveau, chez les Agaricales au sommet de la baside. Dans le type stichobasidié, ces deux fuseaux sont généralement situés l'un au dessus de l'autre. Comme l'a montré M. Boidin, l'orientation du fuseau de première division a moins d'importance pour la distinction des types sticho et chiastobasidiés; il peut arriver que, pour des raisons sans doute variées, l'orientation du fuseau de première division soit plus ou moins éloignée de la transversale chez certains types chiastobasidiés; chez les Agaricales, j'ai signalé un cas où il est même fortement oblique parce que sa longueur est bien plus grande que la largeur de la baside. Il est possible que, dans quelques cas, la distinction sticho- chiastobasidié ait conduit à trop éloigner l'un de l'autre des groupes qui présentent une certaine affinité. J'ai toujours pensé que Omphalia chrysophylla présente de l'affinité avec certaines Chanterelles, et le fait que M. Fiasson ait reconnu que cette espèce est riche en caroténoïdes comme nombre de Chanterelles vient à l'appui de cette opinion car les caroténoides ne sont que rarement responsables d'une coloration chez les Agaricales. Or j'ai démontré que O. chrysophylla est purement chiastobasidié, alors que l'on sait depuis longtemps que les Chanterelles sont stichobasidiées.

SINGER: Il y a beaucoup de temps, vous avez publié une étude sur les nombres de noyaux dans les articles des hyphes dans les pieds des agarics. Avez-vous rencontré quelque relation entre la quantité de noyaux dans les hyphes et dans les spores?

KUEHNER: Aucune relation.

HORAK: Professeur Corner a décrit une espèce de la flore tropicale où on trouve, sur la même lamelle, des basides grandes, petites et normales. Avez-vous trouvé ?

KUEHNER: Non. Je pense que Monsieur Smith va vous en parler parce qu'il a en Amérique du Nord un Hygrophore qui a deux sortes de basides.

SINGER: Il serait toutefois très intéressant d'investiguer ces espèces, principalement tropicales.

KUEHNER: Certainement.

BAS: Avez-vous aussi étudié la structure des hyphes dans le stipe dans le groupe d'<u>Hygrophorus conicus</u>?

KUEHNER: Dans le stipe il y a des hyphes très longues. Dans le stipe on a des hyphes qui ont des articles qui arrivent à atteindre 2,3 ou 4 mm de long. Il sont plus longs encore dans le stipe que dans les lames.

BAS: Mais c'est seulement cet élément long et mince.

KUEHNER: Bien sûr. Comme dans tous les Agaricales il y a deux sortes d'hyphes. Il y a des hyphes fondamentales et ce que Fayod avait appelé les hyphes connectives.

BAS: Mais, vous connaissez les publications de Corner sur les types sarcodimitiques ...

KUEHNER: Je comprends très bien. Mais malheureusement Corner est arrivé à cette notion après avoir étudié les vrais dimitiques, ceux où l'on distingue des hyphes génératrices et des hyphes squelettiques. Chez les Agaricales les hyphes connectives sont caractérisées, non seulement par leur calibre en général plus faible, mais encore, comme je l'ai montré dès 1926 par la présence, à leur intérieur, de petits cristaux ou cristalloïdes (vraisemblablement protéiques), bien visibles déjà sur le vivant, mais absolument évidents, car noirs, sur des préparations colorées à l'hématoxyline ferrique. Un moment j'ai caressé l'espoir que la présence ou l'absence de ces hyphes à cristalloïdes permettrait de délimiter les Agaricales par rapport à d'autres groupes, tels que les Polypores. Aussi ai-je demandé à Madame David, notre spécialiste lyonnaise des Polypores, de rechercher si de telles hyphes à cristalloïdes existent ou non chez ces Champignons; elle en a trouvé, même dans les vrais dimitiques; ceux ci ont donc trois sortes d'hyphes et non deux.

BAS: Mais il est encore possible que cette structure ait une importance taxonomique au niveau des Agaricales. Dans le stipe d'<u>Oudemansiella</u> elle est très évidente.

KUEHNER: Dès 1926 j'ai indiqué que dans le stipe des <u>Oudemansiella</u> la différence entre les hyphes fondamentales et les hyphes connectives à cristalloïdes est particulièrement frappante, à cause d'une énorme différence de calibre, et j'ai même donné un dessin. Si une telle structure est la caractéristique du Sarcodimitisme, alors toutes les Agaricales sont sarcodimitiques. Le dimitisme ou le trimitisme des Polypores résultent de différencientions surajoutées au sarcodimitisme.

BAS: Il y a beaucoup d'Agaricales où l'on trouve dans le stipe seulement des hyphes du même diamètre.

KUEHNER: C'est vrai; les Chanterelles sont dans ce cas; c'est ce que Fayod appelait une structure homomorphe; mais cette structure n'est homomorphe que par le calibre des hyphes; parmi ces hyphes de même calibre, certaines ont des cristalloïdes et d'autres pas.

CLEMENCON: Monsieur Kühner, connaissez-vous des essais de culture des Hygrocybes?

KUEHNER: Oui, mais les essais faits dans mon Laboratoire ont toujours abouti à des échecs alors que l'on réussit à cultiver plusieurs <u>Hygrophorus</u> sensu stricto (c.à.d. Limacium).

WATLING: (I'd just like to make a comment). In the 1960's when Orton was looking at *Hygrocabe* strangulatus and marchii in his notes he separated four taxa: strangulatus, substrangulatus, marchii and submarchii. Two were published, strangulatus and substrangulatus. H. submarchii had only slight morphological differences, so he said "no good" but it would appear that in the two nuclei - one nucleus division indicated by Prof. Kühner the collections with two nuclei in the Edinburgh material apparently refer to the manuscript submarchii.

Herbette Symposium on Species Concept in Hymenomycetes 1976

GENETIC BASIS FOR SPECIATION IN HIGHER BASIDIOMYCETES WITH SPECIAL
REFERENCE TO THE GENUS POLYPORUS

Karl   Esser   &   Peter   Hoffmann

I. Introduction
Each branch of science requires well defined concepts. In botany this
was first attempted by Theophrastus. However, his expressions "genos"
and "eidos" cannot be equated with the present concept of genus and
species. They are relative because a "genos" can be divided into several "eidos" and vice versa an "eidos" can also comprise several
"genos".

It is certainly to the merit of Linné (1737) to have created the
first well defined species concept: "Species tot sunt quod diversas
forma ab initione produxit infinitum ens". Whereas his contemporaries
like Tournefort did not distinguish between species, varieties and
cultural forms, Linné states: "Varietas est planta mutata a causa
accidentali: climate, solo, colore, ventis". This means that Linné in
fact did not think that a species is as invariable as the first quotation suggests. Furthermore, he unintentionally anticipates the control of the expression of the genotype of an individual by environmental factors.

In the period following Linné the species concept ramained fixed
mainly on morphological and anatomical criteria as expressed in a definition presented by Wettstein in 1901: "Man wird daher als Art die
Gesamtheit der Individuen bezeichnen, welche in allen dem Beobachter
wesentlich erscheinenden Merkmalen untereinander und mit ihren Nach-
kommen übereinstimmen". This concept is rather unprecise because it
depends largely on the subjective establishment of appropriate criteria.

The species concept became clearer in the 1920s with the introduction
of cytogenetic criteria. Chromosome numbers and crossing barriers became invaluable tools for the definition of a species. At the beginning of the 1960s biochemical criteria and numerical taxonomy were
introduced for species classification but the so-called g e n e -
t i c   c o n c e p t   o f   a   s p e c i e s   still seems to be

the most precise one because it is based on more objective criteria which are, moreover, reproducible. A s p e c i e s c o n - s i s t s o f : " o n e o r m o r e p o p u l a t i o n s , t h e i n d i v i d u a l s o f w h i c h c a n i n t e r - b r e e d , b u t w h i c h i n n a t u r e c a n n o t e x c h a n g e g e n e s w i t h m e m b e r s b e l o n - g i n g t o o t h e r s p e c i e s " (King 1974). This definition implies that crossability is sufficient to define a species. However, in recent years some examples have become known where organisms, quite clearly belonging to the same species are no longer intercrossable because of genetic barriers (for lit. see Burnett 1975). For this reason some taxonomists have questioned the genetic species concept.

It is one of the aims of this paper to show that a diminution of the genetic species concept is not necessary, provided one understands the genetic parameters which control sexual propagation. In other words: <u>one needs to understand whether the failure of a cross between two morphologically similar isolates or races is due to gross genetic differences, pointing to a species difference, or whether it is only the result of adverse environmental conditions or physiological crossing barriers</u> caused by relatively trivial genetic diversities.

Crosses to test these alternatives can be performed fairly easily with higher plants. In the Fungi however, especially the higher basidiomycetes, crosses must be performed in the laboratory and the criterion "environmental conditions" (e.g. media, light, temperature, humidity) is much more critical than in higher plants.

In addition, the action of the breeding systems, which may create the "physiological barriers", is not as easy to determine as in higher plants where analysis of pollen tube growth and fertility test of seeds is sufficient. We shall use the genus <u>Polyporus</u> as an example to test the validity and practicability of the genetic species concept. But to begin with some general remarks are necessary.

II. Genetic parameters controlling fruit body production

In order to handle fungi in the laboratory one needs not only to know the culture conditions but also to be aware of the genetic parameters which control fruit body production. Since we have pointed out these facts in detail elsewhere (Esser 1974), we need to mention here only those genetic parameters which are specifically required to handle the genus Polyporus. For details see Esser and Stahl 1973, 1974; Stahl and Esser 1976. We can summarize these results by reference to fig. 1:

1) There are two different routes leading to the production of fruit bodies; the first called dikaryotic fruiting is part of the sexual cycle and leads to the "normal" fruit bodies whose morphology serves as a taxonomic criterion (see fig. 3). The dikaryotic fruit bodies, which predominate in nature, contain basidia with four spores. During the development of the basidia karyogamy and meiosis occurs. The second route to fructification is called monokaryotic fruiting and takes place in the vegetative phase. Monokaryotic fruit bodies look rather abnormal (fig. 2) and rarely occur in nature. They contain only two-spored basidia and spore formation is preceded by a mitotic division only.

Monokaryotic fruiting is not restricted to the genus Polyporus, it seems to be universal in the Polyporaceae and also in the Agaricaceae (see list in Stahl and Esser 1976).

2) Both modes of fruit body production are initiated by a single gene called "fruiting initiation" (fi). Monokaryotic fruit bodies are only formed when a monokaryon carries the allele fi^+. In contrast, fi strains do not fruit at all.

For normal fructification of dikaryons it is necessary for one of the nuclei to carry the fi^+ allele. If both nuclei are fi, fruit body production is almost absent. Only very few fruit bodies are formed, some weeks later than normal, and in some combinations they never develop.

In this context, it must be mentioned that in the Agaric Agrocybe aegerita the same mechanism was found to determine both dikaryotic and monokaryotic fruiting. In this fungus the action of the fi genes in dikaryons is even more pronounced because there are no fruit bodies at all if the two nuclei of a dikaryon are fi (Esser et al.1974; Esser and Meinhard 1977).

3) The shape of fruit bodies is determined by additional genes which become effective only in the presence of fi^+. Genes of this kind have been identified in both fungi (Polyporus and Agrocybe) for monokaryotic fruit bodies (see fig. 2). The participation of these genes in the formation of dikaryotic fruit bodies has to be shown by appropriate experiments which are not yet completed.

4) The formation of a dikaryon, which is the prerequisite for dikaryotic fruiting, is controlled by the incompatibility factors A and B. A stable dikaryon can only be formed when the two mycelia differ in both factors (e.g. $A_1B_1 \times A_2B_2$, abbreviated A≠B≠). In other cases, if one or other factor is identical (A=B≠ or A≠B=), no dikaryon is formed and obviously no dikaryotic fruit bodies appear. For further information concerning the action of the AB factors, which is determined by the tetrapolar mechanism of homogenic incompatibility, see reviews of Raper (1966), Esser and Kuenen (1967) and Esser (1971).

5) There is an interaction between the AB factors and the fi locus. In dikaryotic mycelia (A≠B≠) monokaryotic fruiting is suppressed, even in the presence of the fi^+ allele. This is due to the action of the B factor, because in the combination A=B≠, where no dikaryon and no dikaryotic fruit bodies can be formed, monokaryotic fruiting is also suppressed. The heterogeneity of the B factor, initiating the first stages of dikaryosis (nuclear exchange and nuclear migration) is therefore a switch channelling the differentiation of fruit bodies on to the sexual track (dikaryotic fruiting) and blocking the asexual pathway (monokaryotic fruiting). For completion of the sexual pathway, heterogeneity of the A factor is necessary also.

It is now evident that it is not only the incompatibility factors A and B which are involved in the control of fruit body production, but in addition there is at least the fruiting initiation locus, which, depending on its allelic configuration (fi^+ or fi respectively) interferes with dikaryotic fruiting. If one is not aware of this genic interaction, it may cause considerable confusion when natural isolates are brought into the laboratory. On the other hand, these findings explain a phenomenon often recorded by mycologists (Lamoure, pers. comm.) that two monokaryotic isolates, despite their ability to form clamp connections, are not able to produce fruit bodies. This has sometimes been interpreted as an inability to fruit under artificial conditions, but from our genetic data we conclude that this kind of "sterility" may be caused by genes like the fi locus.

Furthermore, the action and interaction of the various genes that are instrumental in fruit body production show clearly the dominance of the dikaryotic (sexual) pathway in the life cycle of basidiomycetes. This favours outbreeding by recombination and thus gives a selective advantage in evolution. Which pathway arose first, the simple monokaryotic or the complex dikaryotic (requiring additional controlling elements such as the AB incompatibility system) is open to speculation.

Last but not least, the morphogenetic aspect has to be seen: one single gene is required to switch from a merely vegetative hyphal growth to the differentiation of rather complicated stromatic structures. This developmental change is not an integrated part of the sexual cycle.

In conclusion: to fulfill the main criterion of the genetic species concept, one must be sure whether or not two crossable strains are able to form fruit bodies and viable spores. This certainly can only be achieved if together with the environmental influences all genetic parameters are fully explor e d   as well.

III. Taxonomic description of the
genus Polyporus

Polypores were already described in the pre-Linnean era by the Italian botanist Micheli (1729). Linné himself (1737) put them together with the boletes into the type "fungus horizontalis, subtus porosus".

In the period following, the family "polyporaceae" contained not only the pore fungi but also species with bracket-shaped fruit bodies (Fomes) or gill fungi such as Pleurotus. According to present taxonomy (Donk 1969, Kreisel 1963) this confusion can be avoided by restricting the family to the genus Polyporus s.s. and including only the synonyms Leucoporus, Favolus and Polyporellus.

These fungi are characterized as follows: Pileus resupinate, cap-like, leathery or tough-fleshy; hymenium concrete with the hymenophore, consisting of pores arranged regularly (Clements and Shear 1964). The taxonomy of many species is far from being settled. The original descriptions are brief and often incomplete and except in a few cases, it is impossible to associate specific names with specimens that are still preserved (Donk 1969). The great variability found in nature is a further problem.

Kreisel (1963) differentiates the main species of European polypores (sensu strictu) on gross morphological characteristics: pore size, fructification period and colour of the fruit bodies (fig. 3).

1. P. arcularius (Batsch ex) Fr.: pores rhombic, radial, - 2 mm diameter; dissepiments irregularly lacerate when dry; pileus 1-3.5 cm, ochre, May - June.
2. P. brumalis (Pers.ex) Fr.: pores oblong, $\pm$ radial, - 2 mm $\emptyset$, pileus 2-7 cm, dark brown; Oct. - Nov. and March - April.
3. P. ciliatus (Fr.ex) Fr.: pores minute, isodiametric, - 0.3 mm, pileus 1-8 cm, grayish - brown, margin ciliated; forma ciliatus: July - August, forma lepideus: April - June.

Concerning the main points of this classification Donk (1969, 1974) agrees with Kreisel, but states that "the correct interpretation is still an open question; further observations are urgently needed".

The main difficulties lie in the nomenclaturial confusion of many authors (for details see Donk 1969). This caused a misclassification of P. brumalis as P. arcularius and P. ciliatus as P. lepideus or P. brumalis. It was one aim of our experiments to settle these open questions of taxonomy on a genetical basis.

IV. Genetic classification of the genus Polyporus*)

For our experiments we used 27 races of Polyporus. Origin, taxonomic classification, conditions for fruit body production and additional data to be discussed later (p....) are summarized in table 1.

In our first attempts to grow fruit bodies of these races under laboratory conditions we found that a change in temperature and humidity during fruiting caused ciliation of the pileus in some races of P. brumalis, making this criterion rather useless for the taxonomic differentiation of P. ciliatus (name!) from the other species. This was the case with three of the P. brumalis races we obtained from abroad (see footnote table 1). Superficial observation may thus lead to incorrect classification.

As a first genetic classification, we checked the breeding system controlling dikaryon formation in these races. We found that in all cases the tetrapolar mechanism of homogenic incompatibility is responsible for the mating relations. In other words we found that for each strain there were 4 mating types determined by the A and B factors.

The ability to produce monokaryotic fruit bodies of various shapes was common to nearly all races. A further genetic analysis of this phenomenon was not carried out (see table 1).

In a second set of experiments we determined the specificities of the mating types for each race by performing intraspecies tests. The results of these experiments are integrated in table 1. Within the 9 strains of P. ciliatus 14 A and 16 B alleles were found, in P. brumalis 23 A and 21 B alleles.

---

*) For methodological and experimental details see: P. Hoffmann: Die genetischen Grundlagen der Artbildung in der Gattung Polyporus, Thesis, in preparation.

Naturally from these data we can deduce nothing about the equivalence of mating type factors between different species, since interspecies tests, as discussed later, failed. The designation of the A and B factors is unlikely to show any correspondence between different species.

After these genetic determinations we were able to perform a basic experiment to obtain g e n e r a l   i n f o r m a t i o n
a b o u t   t h e   c r o s s i n g   a n d   c o m p a t i b i l i -
t y   r e l a t i o n s h i p s   w i t h i n   t h e   g e n u s
P o l y p o r u s : From each of the 27 races we selected two compatible mating types, which in addition were compatible with all other mating types of the same species. These races were crossed in all possible inter- and intraspecific combinations. From the results of these experiments, which are given in table 2, we may draw the following conclusions:

1. I n t r a - s p e c i e s   i n t e r a c t i o n s .
I n t r a s p e c i e s   c r o s s e s   w e r e   f e r t i l e .
In all combinations fruit bodies were produced. In some cases, however, time for fruiting was prolonged for up to three weeks (P. ciliatus) or 12 weeks (P. brumalis) and the number of fruit bodies was reduced. This phenomenon was correlated with the appearance of a distinct line in the contact zone of the two monokaryotic mycelia called "barrage" (fig. 4a).

Since the descriptions of barrage formation by Vandendries (1927, 1933), this phenomenon has been observed in many other fungi (for references see Burnett 1975).

Barrage formation is independent of mating type because it also occurs in incompatible (A=B=) and hemicompatible combinations (A=B≠, A≠B=). Furthermore it is not correlated with any specific configurations of the AB factors. Barrage formation is also independent of the geographical origin of the races because it is found between single spore isolates of a single fruit body taken from nature.

The barrage phenomenon needs some further consideration. C y t o -
l o g i c a l   a n d   g e n e t i c a l   s t u d i e s   p e r -
f o r m e d   w i t h   t h e   b a r r a g e   f o r m i n g   r a -
c e s   Nr. 1 and 4 of P. ciliatus, originating from different geographical regions, have given the following results which were confirmed by extension of the experiments to other races of both P. ciliatus and P. brumalis.

1) The hyphal tips of barrage formers interlace for about 1 mm. In this zone of contact, which is identical with the macroscopically visible barrage, there is less mycelial growth i.e. no aerial hyphae are formed and rather few hyphae grow into the agar medium (compare fig. 4b with e) but the growth rate of the individual hyphae is not reduced.

This change of mycelial appearance which is evidently caused by an interaction of the barrage partners, can also be observed when barrage formers are grown together in liquid culture.

2) The number of hyphal fusions is markedly reduced. Often one may observe that when two compatible hyphae come in contact, they continue to grow parallel to each other instead of forming an anastomosis (fig. 4 h).
Whereas in normal combinations the first anastomoses can be seen 2 d after inoculation, in the barrage zone very few may be seen even after 7 d (fig. 4i).

3) There is no immediate plasmatic disintegration after hyphal fusion as described for barrage formation in the ascomycete Podospora anserina (Blaich and Esser 1971). However, a dissolution of the hyphal walls begins in the barrage zone 3-4 weeks after its establishment (compare fig. 4g, j with d).

4) Nuclear migration seems to be unilateral, because dikaryotization and clamp connection formation occurred only on the side of strain 4. The dikaryotic status spread slowly into the other partner, possibly by the growth of the dikaryotic hyphae formed in the barrage zone (compare fig. 4c with f). Fruit bodies on this side were established after a 10 d delay.

5) The genetic analysis, which is not yet concluded, reflects the complicated situation seen in the cytological experiments. At least 2 or 3 unlinked genes, one of which seems to be epistatic, are involved in the formation of a barrage.

2. Interspecies interactions.
Interspecies matings failed to give a dikaryotic mycelium with clamp connections and dikaryotic fruit bodies. All combination showed a more or less pigmented border line, which is easily distinguishable macroscopically from the barrage reaction (fig. 5).

Hyphal fusions between monokaryons of different species have not been clearly observed, the few anastomoses near the border line were always found to come from hyphae of one monokaryon. No hyphal disintegration was found in the border line, even after a few weeks of contact.

Using this criteria of classification, the strains of P. lepideus do not represent a true species but have to be put as already suggested by David and Romagnesi (1972) into the species P. ciliatus (see table 2) as a forma (see Kreisel 1963).

V. Summary
In using the basidiomycete genus Polyporus as an example we have tried to prove the genetic species concept (p. 190).
As a prerequisite for our experimental studies we have pointed out in some detail the manifold genetic parameters which control fruit body formation in higher basidiomycetes and have therefore to be considered in any evaluation of the success or failure of crosses.

1) The species concept as based on morphological characters of Polyporus has been confirmed by our cytological and genetic experiments. Furthermore another morphological very simply to prove criterion for species delimitation was found. This is the formation of a macroscopically visible border line, which always occurs between cultures of different species, independent of mating type di- or monokaryotic status of the mycelia.

2) The b a r r a g e f o r m a t i o n, which macroscopically can be clearly distinguished from the border line, o c c u r s o n l y w i t h i n a s p e c i e s. Cytological observations revealed that the phenomenon of barrage causes a retardation of all procedures being necessary for the completion of the sexual cycle. It might thus be considered as a s t e p t o w a r d s r e - p r o d u c t i v e i s o l a t i o n because it diminishes fruit body production and provides the non-barrage combinations with a selective advantage in nature.

It differs from the barrage reaction observed in the ascomycete Podospora anserina (for ref. see Blaich and Esser 1973) which does not allow the coexistence of different genetic material in vegetative cells (heterogenic incompatibility) and leads to total isolation.

3) The experiments have demonstrated the v a l i d i t y o f t h e g e n e t i c s p e c i e s c o n c e p t, at least in the present case. Further observations using a greater number of species, are of course needed to give our results a broader basis. The genetic species concept provides then a simple means to test taxonomic problems. However, one must be aware that every factor controlling sexual reproduction, for example the genes for monokaryotic fruiting or interactions of the barrage type, influence the results of crossing experiments.

In considering these phenomona the species concept quoted on p. 190 needs to be modified:
P o p u l a t i o n s (r a c e s) b e l o n g t o d i f - f e r e n t s p e c i e s w h e n t h e f a i l u r e t o i n t e r b r e e d a n d t o p r o d u c e v i a b l e o f f s p r i n g i n n a t u r e i s n o t c a u s e d b y g e n e t i c p a r a m e t e r s o p e r a t i n g i n t h e c o m p l e t i o n o f t h e s e x u a l c y c l e.

A c k n o w l e d g e m e n t s : We are very much indepted to all colleagues and instiutions who have sent us strains. For their technical assistance we wish to express our thanks to the technical staff of the Lehrstuhl especially to Frau G. Lembke. This work was supported by the Deutsche Forschungsgemeinschaft (Bad Godesberg) and the Landesamt für Forschung (Düsseldorf). The junior author was supported by a research grant of the Konrad Adenauer Stiftung (Bonn).

References

Blaich R & K. Esser 1971: The incompatibility relationships between
    geographical races of Podospora anserina. V. Bioche-
    mical characterization of heterogenic incompatibility
    on cellular level. Molec.gen.Genet. 111: 265-272.
Burnett J.H. 1975: Mycogenetics. An introduction to the general gene-
    tics of fungi. New York, London, Sydney, Toronto.
Clements F.E. & C.L. Shear 1964: The genera of fungi. New York and
    London.
David A. & H. Romagnesi 1972: Contribution à l'étude de Leucopores
    français et description d'une espèce nouvelle: Leuco-
    porus meridionalis nov. sp. Bull. Soc. Mycol. France
    88: 293-303.
Donk M.A. 1969: Notes on European polypores. III. Notes on species
    with stalked fruitbody. Persoonia 5: 237-263.
Donk M.A. 1974: Check list of European polypores. Amsterdam, London.
Esser K. 1971: Breeding systems in fungi and their significance for
    genetic recombination. Molec.gen.Genet. 109: 186-192.
Esser K. 1974: Some aspects of basic genetic research fungi and their
    practical implications. Adv.Biochem.Engin. 3: 69-87.
Esser K. & R. Kuenen 1967: Genetik der Pilze. Berlin, Heidelberg,
    New York.
Esser K. & R. Blaich 1973: Heterogenic incompatibility in plants and
    animals. Adv. Genet. 17: 107-152.
Esser K. & U. Stahl 1973: Monokaryotic fruiting in the basidiomycete
    Polyporus ciliatus and its suppression by incompati-
    bility factors. Nature 244: 304-305.
Esser K. & U. Stahl 1974: A genetic correlation between dikaryotic
    and monokaryotic fruiting in Basidiomycetes. I.Inter-
    s.Congr.Intern.Assoc.Microbiol.Soc. Abstr. p.17
    (Tokyo).
Esser K., M. Semerdzieva & U. Stahl 1974: Genetische Untersuchungen
    an dem Basidiomyceten Agrocybe aegerita. I. Eine Kor-
    relation zwischen dem Zeitpunkt der Fruchtkörperbil-
    dung und monokaryotischem Fruchten und ihre Bedeutung
    für Züchtung und Morphogenese. Theor.Appl.Genet. 45:
    77-85.
King R.C. (edt.) 1974: Handbook of genetics. New York and London.

Kreisel H. 1963: Über Polyporus brumalis und verwandte Arten. Feddes
    Repertorium 68: 129-138.
Linné C. 1737: Genera plantarum. Leiden.
Micheli P.A. 1729: Nova Plantarum Genera. Florenz.
Raper J.R. 1966: Genetics of sexuality in higher fungi. New York.
Stahl U. & K. Esser 1976: Genetics of fruit body production in higher
    Basidiomycetes. I. Monokaryotic fruiting and its correlation with dikaryotic fruiting in Polyporus ciliatus. Molec.gen.Genet. in press.
Vandendries R. 1927: Le comportement sexuel du Coprin Micace dans ses
    rapports avec le dispersion de l'espèce. Bull.Soc.Roy.
    Belg. 60: 62-65.
Vandendries R. & H.J. Brodie 1933: Nouvelles investigations dans le
    domaine de la sexualité des Basidiomycètes et l'étude
    experimentale des Barrages sexuels. La Cellule 42:
    165-207.
Wettstein R. 1901: Handbuch der Systematischen Botanik. Leipzig und
    Wien.

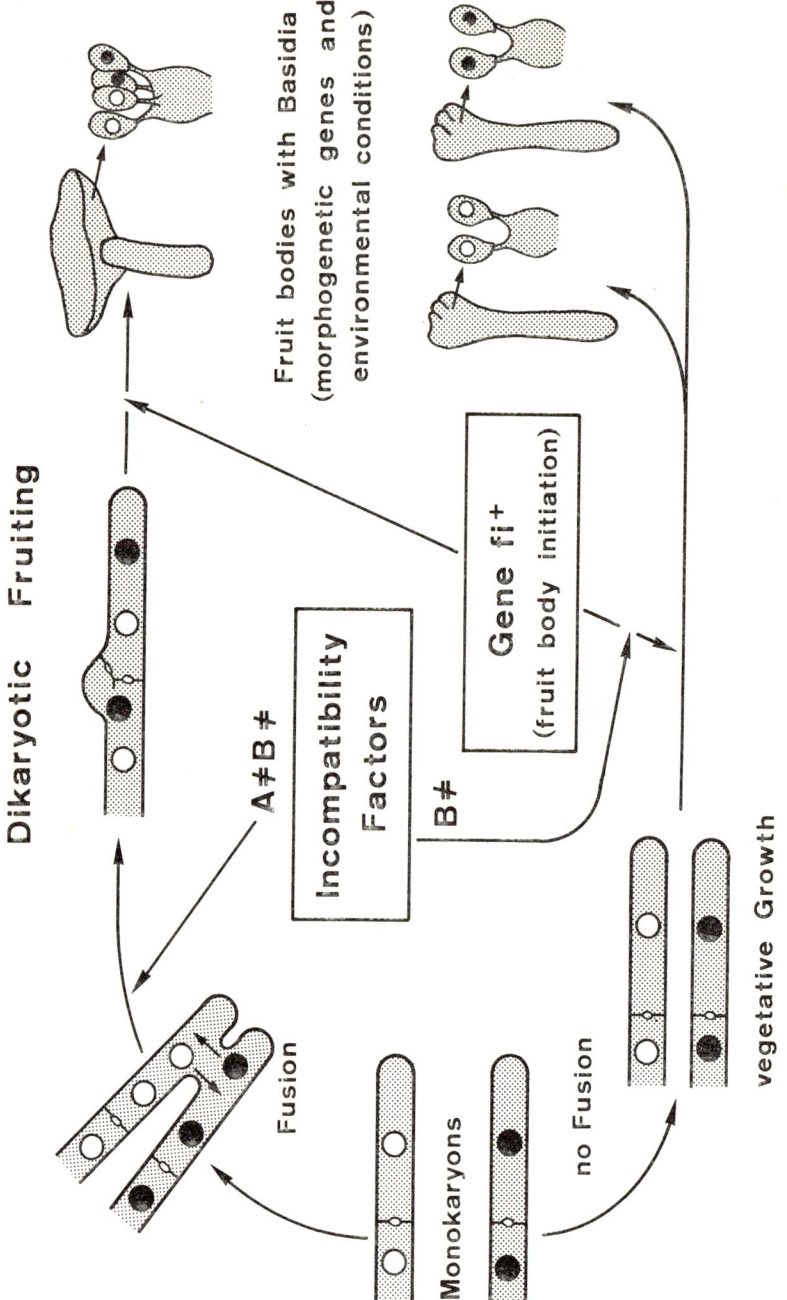

Figure 1: Genetic control of fruiting in higher basidiomycetes. Schematic representation based on Stahl and Esser (1976), for details see text.

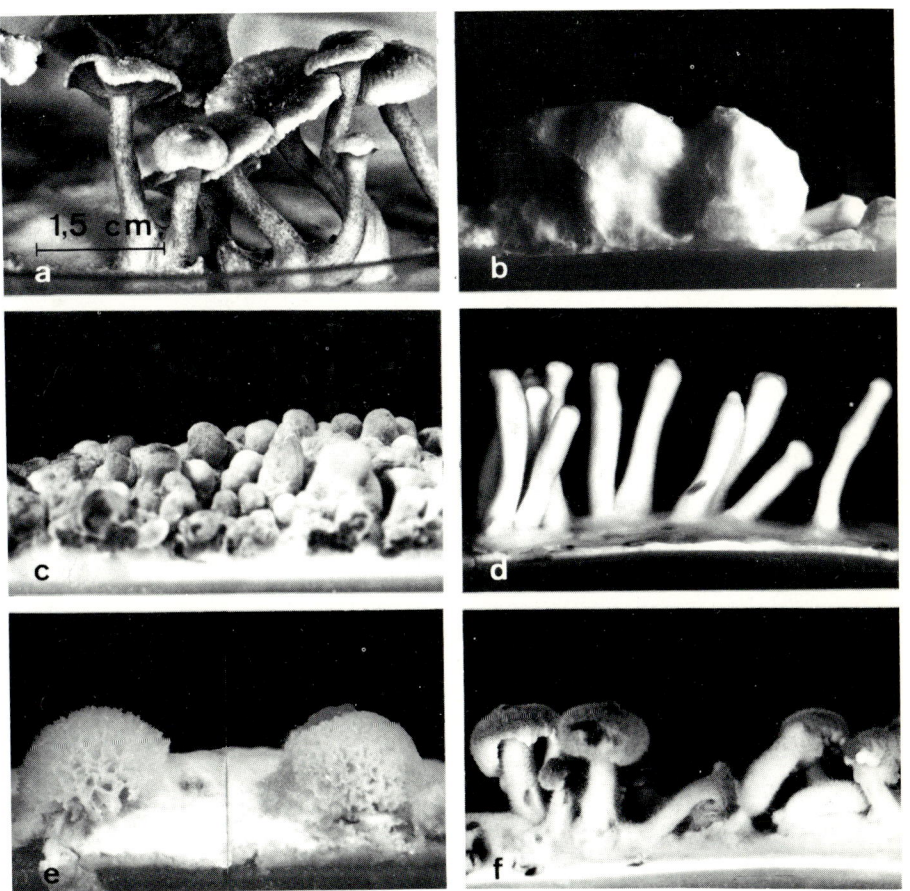

Figure 2: Morphology of fruit bodies of <u>Polyporus</u> <u>ciliatus</u>. (a) Dikaryotic fruit body; (b-f) various types of monokaryotic fruit bodies. (From Stahl and Esser 1976).

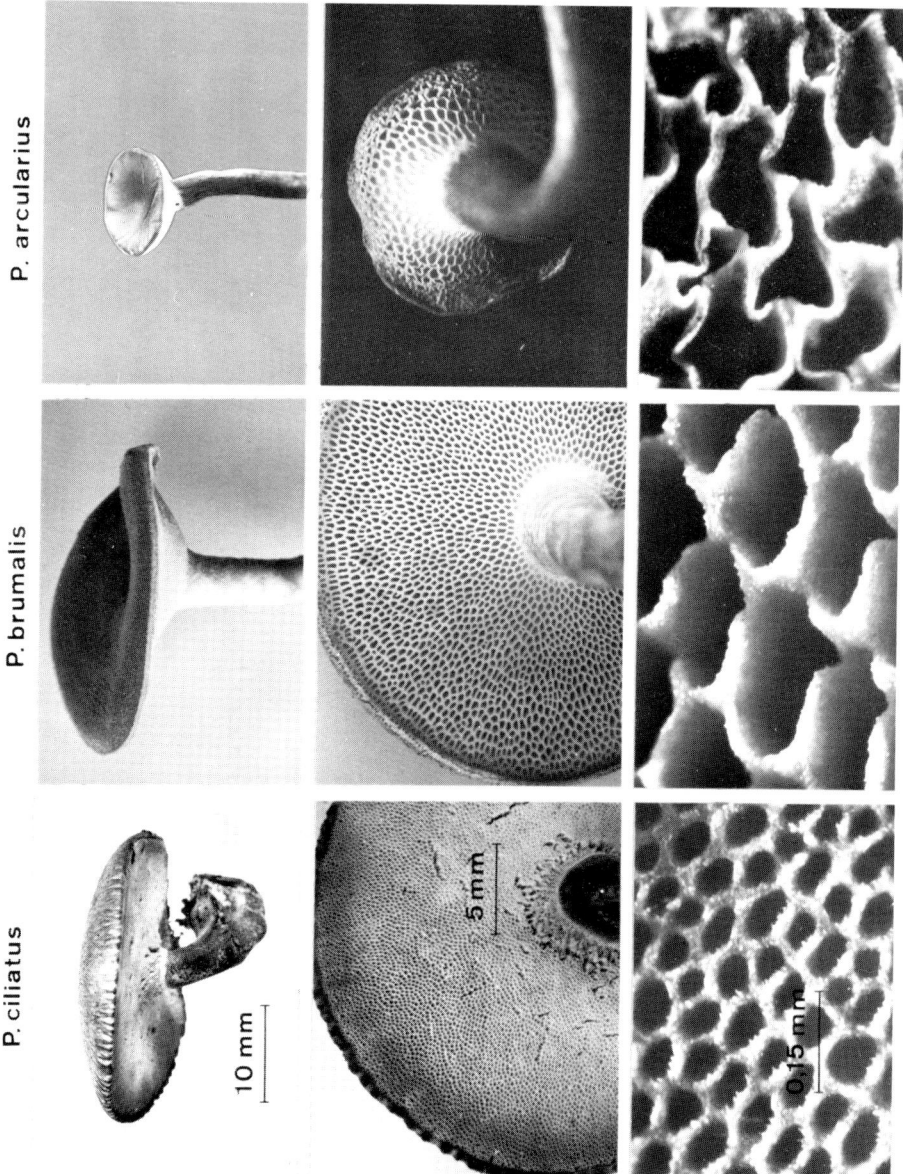

Figure 3: Morphology of fruit bodies from three species of Polyporus. Top: shape of the fruit bodies; middle: view of the hymenophores; bottom: enlarged view of the hymenophores showing pore size.

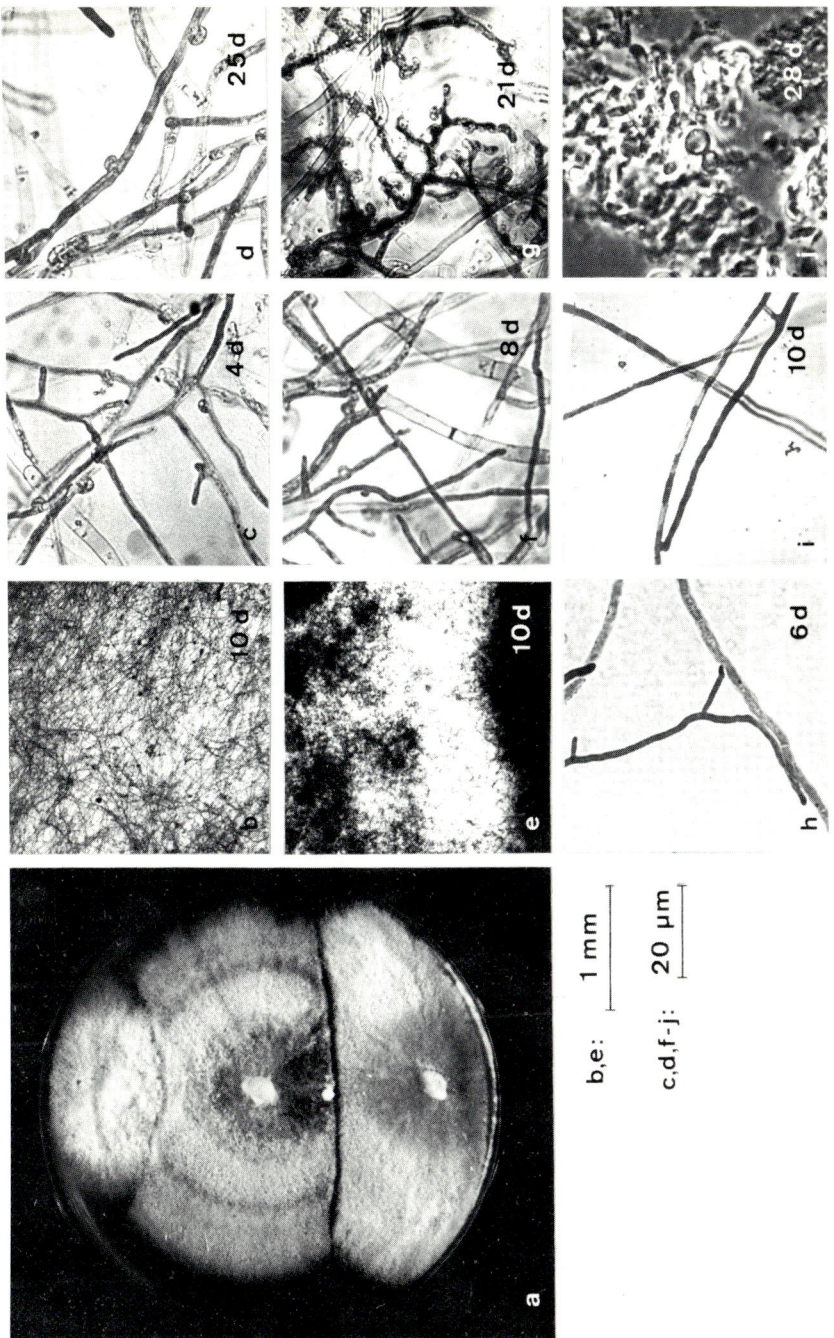

Figure 4: Barrage formation in the intraspecies crosses between monokaryons of <u>Polyporus ciliatus</u> compared with normal contact. (a) Survey, showing petri dish inoculated with three monokaryons; bottom: barrage formation; top: normal contact; (b-d) microscopical view of normal contact; (e-j) stages of hyphal contact between barrage forming monokaryons. Age of mycelia is given in days; for further information see text.

Species	Polyporus Strain Number	Origin	Conditions for Fruiting				Mating Type		Monokaryotic Fruiter	Induction of mf in A=B≠ heterokaryon
			Medium	Temperature °C	Humidity %	Time	A	B		
ciliatus	1	Göttingen	synth.	22°	50-60	8-25 d	1,2	1,2		
	2	Bochum					2,4	3,4	+	+
	3++	Bochum					2,6	5,6	+	+
	4+	Kassel					7,8	7,8	+	+
	5++	Bell's corner, Ont.					7,8	7,8	+	
	6++	Weimar					9,10	9,10	+	
	7++	Weimar					11,12	11,12	+	+
	8++	Jüterbog					13,14	13,14	+	
f.lepideus	24	Lyon LY AD 720					15,16	15,16		
	25	LY AD 727								
brumalis	9	Bochum	natur.	10°	50	3-12 weeks	1,2	1,2	+	
	10	Bochum					3,4	3,4	+	
	11	Bochum					5,6	5,6	−	
	12	Bochum					7,8	7,8	+	+
	13	Bochum					8,10	9,10	+	
	14	Bochum					11,12	11,12	−	
	15	Bursfelde, Weser					5,14	13,14	−	
	16	Hann.Münden					15,16	15,16	+	+
	17	Kingsmere, Quebec					17,18	17,18	−	
	18	ATCC 9385					5,20	12,20	−	
	19	CBS 470.72					21,22	21,22	+	
	20	For.Prod.Res.Lab. 174a					23,24	23,24	+	
	21	For.Res.Inst. 285					25,26	25,26	+	
	22	Paris, LY AD 717					19,28	3,23	+	
	23	Lyon, LY AD 719					29,30	24,26	−	
arcularius	26	Lyon, LY AD 726	like ciliatus							
brumalis ?	27	Kenilworth, Aust.	like ciliatus						−	

+ obtained as P.ciliatus f. lepideus
++ obtained as P.brumalis; redetermined as P.ciliatus

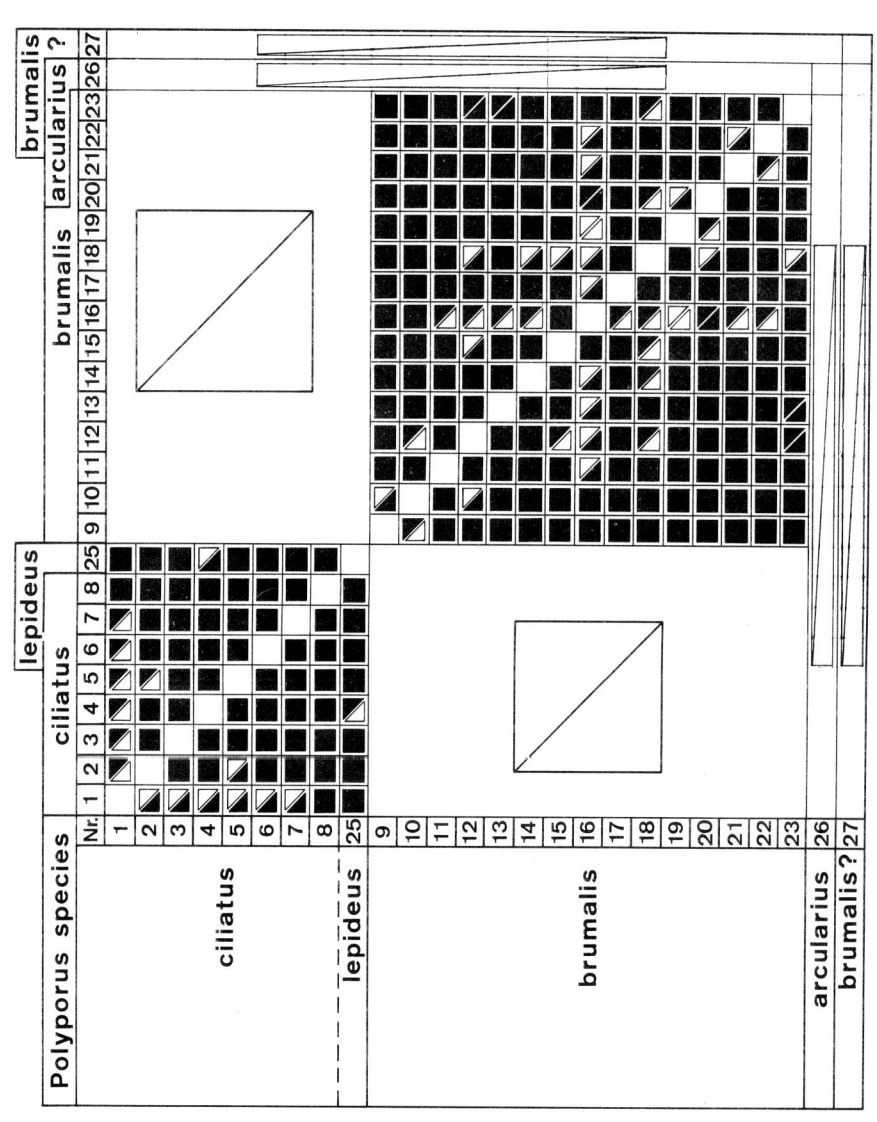

DISCUSSION FOLLOWING DR. ESSER AND MR. HOFFMANN'S PAPER

CLEMENCON: We all know that a newly isolated dicaryon may fruit, but when you take cultures from CBS or ATCC that are a couple of years old, they frequently do not fruit. Is this due to a loss of the fi^+-gene? Do you know something about it from the genetic point of view?

ESSER: Yes, we are confronted with it in our industrial strains. And these strains mutate and they loose the capacity to produce certain antibiotics or certain enzymes. We call these generally senescent. Senescence has been studied or initiated by Rizet and Marcou in Paris some twenty years ago and we must consider that we do not deal with individuals, that we deal with a population of nuclei. And in this population of nuclei one or two mutate, and when there is a slight selective advantage they will overgrow the mycelium and the other nuclei are gone. So there are many reasons for the loss of the capacity to fruit. This might be the case, but there might be other genes, and we only can decide from case to case. In Podospora, an ascomycete, we could show that this senescence can be stopped by the synergistic action of two genes. And we are also able to prevent senescence in Podospora mycelium by treatment with an antibiotic.

CLEMENCON: Why does it not occur in nature?

ESSER: They are eliminated in nature. Color-less spores in the ascomycetes occur very often, but they are eliminated.

SMITH: They do occur in nature. In Lactarius we will have certain characters, like taste, that will drop out. In an acrid species like Lactarius kaufmannii, you can find an occasional fruiting body that is mild. I have another question that I would like to bring up, because in my Mycena work, back in the dark ages, in the 1930ies, I encountered as part of a genetic type young basidioles with single nuclei, but the fruiting bodies were normal. I never did make up my mind as to whether these were haploid parthenogenetic or diploid parthenogenetic. No clamp connections. Would you have a guess?

ESSER: It could well be monokaryotic fruiting, because in Agrocybe aegerita, fruiting bodies in Petri dishes are smaller than in nature.

Genetics of Polyporus

A dicaryotic fruiting body, and a monocaryotic fruiting body, having the genes fi^+ and fb^+ look exactly the same, there is no difference.

SMITH: But is there such a thing as diploid parthenogenetic?

ESSER: Do you mean dikaryotic? Diploid, no!

SMITH: This is what I was wondering. I never could count the chromosomes. But the fruiting bodies were exceptionally large and robust.

ESSER: Well, they are large. You will have to grow them in big bottles. The smaller dikaryotic fruit bodies can have the same size as the biggest monokaryotic fruit bodies.

SMITH: This is true in Mycena megaspora. It is a robust Mycena, and it has one- and two- and three- and fourspored basidia, and it seems to be a very vigorous species, but yet it must be a monokaryotic fruiting.

CLEMENCON: You stated that we must know the genes fi^+ and fb^+ in order to establish relationships between genetic species. Now, does this mean that the formation of a dicaryon and the formation of clamps, simply by confrontation of two monocaryons has limited value?

ESSER: Certainly, but I am dealing with the next step, I have strains of the same species which do form clamp connections but which never fruit. I can now explain at least part of it; because these genes (fi^+ and fb^+) are not present.

CLEMENCON: So, when clamp connections are formed, you would agree that they are the same species.

ESSER: Certainly I would.

SMITH: Going back to clamp connections. I have an incomplete study made on the Pluteus cervinus group. In Pluteus salicinus nearly every septum has a clamp. In Pluteus cervinus very few have clamps. You can count a hundred septa and not find any clamps. And then you can find one or two. And then you have some taxa that are inbetween: 25 out of a hundred, 50 out of a hundred. What can you say about the genetics of clamp connection itself as a taxonomic character? Have you got that far in genetic studies?

ESSER: If you consider Agaricus - no clamp connections. I can only explain positively: if you have strains where clamp connections occur, then I would consider it as a species character. But if they do not have clamp connections then I cannot answer the question.
My species concept says that there must be exchange of genetic material, but they also must have viable offsprings. And that is the reasons because I want the fruiting bodies. I agree basically, but I want to go one step further: have a fruiting body and keep the offsprings viable! An unviable offspring means that there are such gross genetic differences which cannot be put together.

BRESINSKY: Do you think there are species which might be obligatory monocaryotic fruiters?

ESSER: I have no information about this. Monokaryotic fruiting means that there is no sexual cycle involved, and I do not think they are good species.

BRESINSKY: The question was necessary in order to start now my further question. I think we have in fungi real apomictic situations without sexual cycle. And in these apomictic organisms very minor alterations by mutations may occur at a considerable rate. All these alterations are preserved as differences, because there is no exchange and distribution of characters. I think that in this case of apomictic species your species concept would not be workable. Taxonomists would have much to do to come along with the great bulk of apomictic species. I think we should admit exceptions in your species concept.

ESSER: No, no, I do not agree. Look, in an apandric species, like Sordaria microspora, there are normal sex organs, but there is not a sexual cycle. This is an apomictic organism. We have recombination in Sordaria macrospora, and then you get ascospores with one nucleus, and you get haploid monocaryotic mycelia and they are intercrossable.

BRESINSKY: But there are cases of apomixis where you don't have any sexual recombination. There you have caryogamy and meiosis in a closed system of a population of nuclei.

ESSER: It is alright, that is what all self-compatible strains have.

BLAICH: You can cross apomictic species. Make two self-sterile strains and then they are obliged to help each other and they will cross. All fungi which form anastomoses will cross. If they do not cross, there might be two species.

BRESINSKY: I really don't see why you cannot follow me! We have examples in the taxonomy of higher plants, <u>Alchemilla</u>, <u>Hieracium</u> etc.

ESSER: I think we are running on different tracks.

WATLING: Could we not have an example of an apomictic Basidiomycete which we are talking about?

SINGER: Might it be <u>Mycena galericulata</u>?

ESSER: In the ascomycete <u>Sordaria</u> the life cycle is very simple. You have a mycelium, you have the ascogonia and then you have apomictic perithecia. And you can make crosses. The problem is, if you cross self-fertile species, how do we distinguish fruiting bodies from a cross from fruiting bodies from selfings? It is easily done with colored spores. We have marker genes, colored spores for one strain or the other, you have segregation for the spore colors. So this is a means to determine whether they are the same species or not. Another way, which is more complicated, is to make them sterile, or make them auxotrophic. Make one auxotrophic for leucine, the other for adenine, then grow them together, and when you observe segregation you had a cross. So the species concept holds with the self-compatible strains. It is only needed to have appropriate markers which show that this is a compatible strain, and this is an in compatible strain. When they grow together, and when they form fruiting bodies, then they contain both nuclei. So the species concept shall work.

BRESINSKY: It works anyway, but in some cases where you have apomictic situations it causes much trouble. My example would be <u>Morchella</u>. There fusions of hyphae grown from single spores do no occur. You have a great variety of different forms and characters. And if you consider all these different very minute characters you would get a great bulk of different apomictic species.

ESSER: This can easily be overcome. Make protoplasts and they will fuse. Protoplasts from + and + cells from yeasts fuse. So with protoplasts you can overcome all failures of fusion of cells which have been stopped by the barrage of cell walls. And in yeasts, as you know, the substances causing the mating types, or causing the incompatibility of the mating types, are located in the cell wall. So I think if you have a sample of 3 strains, you will find them all incompatible. How many strains did you have?

BRESINSKY: 190 crosses or something less. And they do not fuse.

BLAICH: How has it been proved that they do not fuse?

BRESINSKY: I used an arrangement on slides coated with agar. I put one mycelium (single spore mycelium) on one side, the other one on the other side, and I observed the zone of contact with the microscope.

ESSER: Telling you about the fusion. We never saw any anastomosis in intra-species crosses. But we have fruiting bodies and segregates, there must have been anastomoses. And we took many efforts to see them.

KEMP: A point I would like to make is that in Psathyrella we have found monocaryotic fruit bodies in the wild. I think you said they are likely to be eliminated but it seems that they can develop. It seems clear that in Polyporus and in many other basidiomycetes reaction patterns between mycelia can occur. But in Coprinus it is not the rule that reaction lines are formed. But between just a few species reaction lines are formed.

ESSER: May I just answer?
Well, I am aware that the same phenomena do not happen in all species. E.g. there is a slight difference between the barrage in Polyporus and in Agrocybe. In Agrocybe the barrage is even stronger, quantitatively. In Polyporus we get weak lines, that is the problem.

KEMP: Although reaction lines are not common between species they are sometimes very distinct between matings of strains belonging to the same species, e.g. C. sassii, (2-spored).

ESSER: Did you try monokaryons or dikaryons?

KEMP: Both. Let me come back to anastomoses. We found in Coprinus section Lanatuli that you can get good hyphal anastomosis, and a lethal reaction, between practically all species in the section. Oidial homing shows quite clearly, too, that there is no lack of recognition at the hyphal tip level. After fusion between different species I believe that the cell always dies if both nuclei start to function. So I do not believe that the presence or absence of anastomosis is of any use in distinguishing between species.

ESSER: I would agree with you. If we have anastomosis it does not mean anything, but when we do not have, then it means probably something.

KEMP: May I make a remark about fruiting. You said that in strains where you had unilateral matings fruiting was generally slower. We generally find the converse in Coprinus and Psathyrella. In some species we can only get fruit bodies to develop if we inoculate a plate with macerated fragments or inoculate with dicaryotic and migration blocker monokaryotic strains. In some species we never get fruit bodies if we start with dicaryotic inocula. A dimon mating involving a blocker monocaryon is often a very quick way of getting fruit bodies.

SINGER: In the light of this last discussion I would like to ask now the following question. I think it makes us all very happy to know that you have elaborated an observable difference between intraspecific barrage and interspecific sterility, but to what degree do you now think can your observations be generalized? I mean, could they be generalized for all higher Basidiomycetes?

ESSER: This I cannot answer. I think more research work has to be done.

WATLING: I am very impressed by this picture in which we have a morphological species which we can see in the field. But I would like to ask if Prof. Esser's pretty picture indicated any segregation into different isolating groups, e.g. do we really have a picture in which the <u>ciliatus</u> broke up into two groups, <u>brumalis</u> into three groups etc. but without accompanying morphological differences between the collections. As a taxonomist and a morphologist it is this which perplexes me, because, although in the laboratory one can see differences, in the field we do not see them.

ESSER: I think you brought up a very interesting and very important point which was necessary to be brought up. This intra-species barrage is the beginning of a new specification. And it will give a selective disadvantage to any exchange of genetic material, because there are very few anastomoses between the two strains, very late fruiting, so it will be the beginning of a formation of new species because mutations will not freely exchange, so they will accumulate, and eventually give rise to morphological differences which you can detect in nature.

WATLING: But what happens before that? How can we fit this information into our International Rules of Botanical Nomenclature which in fact we have to work to at this moment in time.

ESSER: At the present time I would not put it into the International Rules, because it is exactly the same case as the histo-incompatibility in human beings of hearts. It is the same thing. You can transplant organs only from identical twins, nothing will happen. But, in sisters, brothers, there are genes, histo-incompatibility genes present. As we could show in <u>Podospora</u>, one is enough to kill each other. But these genes do not infer incompatibility factors, this does not inhibit sexual propagation.

WATLING: I think from work of Professor Romagnesi in <u>Psathyrella</u>, and mine in <u>Conocybe</u> one does have these groups, not always with slight morphological differences, and my problem as a morphologist is where I draw a line and say "that is species A and that is species B".

ESSER: Cross, and test the offspring! That would be my advice.

KEMP: Can I follow on from Dr. Watling's point? In <u>Psathyrella</u> the crossability tests indicate that it is probably quite common to have complete isolation (in what I would call species, because they cannot interbreed) in the absence of any known morphological differences. There are species which do have morphological differences and also have a reasonably high level of interfertility. That is much more common than to have no differences at all. Is'nt that so?

WATLING: Yes, I would say so.

SINGER: Should we necessarily confine ourselves to morphological differences? Could'nt certain physiological or chemical differences also justify separation of species?

THIERS: Professor Esser, in the boletes, as you know, we cannot germinate the spores and therefore cannot do much with them from a genetic point of view. Therefore, in the genus <u>Leccinum</u> in which there is great similarity between species, how would you, as a geneticist, handle this problem where we cannot use genetic characters? We have to use merely morphological features.

ESSER: One could regenerate mycelium out of a fruit body and test it for barrage or any kind of border line formation.

BRESINSKY: You use in your species definition the term population, so I think you should give us your definition of population. Do you mean the totality of nuclei in a mycelium and all the nuclei which may join this mycelium, or do you have another idea?

ESSER: I use population in a sens of a geneticist who goes out in nature and collects fruit bodies in an area and grows the mycelia. I use population and race equivalent, they are taken from nature. However, this may be debatable.

CLEMENCON: I have another question: in a dikaryotic population of nuclei a mixture of genetically different units exists. Do you think that the term parasexuality, as defined and found in Aspergillus etc., can be used to describe the situation in Basidiomycetes? Is it the same thing, can the parasexuality be extended to Basidiomycetes?

ESSER: It is the basis of mitotic recombination, and mitotic recombination, I am sure, is a wide-spread phenomenon, but you must find ways to demonstrate it.

BLAICH: In Schizophyllum it has been proved.

CLEMENCON: I think this would be quite interesting. When you have an established dikaryon.....

ESSER: Oh, I see what you mean, I am sorry. You mean: that in nature somatic recombination occurs very often, and the diversity and the heterogeneity of the dikaryons may be caused by parasexual procedures.

CLEMENCON: Yes.

ESSER: I cannot answer this question, because it is not known.

CLEMENCON: Dr. Kühner showed several years ago, that the fungal cells in the stipes of many agarics have many nuclei. So we have a population of nuclei there, and that would allow parasexuality in the fruiting body.

ESSER: Certainly, but the frequency of somatic recombination is very low.

KEMP: Can I follow up on this aspect of somatic recombination. I think that in Coprinus and Schizophyllum and probably in most species of basidiomycetes if you have suitable markers then somatic recombination will be found. I do not think it makes any difference to the species concept because we still must have plasmogamy to get the nuclei into a common cytoplasm. The only event which is modified is the timing of karyogamy. Having it in the hyphae you can get diploid nuclei without having a fruit body. Instead you can have haploidisation and mitotic recombination in the hyphae. So genetic recombination is still there but we do not have the fruit body. The hyphae still have to recognise each other and somatic recombination is still within the species concept.

CLEMENCON: So from your experience you do not expect that genetic information coming along with a spore from a different species can get into an established mycelium.

KEMP: From another strain of the species, yes.

CLEMENCON: No! Different species!

KEMP: Impossible. Because the recognition of self and non-self is an internal cytoplasmic event.

CLEMENCON: of course, if you are convinced that this is impossible, that has great bearing on our species concept.

PETERSEN: I am afraid that my question may overlap considerably with what has already been discussed, but I am interested in this kind of an idea: in the last two days and in the literature there are "species" which have not yet been explored enough to fulfill your definition. What kind of terminology would you use to describe the kinds of "species" for which proof of interbreeding and viable offspring has not been forthcoming. Are these "provisional" species?

ESSER: No. I would call them species, because, in many things we have ideals which we want to attain, but in life we never will attain all ideals. We should have in mind that we want to come to a species concept, so we should have one which we can fulfill.
But there is another phenomenon which must be taken in consideration. Some years ago EGER (TAG 38, 23-27, 1968) observed in my laboratory, that pieces from fruiting bodies of Pleurotus when brought on a Pleurotus mycelium, enhanced the fructification. This phenomenon could be misinterpreted as interbreeding. But a more detailed analysis proved that the formation of fruit bodies was merely caused by an unspecific induction, because the same phenomenon occurred when a piece of a Agaricus fruit body or fruit body extracts were used and even when simply L-asparagine or urea were applied. The induction phenomenon could be confirmed by the fact that the spores of all fruit bodies were homogeneous. This again points the necessity for genetic analyses of all different kinds of inter-race, inter-strain, or even "inter-species" crosses.

SMITH: Dr. Esser, back in the dark ages again, when I was culturing Mycena leaiana, we often had contamination by Penicillium in a Petri dish. The _Penicillium_ would stimulate production of fruiting bodies on monocaryotic mycelia. The fruiting bodies were not mature, but they would start. We explained this as a chemical stimulus, just off the cuff, we never proved it.

ESSER: You are right, that must be a chemical stimulant, but chemical stimulants can be strictly related to the B-factors. The B-factors block fruiting bodies, and the blocking effect of the B-factors is released by the presence of an other species.

OBERWINKLER: What sort of fruiting bodies do you have with monokaryotic strains of different species?

ESSER: If we have fi^+ present you get the stipe-type, if you have fi^+fb^+ present you have the cap-type.

BLAICH: Is this induction dependent on the type of the monokaryotic inducer?

ESSER: So far no. These are just preliminary experiments.

KEMP: Do you know the type of anastomoses?

ESSER: According to our information there should be no hyphal anastomoses. What we do now, we are putting a membrane between.

Herbette Symposium on Species Concept in Hymenomycetes 1976

ENZYMES AS AN AID IN TAXONOMY OF HIGHER BASIDIOMYCETES

Rolf   Blaich

1) Introduction

One of the main difficulties for a clear delimitation of species in fungi is the large variability of their morphological characters, as for instance, colour, shape and size of fruiting bodies. These may vary to a large extent without affecting the reproduction of the species.

A delimitation of species is facilitated by the analysis of characters which are under a high selective pressure and thus not easily changed by mutations under natural conditions. Such additional characters may be found among proteins and secondary metabolites, whose analysis is therefore one of the cardinal tasks of chemotaxonomy.

The use of chemotaxonomic methods by systematic mycologists is very often limited by the lack of appropriate equipment, of time, and sometimes also of experience with biochemical methods.

The main problem, however, seems to be a certain incompatibility between mycologists with biochemical interests on one hand and taxonomists on the other hand, many biochemists thinking taxonomic work to be hopeless by out of date, and taxonomists being convinced that the organism a biochemist works with is known to him only in the form of extracts.

Some of these difficulties may be overcome by new methods which were developed in the last decade and which may be used by mycologists not acquainted with biochemical procedures. One of these methods is thin layer chromatography of pigments and other easily detectable metabolites. This subject will not be discussed here; detailed reviews were given by BRESINSKY (1974). Another possibility is the comparison of protein electropherograms.

Due to the good separations obtained and the possibility to analyze relatively dilute crude extracts without critical concentrations procedures, disc electrophoresis is widely used for that purpose.

Though analytic electrofocusing has the same advantages, the evaluation and interpretation of the results is sometimes difficult and needs more equipment. It is recommended therefore only for experienced laboratories (for literature see RIGHETTI and DRYSDALE 1974).

2) Enzymes in Taxonomy

A number of papers have been published using electropherograms of proteins as criteria for taxonomy. In fungi mainly ascomycetes and imperfect forms were studied. The conclusions of the authors concerning the taxonomic value of protein patterns have, however, been contradictory.

For literature see ABBOTT and HOLLAND, 1975; DURBIN, 1966; GAIROLA, 1971; GLYNN and REID, 1969; HALL, 1967; KULIK and BROOKS, 1970; MEYER and RENARD, 1969; MILTON et al., 1971; NASUNO, 1972 a, b; NEALSON and GARBER, 1967; PELLETIER and HALL, 1971; REDDY and THRELFELD, 1971; SELVARAY and MEYER, 1974; SORENSON et al., 1971; STIPES, 1970; STOUT and SHAW, 1973; WHITNEY et al., 1968.

From this work it is obvious that the suitability of new methods must be studied individually for each group of fungi. Our own research indicates that the taxonomy of higher basidiomycetes is accessible through electrophoretic studies. For the following reasons, we prefer the use of enzyme patterns rather than examinations of protein spectra:

1) Most of the stains used to dye enzymes in electrophoresis gels are less time consuming than the unspecific staining of proteins which requires a destaining procedure.

2) There is an enormous number of proteins present in crude extracts. Though it is sometimes possible to separate 20 or 30 bands in one gel, most proteins will overlap, or strong bands will mask the weaker ones. The staining of enzymes simplifies the pattern and, in addition, introduces a "second dimension" by allowing the separation of different enzymes by specific staining methods in parallel runs.

3) In most cases enzyme tests in gels are far more sensitive than protein stainings, allowing the analysis of less concentrated protein solutions. This is of considerable practical importance because it is sometimes difficult to get concentrated protein extracts out of fungal mycelia.

4) According to genetic studies with single mutants of the ascomycete Podospora anserina (unpublished) it seems that enzyme patterns are - probably due to statistical reasons - less influenced by mutations than protein spectra.

Enzymes   217

The first to use enzymes for taxonomic purposes was BAVENDAMM in
1927. His test was based on colour reactions of fungal phenoloxidases
with phenolics added to the substrate (for literature see KÄÄRIK,
1965; NOBLES, 1958; BLAICH and ESSER, 1975). The determination of enzymes in crude extracts is, however, not sufficient to yield specific
characters for the delimitation of species, because in most cases enzymatic activities are not due to the action of one single enzyme,
but to the joint action of a number of multiple forms with similar
specificity.

To characterise a species, it is therefore necessary to use the above
cited methods by which such isoenzymes can be demonstrated separately.

There are no rules, concerning the   s e l e c t i o n   o f   e n -
z y m e s   f o r   t a x o n o m y . It is, however, convenient to
use types which are revealed by simple procedures and cheap chemicals.
An enzyme pattern should not be too simple in order to offer more
possibilities for variation. On the other hand too many enzyme bands
are difficult to handle. From over 10 types of enzymes tested by us,
esterases (more exact: ʌ -naphtylamidases), aminopeptidases (more
exact: aminoacid-ɔ-naphtylamidases) and phenoloxidases (laccases)
were best suited for taxonomic purposes.

Like morphological characters, enzyme patterns of fungi are also submitted to changes. If fungal mycelia are cultured under various conditions, especially in different culture media, if they are harvested
at different ages and if the mode of preparation of crude extracts is
changed, the enzyme patterns obtained are not identical. There are
several reasons for this behaviour, as for instance the induction of
various enzymes by different substrates, differential gene activity
during growth of the fungal thallus, reversible formation of enzyme
polymers, etc. One might conclude that enzyme patterns therefore are
of doubtful value for taxonomy. In fact it is difficult to compare
results from different laboratories. However, our own studies have
shown that patterns are reproduciable over the years, if conditions
are kept constant.

To obtain optimal results it is necessary to grow the fungi in question at the same time and under identical conditions, or to use sporophores of the same age. The analyses should be carried out in parallel runs with identical equipment. If there are many species and/
or enzymes to be compared, it is recommended to select one standard
species (e.g. the well-analyzed Trametes versicolor), which is grown
and analyzed parallel to the species in question. The growth rate of
mycelia may be quite different. It is recommended to compare thalli
not of the same age but at the same stage of development.

3) Methods

a) Mycelia

Both fruit bodies and vegetative mycelia of fungi may be used to prepare crude extracts. Their enzyme patterns are, however, only partly identical and may be compared only if obtained from the same structures (Fig. 1). Though the use of fruit bodies is very convenient and both fresh and deep freezed specimens may be used, protein extraction is only possible in species with fleshy caps. Fortunately the mycelia of fungi with tough fruit bodies are growing fast and are easily cultured, whereas slowly growing species, e.g. mycorrhizal ones, usually have soft sporophores.

For the growing the mycelia any medium will do, if the fungi accept it. However, it should be kept in mind that all fungi which are to be compared must be grown under identical conditions, so sometimes the medium will not be optimal but a compromise for all species.

A culture medium which has proved suitable for most higher basidiomycetes (with the exeption of mycorrhizal fungi) is cornmeal-extract (10 g per 100 ml water) with 3% malt extract added. Its pH-value should be adjusted to 5 by adding KOH.

Mycelia are isolated by cutting out pieces of the trama of a sporophore under sterile conditions and by placing them on agar at room temperature. These pieces will form colonies which may be subcultured if necessary. The mycelial colonies are then cut in small fragments (2 x 2 mm) and transferred in flasks containing a 1 cm layer of liquid medium. The best results are obtained with still surface cultures using FERNBACH or ERLENMEYER flasks. To get enough mycelium for analysis a surface of about 10 dm^2 is needed.

There should be a constant number of pieces in proportion to the surface of liquid. We use one block per 5 cm^2. The number is, however, not critical.
The mycelium is harvested when a closed mat has formed; the time depends on the growth rate of the species in question. Most polypore fungi may be harvested after one week at room temperature, whereas the growth of Agaricales is usually rather slow. Mycelia are separated by filtration on cheese cloth and squeezed out between filter paper. They may be stored some days at -30°C without changing the enzyme content. Freeze drying has proved harmful.

b) Preparation of Crude Extracts

The mycelia, or pieces of sporophores, are ground with twice their weight of sand and the same weight of water in a mortar until a creamy mass is obtained. It is usually not necessary to use a buffer because the pH-value of the crude extract is nearly neutral and buffered by the amount of ampholytes it contains. Sand and cell debris are then centrifuged off and the supernatant is analyzed by disc electrophoresis. Usually 0.1-0.4 ml of crude extract is sufficient to yield good results in a gel column.

Enzymes

c) Electrophoresis

The general techniques for disc electrophoresis on polyacrylamide gels shall not be discussed here. (For literature see MAURER 1971). Some additional remarks seem, however, necessary.

In disc electrophoresis the most important parameter for the characterization of enzymes is their Rf-value. Theoretically Rf-values are constant irrespective of the length of the gels used. In practice, however, the separation distance should be held constant.

The use of gel slabs facilitates the comparison of the Rf-values of enzyme patterns, but the liquid volume of the sample to be applied is rather small, so it is more convenient to use glass tubes with cylindrical gels, which allow the application of samples up to 0.5 ml, thus omitting tedious concentration procedures. The length of gels should be 50 mm. The whole length is used for separation. Since the velocity of separation differs by 10-20% between single gels, it is necessary to stop the run separately for each gel when the marker dye has reached its lower end. For this purpose a small rubber stopper is introduced into the top of the glass tube. Be sure to interrupt the electric circuit for this procedure. The electric current, should not exceed 2-3 mA per gel, to avoid distortions of the enzyme bands.

d) Enzyme Tests

After the electrophoretic run the gels are removed from the tubes and immersed in the test solution using small test tubes without rims, in order to facilitate the photographic record. The test solution is replaced by water after the enzyme bands have developed. Since some patterns may change due to the oxidation of the stains by laccases or tyrosinases, it is necessary to take the pictures within a few hours.

Esterases: 1 mg alpha-naphtylamide and 1 mg Fast Red TR;
Aminopeptidases: 1 mg aminoacid-ß-naphtylamide (over 10 types are available) and 1 mg Fast Blue BT are dissolved in a drop of acetone and 3 ml 0.1 m phosphate-buffer pH 5 are added. Solutions must be used immediately after preparations.
Phenoloxidases: 5 mM solutions of guaiacol or paraphenylenediamine are prepared in the same buffer.

4) Results

If enzyme patterns are to be used as taxonomic characters for the delimitation of species, it is necessary to study a number of related species to learn the variability of the patterns.

Testing ß - g l u c o s i d a s e s of over 50 species, we found that this enzyme pattern is always similar and crosses specific and generic lines and even the limits of orders and classes. In addition the enzyme produces oligomers of different molecular size. The resulting pattern is inconstant and useless for taxonomic studies.

In contrast, e s t e r a s e   p a t t e r n s are constant and characteristic. Similarities between species being accidental, electropherograms of these enzymes may be used as "fingerprints" for the identification of mycelia without fruit bodies (fig. 2).

In fig. 3 the a m i n o p e p t i d a s e   p a t t e r n s of 38 species are summarized. In this case not only Rf-values may be used to characterize single enzyme bands, but also their substrate specificities. In fig. 3 the abbreviations within the bands indicate the aminopeptidase-$\mathcal{S}$-naphtylamides cleaved by them. In all species where geographical races were tested, this pattern was nearly identical, even if races were crossincompatible due to heterogenic incompatibility. Obviously, aminopeptidases offer more interspecific similarities than esterases, thus not all species may be separated by these patterns.

On the other hand they seem suited for the delimitation of orders of basidiomycetes which is not possible through the use of esterases. Though this peculiarity does not concern the species problem, it should be mentioned here: In all Agaricales tested there is a very characteristic enzyme band at Rf 0.5 which stains only with gly- ala- or proline-naphtylamide. Since it is not present in other taxa, it seems to allow a classification of doubtful species. According to their aminopeptidase pattern Flammulina and Lentinus seem to belong to the Polyporaceae, a finding which supports the systematic concept of KREISEL.

The application of our method shall be demonstrated by two examples: Fig. 4a shows the esterase patterns of geographical races of Pleurotus ostreatus, isolated in Czechoslovakia, the Netherlands and Germany. They comprise species with grey, brown, and pale fruit bodies, and there are also morphological differences between their mycelia (fig. 4b). The single column on the left of fig. 4a represents the pattern of two races of Pleurotus cornucopiae, which according to PILAT and also to QUELET *) is only a variety of P. ostreatus. The patterns of the P. ostreatus species show only a few differences. It is evident that, according to the esterase-profile, P. cornucopiae is a distinct species.

---

*) cited after MICHAEL and HENNIG (1968)

This result is corroborated by studies of aminopeptidase patterns
(fig. 5). There is no similarity between the profiles of both species.
Aminopeptidase patterns were also helpful in our second example:
According to BOURDOT and GALZIN, and also to PILAT, Phellinus pomaceus is only a subspecies of Phellinus igniarius, whereas DONK takes
it for an independent species *). According to our species concept,
based on the dissimilarity of enzyme profiles, it seems to be a variety (fig.6).

5) Concluding remarks

These few examples show the utility of enzyme spectra as a tool in
taxonomy of higher basidiomycetes. Obviously it is not necessary to
know the function of these "isoenzymes". Very probably they are partly due to artifacts, a fact which does not diminish their value as
reproducible characters of a fungal species. On the other hand the
field of interest should be restricted to the taxonomic significance
of an isoenzyme pattern. Disc electrophoresis and electrofocusing are
valuable tools in the study of enzymes - if these methods are not the
sole sources of knowledge. One should therefore have reservations
about drawing any conclusions concerning function and specificities
of enzymes which are known only as coloured bands in electropherograms of crude extracts.

Acknowledgements

This work was supported by the Deutsche Forschungsgemeinschaft
Bad Godesberg. We are very much indebted to the direction of the
Parco Nazionale Gran Paradiso, Torino (Italy) for the opportunity
to collect geographical races of wood rotting fungi.

---

*) cited after MICHAEL and HENNIG (1968)

6) Literature

Abbott, L.K., Holland, A.A. 1975: Electrophoretic patterns of soluble
 proteins and isoenzymes of Gaeumannomyces graminis.
 Aust.J.Bot. 23, 1-12.
Bavendamm, W. 1928: Über das Vorkommen und Nachweis von Oxydasen bei
 holzzerstörenden Pilzen. Z.Pflanzenkrankh.Pflanzen-
 schutz 38, 257-276.
Blaich, R., Esser, K. 1975: Function of enzymes in wood destroying
 fungi. II. Multiple forms of laccase in white rot
 fungi. Arch.Microbiol. 103, 271-277.
Bresinsky, A. 1974: Zur Frage der taxonomischen Relevanz chemischer
 Merkmale bei höheren Pilzen. Num.spéc.Bull.Soc.Lin-
 néenne Lyon, 61-84.
Durbin, R.D. 1966: Comparative gel-electrophoretic investigation of
 the protein patterns of Septoria species.
 Nature 210, 1186-1187.
Gairola, G., Powell, D. 1971: Electrophoretic protein patterns of
 Cytospora fungi. Phytopathol.Z. 71, 135-140.
Glynn, A.N. Reid, J. 1969: Electrophoretic patterns of soluble fungal
 proteins and their possible use as taxonomic crite-
 ria in the genus Fusarium. Can.J.Bot. 47, 1823-1831
Hall, R. 1967: Proteins and catalase isoenzymes from Fusarium solani
 and their taxonomic significance. Aust.J.Biol.Sci.
 20, 419-428.
Käärik, A. 1965: The identification of the mycelia of wood decay fungi by their oxidation reactions with phenolic compounds. Studia forestalia Suecica 31, 1-80.
Kulik, M.M. Brooks, A.G. 1970: Electrophoretic studies of soluble
 proteins from Aspergillus spp. Mycologia 62,
 365-376.
Maurer, H.R. 1971: Disc electrophoresis and related techniques of po-
 lyacrylamid gel electrophoresis. W. de Gruyter,
 Berlin.
Meyer, J.A., Renard J.L. 1969: Protein and esterase patterns of two
 formae speciales of Fusarium oxysporum. Phytopatho-
 logy 59, 1409-1411.
Milton, J.M., Rogers, W.G., Isaac, I. 1971: Application of acrylamide
 gel electrophoresis of soluble fungal proteins to
 taxonomy of Verticillium species. Trans.Brit.Mycol.

Soc. 56, 61-65.
Nasuno, S. 1972: Differentiation of Aspergillus sojae from Aspergillus oryzae by polyacrylamide gel disc electrophoresis. J.Gen.Microbiol. 71, 29-33.
Nasuno, S. 1972: Electrophoretic studies of alkaline proteinases from strains of Aspergillus flavus group. Agric.Biol. Chem. 36, 684-689.
Nealson, K.H., Garber, E.D. 1967: An electrophoretic survey of esterases, phosphatases, and leucine aminopeptidases in mycelial extracts of species of Aspergillus, Mycologia 59, 330-336.
Nobles, M.K. 1958: Cultural characteristics as a guide to the taxonomy and phylogenie of the polyporaceae. Can.J.Bot. 36, 883-930.
Pelletier, G., Hall, R. 1971: Relationships among species of Verticillium: Protein composition of spores and mycelium. Can.J.Bot. 49, 1293-1297.
Reddy, M.M., Threlefeld, S.F.H. 1971: Genetic studies of isoenzymes in Neurospora: I. A study of eight species. Can.J. Genet.Cytol. 13, 298-305.
Righetti, P.G., Drysdale, J.W. 1974: Isoelectric focusing in gels. J.Chromatogr. 98, 271-321.
Selvaraj, J.C., Meyer, J.A. 1974: Electrophoretic protein and enzyme patterns and antigenic structure in Verticillium dahliae and V. alboatrum. Mycopathol.Mycol.Appl. 54 549-558.
Sorenson, W.G., Larsh, H.W., Hamp, Susan 1971: Acrylamide gel electrophoresis of proteins from Aspergillus species. Amer.J.Bot. 58, 588-590.
Stibes, R.J. 1970: Comparative mycelial protein and enzyme patterns and four species of Ceratocystis. Mycologia 62, 987-995.
Stout, D.L., Shaw, C.R. 1973: Comparative enzyme patterns in Thamnidium elegans and T. anomalum.Mycologia 65, 803-808.
Whithey, P.I., Vaughan, J.G., Heale, J.B. 1968: A disc electrophoretic study of the proteins of Verticillium albo-atrum, Verticillium dahliae and Fusarium oxysporum with reference to their taxonomy. J.Exp.Bot. 19, 415-426.

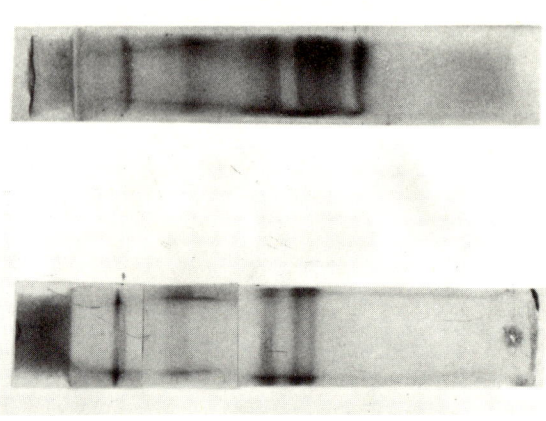

figure 1   Intracellular esterases of Piptoporus betulinus, as revealed by disc electrophoresis on 10% polyacrylamide gels. Staining with 1 mg α-naphtyl-acetate and 2 mg Fast Red TR in 3 ml phosphate buffer (pH 5, 0,05m)

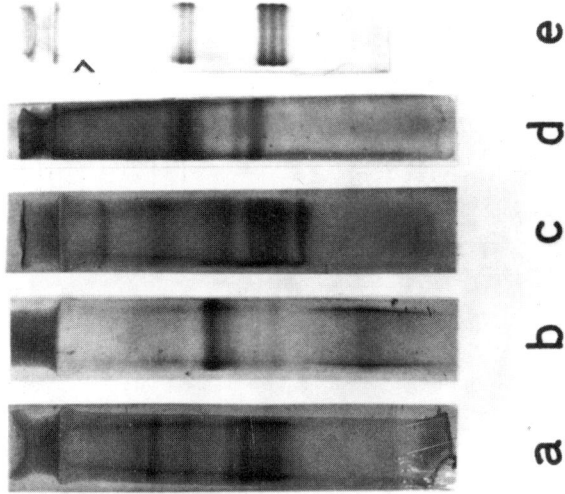

figure 2 Esterase patterns of fungal mycelia

a) Pholiota aurivella
b) Pholiota squarrosa
c) Piptoporus betulinus
d) Armillaria mellea
e) Pleurotus dryinus

Agaricales

Rf-value	Agrocybe aegerita	Pholiotas aurivella	Hypholoma fasciculare	Kuehneromyces mutabilis	Coprinus plicatilis	Pholiota squarrosa
0,1	Glu Asp		Glu Asp	Glu Asp		
0,2	Pro / Leu Lys Arg / Ala PTT	(Pro) / Leu	Leu / Pro	Leu PTT Pro / Arg Lys	Leu PTT Pro / Arg Lys	Leu PTT (Arg Pro)
0,3				similar: Cortinarius glaucopus Cortinarius hasei		
0,4			Leu Lys Arg			Leu Lys Arg
0,5		PTT double band	PTT			
0,6	Ala Gly Pro	Ala Gly Pro	Ala Gly Pro	Ala Gly Pro	Ala Gly Pro	Ala Gly Pro

Poriales

Rf-value	Gloeophyllum abietinum	Heterobasidion annosum	Ganoderma applanatum	Coniophora cerebella	Fomes fomentarius	Trametes gibbosa
0,1	Glu Asp	Glu Asp	Glu Asp	(Glu)	Asp Glu	Glu Asp
0,2		Lys Leu Arg (PTT)		Leu Lys Arg (PTT)		(Gly Pro)
0,3	Pro Leu Ala		(Leu)		Gly / Leu Arg Lys	Leu Lys Arg
0,4				Gly / PTT		Leu PTT
0,5	(Phe Arg) / Ala Gly Pro (PTT)		Ala Gly Pro	PTT	Leu PTT Ala (Gly)	Leu PTT Pro / Ala Gly Pro
0,6	Phe Tyr Leu (Trp)	Phe Tyr Leu (Trp Ala)	PTT Leu		PTT	Leu PTT

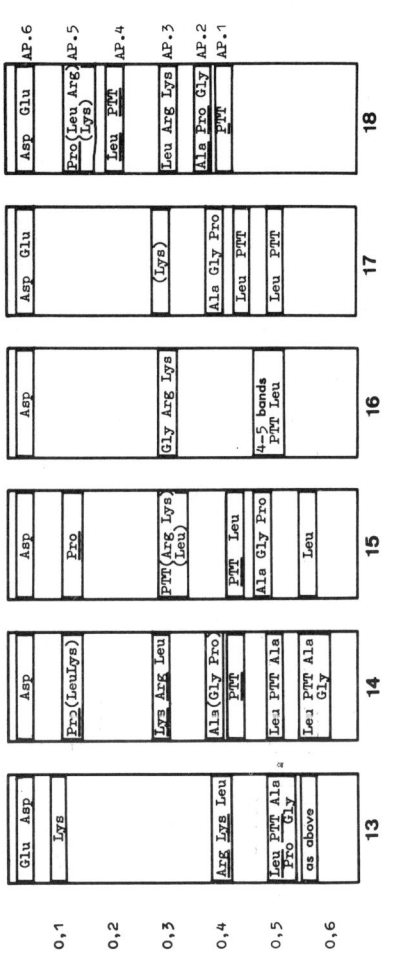

figure 3 Aminopeptidase patterns of higher fungi, as revealed by disc electrophoresis on 10% polyacrylamide gels, stained with 1mg aminoacid-ß-naphtylamide and 2 mg Fast Blue BT in 3 ml phosphate buffer (pH5, 0,05m). The abbreviations indicate the substrate-

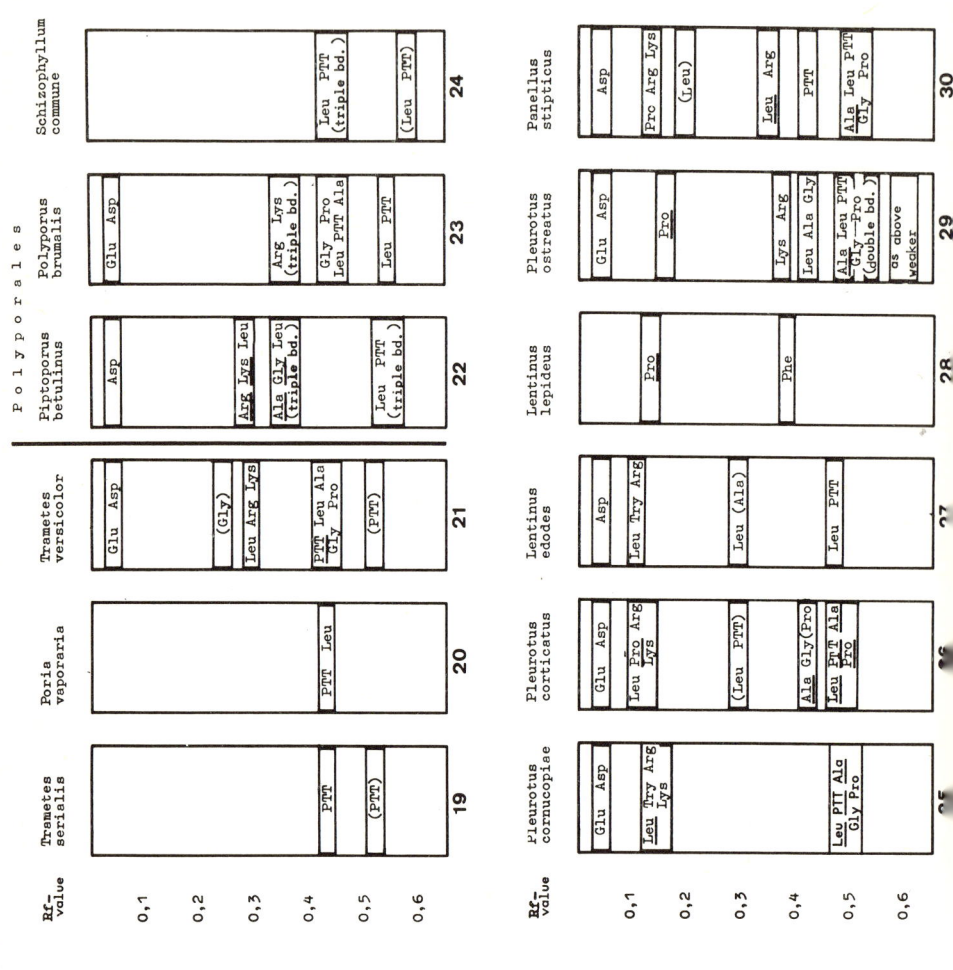

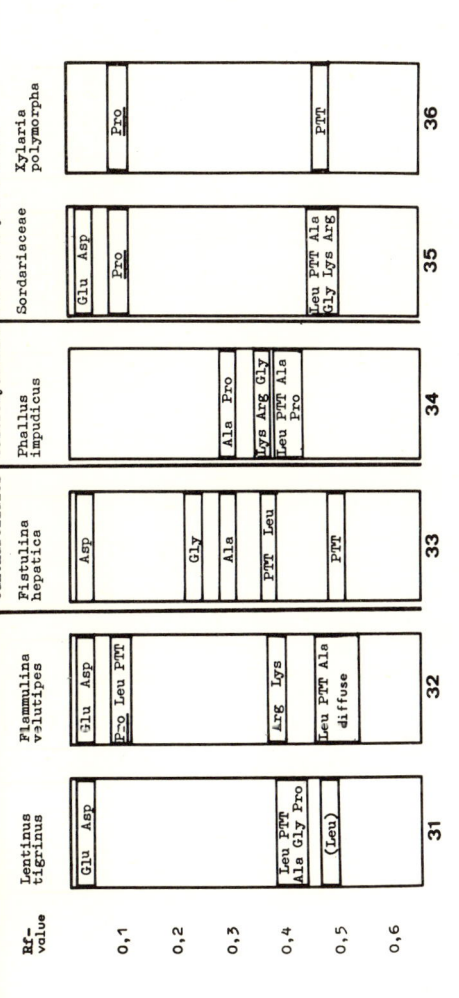

(figure 3, continued) specificity of the bands. PTT = Phe, Tyr, Trp. Example: Leu, Try (Arg) indicates an enzyme band which cleaves leucine-ß-naphtylamide, tryptophane-ß-naphtylamide and arginine-ß-naphtylamide. The relative intensity is indicated by underlining or bracketing respectively.

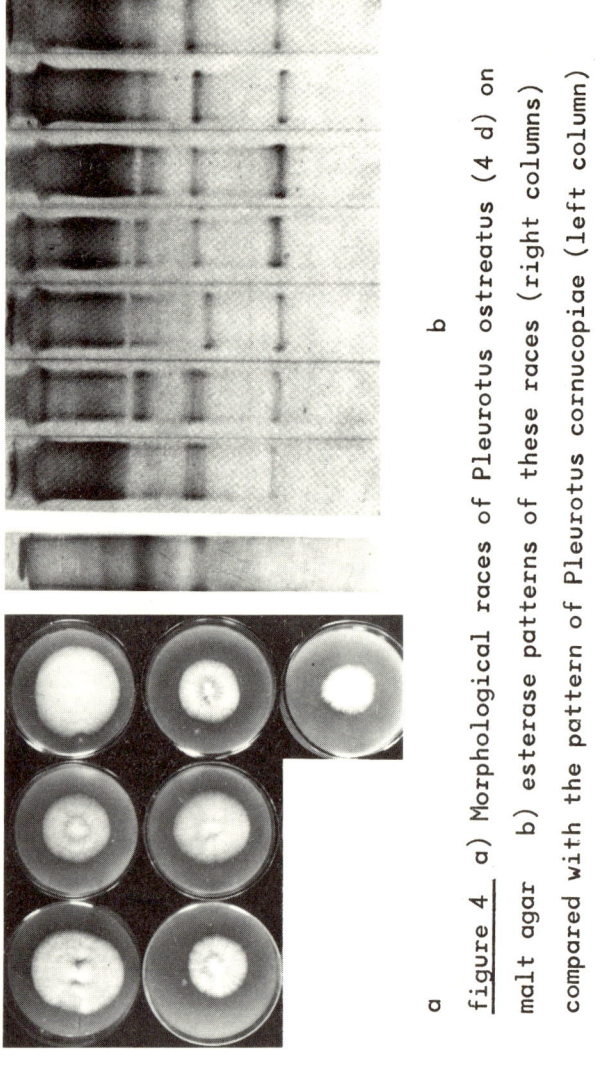

figure 4 a) Morphological races of Pleurotus ostreatus (4 d) on malt agar   b) esterase patterns of these races (right columns) compared with the pattern of Pleurotus cornucopiae (left column)

P. ostreatus   P.cornucopiae

Leu Phe Tyr Try Ala   Leu Phe Tyr Try Ala Gly
Gly

figure 5 Intracellular aminopeptidases of Pleurotus ostreatus and P. cornucopiae, revealed by incubation of disc-gels with different aminoacid-ß-naphtylamides (indicated at the bottom of the gels). See also legend of figure 3.

Aminopeptidases of

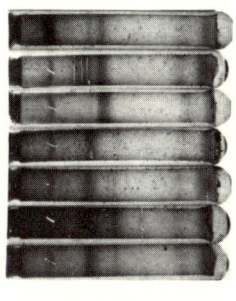

Phellinus igniarius

Phellinus igniarius var. fulvus (Phellinus pomaceus)

Leu Phe Tyr Trp Ala Gly Pro

figure 6 Intracellular aminopeptidases of Phellinus spp. (see also figures 3 and 5)

DISCUSSION FOLLOWING DR. BLAICH'S PAPER

CLEMENCON: I wanted to ask you, where in your classification of methods you put the analysis of other substances like pigments, toxins, etc...?

BLAICH: Oh yes, I forgot that. Of course between morphological characters and enzymes.

CLEMENCON: I think that many taxonomists really concerned with taxonomy have, at least mentally, made the bridge between purely morphological and some more chemical characters. This is really coming into our laboratories more and more.

BRESINSKY: Looking on your material I would be inclined to say that your patterns are fairly good, however the conclusions are keen. Let me explain this. In case of Pleurotus cornucopiae, I think you have had good luck that the enzyme pattern falls together with the morphological species concept, in the other case of Phellinus you just selected one or two species from a great variety of species and what you missed to do, was to look at other species, as far as their enzymes are concerned. It might be, and this you should consider, that all species within the genus Phellinus have very similar enzyme patterns. From the observation of two species of Phellinus having so similar enzyme patterns you are not able to draw the conclusion that they are only subspecies or varieties. I should remind you in this context, that there are existing more taxonomic treatments of fungi and even more modern ones than Michael-Hennig. In Phellinus, a very good and modern systematic revision by Nimelä in Finland gives many morphological characters separating Phellinus pomaceus from Phellinus igniarius on the level of species.

BLAICH: Of course I did not present all the species investigated. We did not restrict ourselves to these two examples. I simply took examples where we could test what we call geographical races. We first tried to show the variability of the enzymes within one species, and then we went farther. And, of course, I do not say that one could draw any conclusions bases on these enzymes only. It is just a new morphological character which is nearer to the genes.

Enzymes   225

SMITH: It seems to me that we are approaching a problem that is very
important to this symposium. But I think also, when we start with individuals, like the individuals of the human race, we find that each
individual is different, and we find that each family, in the sense of
a human family, has characters different from neighbouring families,
and you go on down the line, the whole progression of the living organisms is one of diversity   down to the individual. And our
problem here is to try to define where the species line is a cut-off
point separating subunits down to the level of individuals. I think
here we are getting up to some characters that probably will en up at
the individual bases, like blood types in the human race.

BLAICH: I do not think so. Because, it is a quite common method to
make enzyme spectra. And when an enzyme spectrum is not identical
with the normal one one thinks that this individual is ill. These
patterns are constant and characteristic for the human species.

CLEMENCON: You should not confuse enzyme patterns with protein patterns. I did some protein analyses on electrophoretic gels, a slightly different method than Dr. Blaich is using, and did it with Boletes, some Amanitas, and I got some very interesting results. But, as
Dr. Blaich said, much to many proteins to deal with. My method allows
to separate up to 70 bands, so it became impractical and in reading
Dr. Blaichs paper I became quite convinced that enzymes are much better than proteins. With proteins we run into the problems of individuals, with enzymes this is much less the case.

KEMP: So far from Monday until now we have been looking at different
ways of detecting patterns and variation. As Prof. Smith said we now
come to the point of deciding which taxa we can call species and
which ones are below the species level. To me the only way of valida-
ting whether a pattern is characteristic of a species is by doing
crossability tests between the isolates involved. Basically, if viable dicaryons are formed then the strains belong to the same species. We always come back to the sexual life-cycle, to the sexual
species concept.

BOIDIN: Vous avez souligné que les protéines, et parmi elles les enzymes, sont extrèmement nombreuses. Les chimiotaxonomistes qui utilisent les pigments pensent que les substances taxonomiquement les plus

intéressantes sont les pigments physiologiquement secondaires ou inutiles. En est-il de même pour les enzymes? Lesquelles faut-il choisir?

BLAICH: Il faut essayer si les enzymes soient constant ou non, s'ils soient typiques ou non, etc. Il faut choisir.

BOIDIN: Vous faites alors comme le morphologiste. Vous essayez et faites confiance à votre flair!

BLAICH: Oui.

BOIDIN: Il est vrai que vous avez comparé des groupes très divers. Est-ce intéressant, ou ne faudrait-il pas comparer les espèces d'un même genre ou de genres voisins?

BLAICH: On peut faire les deux.

BOIDIN: Les comparaisons protéîques d'espèces trop éloignées ne sont-elles pas dénuées de sens?

BLAICH: Il a une certaine variation maximale, et dans les espèces très éloignées les variations ne peuvent être encore plus variées. Il y a des coincidences. Par exemple, le Fistulina hepatica et Fomes fomentarius. Il semblerait que ces deux espèces seraient relativement proches.

CLEMENCON: Cela veut dire qu'on ne peut pas comparer des taxa très éloignés.

BOIDIN: Je le pense, oui.

BRSINSKY: Lawson, Harris et Balla (1975) *, working with different polypores, found less resemblance of enzyme bands between Pycnoporus cinnabarinus and Pycnoporus sanguineus, two very closely related species, than between other species that are far away from each other.

BLAICH: That's right! You can use such a pattern only as a finger-

---

*) Economic Botany 29: 117-125, 1975

print. The resemblances may be purely accidental. But if you take these patterns from one species, they are fairly constant.

CLEMENCON: This is in line with what a friend told me. He works with proteins obtained from ribosomes, and he says it works best with species closely related and degrades rapidly in significance as the species get farther away taxonomically. He is working with yeasts.

BLAICH: We analysed about 50 species by now, and I am able to recognize each individual species according to its esterase pattern, but it is not possible to group them. Even Basidiomycetes and Ascomycetes are not more different than two species,

CLEMENCON: That means that any formula of numerical taxonomy is useless.

BLAICH: Yes.

HORAK: Did you try to culture the same species on different media, and what was the influence?

BLAICH: Yes, I tried. The growth rate is different and the morphology of the mycelium is different, even the fruit bodies may differ a little bit, but we did this only in order to find the most suitable medium.

HORAK: In connection to the enzyme patterns, any differences?

BLAICH: Different enzyme types react differently. Again trial and error, but if you compare fungi, of course you must use the same media.

KUEHNER: Je pense que les travaux préliminaires de Mr. Blaich sont intéressants et prometteurs. Il est bien évident, comme il l'a dit, qu'une identité dans les stocks enzymatiques de deux souches n'indique pas qu'elles appartiennent à une même espèce, pas plus qu'une identité dans le nombre de noyaux des spores de deux souches d'<u>Hygrocybe</u>. Il n'en pas moins vrai que de tels caractères, négligés jusqu'ici par des Systématiciens, peuvent aider ceux ci car ils s'ajoutent à l'arsenal des caractères morphologiques et anatomiques et que l'on ne possédera jamais trop de caractères pour apprécier les affinités des champignons. Il me semble que c'est dans cet esprit que Mr. Blaich a conclu tout à l'heure.

BLAICH: Exactement.

THE GENUS PLEUROTUS AS AN AID FOR UNDERSTANDING THE CONCEPT OF SPECIES IN BASIDIOMYCETES

A. Bresinsky, O. Hilber*) and
H.P. Molitoris**)

1. Introduction

The species as basic unit of systematic diversity has been defined for the higher fungi by generations of mycologists on the basis of observations upon material collected in the field. In many cases, however, cultural experiments have been omitted, due to difficulties in obtaining fruit bodies. Careful observations and descriptions are important elements to recognize the evolutionary patterns, they are, however, not sufficient per se to understand the biology of species. Observations and descriptions, therefore, should be supplemented by experiments for delimitation and definability of species. In addition, questions about the evolutionary process of speciation must be solved experimentally.

Before dealing with the special suitability of the genus Pleurotus for the understanding of the species concept in higher fungi, we first would like to discuss some relevant questions and exemplary investigations.

Concerning constancy and stability of the observed taxonomic features against environmental conditions, Gäumann (1923) has made important contributions by a number of classic experiments. They are based on the consideration that phenetic differences may be modified by superposition of environmental influences which makes it difficult to recognize the differences due to genetic factors. In fact, Gäumann's experiments on Peronospora have shown that extreme fluctuations of temperature and humidity in connection with great differences in age result in deviations of maximal 11 microns in the diameter of the conidia. This equals after all half the diameter of a normal conidium

---

*) Cultivation of carpophores, crossings and their evaluation by O. Hilber with technical assistance of Mrs. R. Maier.
**) Growth experiments and enzyme spectra by H.P. Molitoris with cooperation of A. Schärtl.

of Peronospora. In the field, however, character-shifting environmental conditions will hardly play together in that way that extreme deviations result from these.

It is an other question to what extent different substrates are to be made responsible for differences in the features of two compared fungi. The role of substrate-induced modifications and the extent of difficulties which arise by them for the delimitation of species is controversial. Concerning the powdery mildews (Erysiphales) and downy mildews (Peronospora) it was possible by crossinoculation and by observation on mutual hosts to exclude a hypothetically possible influence by different hosts (Blumer 1967, Gustavsson 1959). Contrary to that, other authors (Schweizer, 1919) observed substrate induced modifications of characters used as taxonomic criteria in other parasitic fungi (e.g. Bremia).

A profound understanding of the species as taxonomic unit is impossible without information about the evolutionary processes which result in speciation. Variation and selection are nowadays commonly accepted factors in evolution. Furthermore, for the taxonomist whose job is to look for discontinuities the i s o l a t i o n of species is another very important, almost predominant complex of factors. This, because due to the different isolating mechanisms developing divergencies are stabilized and therefore the distinction of species becomes possible.

In the fungi, inhibition or prevention of free mating by genetic isolation is the main isolating factor in the diversification of species. Ecological isolation, e.g. by specialization for a certain substrate or host, is in this context merely an additional factor. Only seldom, like in the case of yeasts, it seems to be an important one. Factors preventing mating become manifest in fungi partially already before a morphological or ecological differentiation may be observed (e.g. between strains of Podospora; heterogenic incompatibility, Esser, 1967). In this context the fungi differ from other organisms. Kemp (1975) even considers genetic isolation to be the first step in the evolution of new species of fungi and proposes the following sequence for the single steps of this process: Mutation of the genome - cytoplasmatic incompatibility - prevention of mating - morphological and ecological differentiation. It agrees well with the importance of genetic isolation in fungi that interspecies hybrids - beside a few exceptions (Blastocladiales: Allomyces; Saccharomycetales; Sphaeriales: Neurospora; Ustilaginales) - have been never obtained. This fact is not only in favour of the efficiency of genetic isolation, but also supports the species concept of the taxonomist not being too narrow. How well the minutious and for the extern observer often suspicious way of the taxonomic work corresponds with the phenomenon of genetic isolation shall be shown by the studies of Lamour (1965 and 1972) on alpine species of Clitocybe (Fig.1).

V a r i a t i o n includes process and result of any change within the genome and leads to differences within a series or a population. Variation includes also the pattern of character distribution among the individuals of a population or a species. Changes may be caused by gene mutations. Within the context of speciation, however, the frequency of mutations during experimental work with fungi is often overestimated, since the process of natural selection in its eliminating and therefore stabilizing function tends to be underestimated. However it may be possible that changes by gene mutation (spontaneous rate of mutation in fungi 1×10^{-7} to 1×10^{-5} according to Esser and Kuenen, 1965) become decisive for evolution in the long run, since fungi show such a high number of nuclei and nucleic divisions before undergoing caryogamy. Unfortunately, changes in chromosome number, e.g. by polyploidization, in fungi can hardly be observed or indirectly via differences in nucleus volume. A change in the genome may also be accomplished by recombination during the sexual cycle according to the possibilities given by the respective mating system. (Esser, 1974). Finally, one has to keep in mind the changes by mitotic recombination following parasexual processes, especially since such processes have been shown to exist also in basidiomycetes (Coprinus: Prud'Homme, 1963; Swiezynski, 1963).

S e l e c t i o n may have as well a stabilizing (under fairly constant environmental conditions) as also a promoting function in the evolution of new species. Promoting factor for evolution means in this context that differences which occur in populations are enhanced by selection under variable and changing environmental conditions. The extended dicaryophase of higher basidiomycetes results in a specific mechanism of selection. According to our present knowledge in a dicaryon, not only the two nuclei within one dicaryotic hyphal compartment but also the pairs of nuclei in different hyphal compartments may differ genetically. A dicaryon therefore may represent a genotypic mosaic (Burnett and Partington, 1957). As indicated above, different parts of a mycelium may be genetically different and if reaching different substrates they are exposed to different environmental conditions. Forces of selection acting upon one single mycelium therefore may be heterogeneous and may lead to different results by virtue of the above mentioned mosaic structure of the mycelium. The diploid nuclei which are formed in the hymenium are already optimized by the selective processes before fusion of the nuclei takes place. Depending on different conditions of selection, genetically different fruit bodies may arise from one mycelium. The extension of the dicaryotic phase in the life cycle of basidiomycetes enhances the efficiency of selection and reduces the loss of individual mycelia.

In the formation, limitation and stabilization of species all three factors, variation, selection and isolation participate, often even in antagonistic interaction. Knowledge of the interaction of the various evolutionary factors should form the necessary guideline for the taxonomists work which often stresses more practical aspects. Gäumann (1923) during his work on the downy mildews has recognized the importance of experimental work for the solution of taxonomic problems. The results thus obtained for one taxonomic group, however, do not necessarily apply to others. It is therefore necessary to look for each group of fungi for the suitable representatives to work with. It will be shown here to what extent the genus *Pleurotus* is specially suited to reach an understanding of the species within the higher basidiomycetes.

The genus *Pleurotus* comprises mostly wood-decaying species, more seldom saprophytes on other plant material. *Pleurotus* species may be cultivated in the laboratory and fruit bodies may be obtained under certain conditions within 4 (3-6) weeks. Monocaryotic mycelia for mating experiments may be produced from spores. The genus *Pleurotus* includes several complex groups whose members occur in Europe under very different ecological conditions from Lapland in the north to Sicily in the south and which also show extensive variation of characteristics. Since fruit bodies may be obtained in the laboratory within a relatively short time one is able to study the influence of environmental conditions upon the expression of taxonomically rele-

vant characters. The prerequisites are given to study the presence
or even the development of mating barriers in comparison to the various degrees of morphological and ecological divergency and to obtain
insight into the process leading to speciation.

2. S u r v e y   o f   t h e   P l e u r o t u s - o s t r e a -
   t u s - c o m p l e x   i n   E u r o p e

A proper taxonomic treatment of the complex of P. ostreatus is not
yet reached. On the one side attempts to divide the complex and to
evaluate the rank of the units are still controversial, on the other
side the total diversity has not yet been recorded completely. A
great step forward in the taxonomy of this group was done by
Romagnesi (1969) through consideration of a number of microscopical
characters which had been neglected hitherto. The provisional key given at the end of this paper presents a survey of the species which
can be distinguished at the moment.

Unlike the group of P. eryngii, the species complex around P. ostreatus does not show a narrow specificty for different substrates.

3. R e a l i z a t i o n   a n d   m o d i f i c a t i o n   o f
   t a x o n o m i c   c r i t e r i a   u n d e r   l a b o r a -
   t o r y   c o n d i t i o n s

Our own investigations about the realization of characteristics under
various environmental factors are only at an initial stage. Utmost
care must be applied for interpretation. The preliminary results, reported here, show merely the tendency of further studies. Experiments
and observations are discussed separately for some features which are
used in the key for the distinction of species.

C o l o r   o f   c a p : In our experiments different substrates
did not effect the color of the cap. We used strain 2y, originating
from fruit bodies grown an Salix. The strain was identified as P. ostreatus and produced in culture also fruit bodies on stumps of Picea
and Betula. Under any circumstances the color of the cap was grey
brown and was rated as 5F3 according to Methuen *). This color was

---

*) Kornerup and Wanscher 1967.

identical with that of the fruit bodies on Salix at the original
site. Since this fungus, originally fruiting on Salix, produced fruit
bodies also on Picea and Betula it becomes evident that it is doubtful to base P. salignus and P. ostreatus in the key only on different
substrates.

If fruit bodies were produced in the laboratory under artificial
light of relatively low intensity or on wheat straw or wheat grains,
they showed paler colors. When fruit bodies were grown in open petri-
dishes with wheat straw in a humid chamber, they had a yellowish grey
color (Methuen 4B2), whereas the fruit bodies extending from the rim
of Erlenmeyer-flasks filled with wheat grains were brownish grey
(Methuen 5D4). This agrees with observations of Eger, Eden and Wissig
(1976), who in addition determined the influence of temperature. They
noted a darker color of the fruit bodies with lower temperatures and
with higher light intensities.

The colors used in the key for distinction of species remain also under laboratory conditions fairly constant if fruiting occurs under
constant and sufficient light (minimum 500 Lux). Strains of P. pulmonarius retain then their light, often almost whitish color and P.
columbinus will always be distinguished from P. ostreatus by the color of its cap. If, however, fruit bodies of the different species
were produced under different conditions the color may become similar. Changing environmental conditions therefore modify the color
differences between the species. As long as the natural condition of
fruiting is fairly well in concordance with the natural situation,
there is only a shift, not a loss of the specific differences in color.

T h e   s h a p e   o f   c a p may be different according to the
culture method. By higher humidity, as is given within an Erlenmeyer-
flask with substrate, fruit bodies are produced which are funnel-sha-
ped and whose caps are more lobated, the margin being transparently
striate and turned upward. The caps under these conditions are not
seldom hygrophanous. Experiments where the fruit bodies are produced
laterally on Erlenmeyer-flasks simulate most closely natural conditions. Here the shape of the fruit body, the central depression of
the cap, the involution of the margin and other taxonomic criteria
are not changed (e.g. shape of cap of P. cornucopiae).

L e n g t h   o f   s t i p e : Light intensity proved to be important for the length of stipe. Zadrazil and Schneidereit (1972) showed that fruit bodies grown under low light intensities produce long stipes and the fruit bodies finally assume the habitus of P. cornucopiae. Jablonsky (1975) could correlate length of stipe with light intensities. The fruit bodies show under defined light intensities a characteristic length of stipe for each strain, however, if this property is compared under variable light intensities, the overlapping of the length of stipes is rather large. From these and our own observations follows, that length of stipes can be used as taxonomic criterion only under defined and limited conditions. A genotypic differentiation is here - as it is often the case - obscured by modifications.

S c l e r i f i e d   h y p h a e : Romagnesi (1969) paid special attention also to the presence of sclerified hyphae in characterizing P. pulmonarius and partially also P. cornucopiae. According to our own observations, however, this criterion is only of limited value, since it is rather easily modified. Depending on culture conditions sclerified hyphae in fruit bodies may be present or may be totally absent. This phenomenon was already demonstrated by investigation of one single strain of P. ostreatus (2 y), whose fruit bodies if grown on sawdust of Fagus produced many sclerified hyphae, whereas fruit bodies from wheat straw, stumps of Betula and Picea as well as the original collection from Salix did not contain thickwalled hyphae in the context.

In strain 1 w of P. ostreatus we found sclerified hyphae in fruit bodies from wheat straw, but not in those grown on wheat grains (Fig. 2). Possibly also other factors than substrate may be important for the occurrence of sclerified hyphae. Stankowicova (1973) e.g. states that especially older fruit bodies of P. ostreatus contain thickwalled hyphae (Fig. 2).

D i m e n s i o n   o f   s p o r e s : Until now dimension of spores has not been considered for distinguishing species within the Pleurotus-ostreatus-complex, although different maxima of spore length indicate that statistical treatment of spore measurements could result in relevant differences of spore size. To see whether spore size could be used for taxonomic purposes in Pleurotus, we also

looked for direct or indirect correlations between substrate and
spore dimension, a question of general importance. Although at the
moment only of preliminary nature, our data, obtained from a well
suited organism, allow us to take up again the discussion of substrate modification of spores, a problem which is important especially in lower parasitic fungi.

Fruit bodies of one single strain of P. ostreatus (strain 2y) were
grown on different substrates and the distribution of spore dimensions was determined. The length of spores showed greater variation
than the diameter of spores and was therefore used for presentation
of results (Fig. 3). 100 spores each for every experiment were measured. Greatest variation in spore length was found in the fruit bodies
from P. ostreatus on Salix, the maximum of frequency being around 9,5
microns; the shoulder around 11 microns could represent a further,
hidden maximum. With exception of the fruit bodies on stumps of Picea, in all other experiments variation of spore diameters was narrower. The average length of spores from wheat straw was smaller, the
peak of frequency beeing around 8,5 microns. Clearly higher were the
values for spores from fruit bodies on stumps of Picea where the maximum frequency was at 11,5 microns. Different substrates therefore
have a strong influence on the average dimension of spores (8.5 -
10.95 microns) and the maxima of frequency may be in extreme cases as
far apart as 3 microns.

A final interpretation of the results cannot be given at the moment.
The different substrates either could modify directly or act indirectly on the dicaryons through selection, resulting in the observed
picture.

From all this follows that taxonomic characters may exhibit considerable fluctuation due to different environmental conditions. However,
the average environmental situation which induces fruiting in the
field limits the range of phenetic plasticity. Thus overlapping of
features due to unequal shift of environmental conditions as possible
in the laboratory should not be overemphasized in regard to the problem of finding proper phenetic differences for the delimitation of
species.

4. Intersterility barriers between
   species of Pleurotus

In the following two chapters inability of gene exchange between different species is contrasted with internal fertility within species. Lack of external interbreeding combined with the presence of morphological differences between the units supports the validity of an adopted species concept. Regarding the "bona fide" - species of Pleurotus treated in the key this means necessity of validation by looking for the established mating barriers. This procedure was followed in some of the possible combinations.

a) P. pulmonarius x P. columbinus
For both species (P. columbinus, strain 1n; P.pulmonarius, strain 1r) the mating types were determined and the expected scheme of tetrapolar incompatibility has been confirmed. The monocaryotic mycelia were obtained from a single sporocarp of each of both species. For each species matings of the monocaryotic mycelia resulted in 25% formation of clamp connections (table 1a and 1b). However, in no case clamp connections occurred in confrontations between monocaryotic mycelia of P. pulmonarius and P. columbinus (table 1c). Using different strains of both species the crosses were repeated with the same result: in no case formation of clamp connections could be observed.

b) P. pulmonarius x P. ostreatus
Not any combination of monocaryons of these two species led to clamp forming mycelia (1d x 1u and 1r x 1s).

c) P. pulmonarius x P. cornucopiae
Crosses of monocaryons of P. pulmonarius (strain 1r) with monocaryons of P. cornucopiae (strain 4r) did not yield any clamp connections. From these observations follows that P. pulmonarius represents an independent species, which is genetically isolated from P. columbinus and P. ostreatus in spite of only minute differences in macroscopical and in microscopical features.
Further interspecies crosses are listed in table 2.

5. Internal fertility within the
   species of P l e u r o t u s

According to multiple allelism of incompatibility genes the percentage of dicaryotization between strains of one species increases at a considerable rate if these strains are of different origin. This phenomenon is demonstrated with different strains of P. pulmonarius and P. ostreatus (table 3a-3d). Dicaryotization occurred between strains which showed clear deviation of several properties. Since recombination of varying characters is favored by multiple allelism of incompatibility genes this means increase of infraspecific variation. New species may predominantly originate from such units with a high degree of internal variation.

Crosses between monocaryons with throughout different A- and B-factors possess according to Eugenio and Anderson (1968) an increased chance to succeed, since the total weight of fruit bodies increases and by that the number of spores is also higher.

Some remarks on the results presented in table 3 are necessary in order to demonstrate exchange of differing features in intraspecies crosses. The fruit bodies of P. pulmonarius 4h are lighter (nearly white) than those of P. pulmonarius 1r (table 3b: 1r x 4h).

The so-called strain "Florida", originally obtained by Eger from Florida, became important for commercial cultivation, because it does not need a cold-phase for induction of fruiting. The fact that this strain "Florida" interbreeds with P. pulmonarius 1r from Austria shows that the former cannot been taken as an independant species (table 3c: 1r x 4b). Eger, Eden and Wissig (1976) describe the mating of a strain from Florida with some strains from America and Germany. The monocaryotic mycelia of these strains were in the crossings 100% compatible and largely fertile (regarding production of hybrid sporocarps and spores). Unfortunately it is impossible to identify these strains from Germany used by Eger et al. by the given general remarks about color changes due to different light intensities. The assumption of these authors, having crossed Pleurotus "strain Florida" with P. ostreatus, therefore cannot be verified. For our problem, however, it is important that Eger et al. were able to prove that interbreeding was linked with combination of different properties, e.g. temperature optima of fruiting.

Monocaryotic isolates of P. ostreatus from Germany, Westfalen, are 100% compatible with two different strains of P. ostreatus from Japan (table 3d: 1s x 1w; 1s x 1u). Recombination takes place between strains which differ in the color of the cap. The colors of the fungi from Japan are lighter, especially in older stages, than those of the strain obtained from Westfalen. The F_1-offspring of hybrid fruit bodies, which were fertile in regard to spore production, was split into darker and lighter colored individuals.

As a result of these observations we may state, that differences in features between two compared units are not per se reliable criteria to establish a species concept. In every case it is necessary to be sure that the differences are maintained by proper isolation mechanisms.

6. Enzyme spectra for characterization of species and strains of Pleurotus

Chemical analysis as tool of "biochemical systematics" (Heywood, 1973a) has been used successfully and increasingly within the last years in systematics and taxonomy of higher plants and fungi (Davis and Hoywood, 1963; Hall, 1973; Vaughan, 1975; but see Heywood, 1973b for critical comments).

Proteins are - according to the one-gene-one-enzyme-hypotheses (Beadle and Tatum, 1941) - direct products of genes and therefore also indicative for the genetic material. For that reason proteins, together with pigments, have been analyzed for presence, quality and distribution by electrophoretical and in the case of proteins also serological methods. A number of good correlations with morphological and other data has been found (see Hall, 1969; Tyrell, 1969; Hall, 1973; Bucher, 1974).

General protein profiles are often obscured and unspecific by too many proteinbands. It therefore proved advantageous either only to investigate certain fractions of total protein or - even better - only specific enzymes, characteristic for a given organism. With due care of interpretation such isoenzyme spectra appear to be useful for taxonomic and systematic purposes also for fungi (Nealson and Garber, 1967; Wang and Raper, 1969; Reddy and Threlkeld, 1971; Snyder and Kramer, 1974).

For these reasons it seemed to be worthwhile within the scope of our problem to investigate not only the morphologic criteria of the woodrotting genus Pleurotus (white rot) but also the enzymes believed to be responsible for the decay of wood. We tested therefore the phenoloxidases tyrosinase and laccase and in addition peroxidase. It was hoped to find out, whether enzyme profiles in addition to the established morphological criteria could provide useful information for the delimitation of species and characterization of strains in fungi.

For this purpose several strains of Pleurotus (P. columbinus, P. ostreatus, P. pulmonarius) were tested. In order to see the effect of the nuclear status on growth and enzyme spectra, we furthermore analyzed of one species, P. ostreatus, monocaryotic strains (1w8, 1s4, dicaryotic strains (1w, 1s) and the dicaryotic hybrid of the cross 1w x 1s.

At first g r o w t h   r a t e   and   y i e l d   o f   m y c e l i u m on different solid and liquid media at different temperatures were determined in order to characterize the strains and to obtain the optimal time of harvest for the enzyme determinations. Table 4 presents the results for growth on malt-agar and malt-broth at $27^{o}C$. The Pleurotus species and strains showed characteristic and constant growth rates on malt-agar, the growth rates being in the range of values given by Nobles (1965), Macaya - Lizano (1974-1975) and Blaich (1972). Only small differences in growth rate existed between the strains on solid media, whereas in liquid medium mycelial yield showed wider variation. It is striking that on solid medium the growth rate of the hybrid 1w x 1s is about twice as high as that of the monocaryotic crossing partners 1w8 and 1s4, whereas in liquid medium mycelial yield of the hybrid is only about one quarter of that of the monocaryotic strains.

Looking at the   e n z y m e   a c t i v i t i e s   in solid and liquid medium, peroxidase could not be found in any strain with the methods used, neither on solid medium nor colorimetrically in culture filtrates and mycelial extracts nor by staining of gels after electrophoresis. This confirms the results of Lyr (1958) for P. ostreatus.

After 2 weeks of incubation tyrosinase also was absent in malt-agar, in culture filtrates and in mycelial extracts of liquid culture. This enzyme appeared, however, in small amounts after 3 weeks on solid medium in cultures of P. columbinus and P. ostreatus. Only the monocaryotic strains and the dicaryotic hybrid of P. ostreatus did not show tyrosinase activity at this time. Presence of tyrosinase in P. columbinus and P. ostreatus agrees with the data of Hackl (1975). Our results show further the dependency of tyrosinase production from cultural age.

Laccase was always found in every strain tested in solid media with the drop-test (extracellular enzyme), in the culture filtrates (extracellular enzyme) and in mycelial extracts (intracellular enzyme) from liquid culture. This corresponds well with the data of Lyr (1958), Harkin et al. (1974), Blaich and Esser (1975), Hackl (1975) and Leonowicz and Trojanowski (1975a, 1975b).

The discrepancy between the semiquantitative data of the drop-test and the quantitative colorimetric determinations of laccase activity suggests that the "Bavendamm-test" (Lyr, 1958) should be used only for qualitative or preliminary determinations and shows that this test indicates only presence of extracellular enzymes.

In order to check also the presence of intracellular enzymes mycelial extracts in addition to the culture filtrates were investigated. From table 4 appears that laccase is here predominantly an extracellular enzyme. The monocaryons of P. ostreatus synthesize only about 1/3 of the laccase produced by the dicaryons and the hybrid. The inverse relationship between phenoloxidase- and mycelium production (see above) has been observed repeatedly (Bergmann, Molitoris, unpublished results).

As could be shown, a more specific and better reproducible picture of the enzymatic differences between the strains was obtained from the isoenzyme spectra of disc-electrophoresis and isoelectric focusing.

As figure 5 shows, in the culture filtrates generally fewer laccase bands were found, an indication for prevention or retardation of laccase secretion through the cell membrane into the medium (Molitoris and Esser, 1971).

In d i s c - e l e c t r o p h o r e s i s at pH 8.3 in all cases a laccase band remains in the spacer gel, indicating a laccase of high isoelectric point like the B-type laccases with an isoelectric point between pH 4 to 8 (Jonsson et al., 1968) or laccase II of Podospora anserina (Molitoris and Esser, 1971). In addition, all strains show a strong laccase band near the moving boundary, indicating a laccase of low molecular weight and/or low isoelectric point corresponding with the A-type laccases of Jonsson et al. (1968). Additional bands appeared in the mycelial extracts of P. columbinus and P. ostreatus and showed characteristic differences between the strains. In P. ostreatus such additional bands were found only in the mycelial extracts of the mono- and dicaryotic "s"-strains (1s4 and 1s) but not in the mono- and dicaryotic "w"-strains (1w7 and 1w) nor in the hybrid 1w x 1s. This hybrid strain therefore is not, as could be expected, intermediate in its laccase spectrum but is identical with one of the parental strains (1w8).

I s o e l e c t r i c   f o c u s i n g gives considerably better separation than discelectrophoresis. It is therefore not surprising that this method produced more distinct laccase bands. The danger of artifacts, however, is in isoelectric focusing higher. Again, A- and B-type laccases are present and both are secreted into the culture medium. All species and strains of Pleurotus showed different laccase spectra. The similarity of laccase spectra in isoelectric focusing within a species was generally not higher than between species. An exception are the spectra of the culture filtrates of P. ostreatus. In the case of the spectra from isoelectric focusing interpretation of results should await further experimentation.

As a result of these preliminary investigations it can be stated already: The analyzed strains of Pleurotus all produce intra- and extracellular laccase and exhibit characteristic, reproducible isoenzyme spectra in electrophoresis. The spectra of the hybrid (1w x 1s) of P. ostreatus show in disc-electrophoresis identity with one of the crossing partners (1w), in isoelectric focusing the spectra of the culture filtrate are at least very similar to both monocaryotic crossing partners.

Before we finally can assess the suitability of these methods for characterization of strains and species of Pleurotus a number of additional enzymes should be tested as well as the influence on the enzyme spectra of factors like mycelial age, medium and environmental conditions like temperature and light.

7. Species concept

A species comprises all individuals with indentical or continously varying characters. Individuals belong to different species if the variation of characters shows discontinuities. Species are recognized and defined by such discontinuities (Davis and Heywood, 1963). A proper delimitation of species is not influenced by the question how many genes participate in an established discontinuity if stability of differences is warranted on one side, and continuity of variation is maintained within the defined species on the other side. Genetic isolation is the most important mechanism for establishing discontinuities in the variation of fungi.

In order to limit taxonomic work with basidiomycetes on the species level the species concept should be based primarily on macro- and microscopic, macro- and microchemical and finally on ecological criteria. Physiological and biochemical characters would only be useful if the inbreeding units (i.e. the species) are characterized as a whole by such criteria. Otherwise these features would only serve to describe the internal variation and structure of species.

The species concept for the basidiomycetes seems to be comprehensible in an almost objective manner, although there are difficulties in application. In many groups genetic isolation cannot be assayed by mating experiments, and taxonomy of basidiomycetes is therefore based on a "bona-fide"-species concept. Environmental conditions may influence features to a high degree and this inhibits precise determination of specific differences.

Recognition of discontinuities and thereby of species is based on long experience, much patience and on the observation of often minute differences. This demands aptitude for carrying out such observations, which is not given everyone. Finally, species are dynamic structures which can be altered by evolutionary processes. To recognize such changes is for the systematist - contrary to common view - not annoyance but object of his studies.

8. M a t e r i a l   a n d   m e t h o d s
Isolates, pure cultures and monocaryotic mycelia were obtained and propagated using conventional methods.

F r u i t   b o d i e s , very similar to those grown in nature, were produced by filling Erlenmeyer-flasks (300 ml, wide mouth) halfway with wheat grain. After addition of distilled water the grain swelled and filled the flasks up to the rim. The flasks then were inoculated, closed with cotton plugs, covered with aluminium foil and incubated at $23^{o}C$. After 14 days the flasks were transferred to $11^{o}C$ and 500 - 600 lux constant light. The foil was removed on one side to allow lateral growth of the fungus, simulating natural conditions. Fruit bodies were produced after 4 (3 - 6) weeks, depending on strain and species. This method was superior to fruit body production inside flasks and petri-dishes.

F l a s k s : The flasks were filled with 30 g of substrate:
a) straw of wheat; sawdust of b) Fagus, c) Picea, d) Abies,
e) Betula. The substrate was covered by a thin film of the following composition: Inositol 50 mg; KH_2PO_4 0,5 g; $MgSO_4$ 0,5 g; $ZnSO_4$ 0,001g, $FeCl_3$ 0,01 g, $CaCl_2$ 0,055 g, $MnSO_4$ 0,005 g, maltose 20 g, glucose 10 g, agar-agar 20 g, asparagine 1.2 g, alanine 0,8 g, dest.water ad 1000 ml.

The flasks were inoculated and incubated for 14 days in the dark at $23^{o}C$, then transferred to $11^{o}C$ and 500 - 600 lux light intensity; removal of the plug, 80% relative air humidity. Fruit bodies were produced after 8 (6 - 10) weeks.

P e t r i - d i s h e s : 20 g of substrate, covered with a film of medium as above. The petri-dishes were kept closed in the first incubation period. During the $11^{o}C$-period they were opened and kept under a tent of polyethylene foil. CO_2, inhibiting fruiting body production, was removed by absorption by a solution of $Ca(OH)_2$.

M y c e l i a l   g r o w t h : Growth in connection with the enzyme determinations was observed on malt-broth (1,5% malt), solidified if needed by 1,5% agar. The liquid cultures (surface culture, 250 ml Erlenmeyer-flasks containing 90 ml of medium) were inoculated with 10 ml of a homogenate of a 14 day old liquid pre-culture. After 14 days of incubation at $27°C$ and diffuse light, the mycelium was harvested and separated from the culture filtrate by suction through filterpaper on a Buchner-funnel. Wet weight, dry weight ($80°C$) and pH were determined for each experiment.

C u l t u r a l   f i l t r a t e   a n d   m y c e l i a l   e x -
t r a c t s . 5 g mycelium (wet weight) were homogenized for 3 min in a precooled mortar with 10 g of washed seasand. After addition of 30 ml 0,05 M phosphate buffer, pH 6.0, the mycelium was ground for another 5 min. Mycelial fragments were separated from the mycelial extract by centrifugation (20 min, $4°C$, 10,000 x g). If necessary, mycelial extracts and cultural filtrates were concentrated for the enzyme assays in disc-electrophoresis and isoelectric focusing by ultrafiltration.

E l e c t r o p h o r e t i c   s e p a r a t i o n   o f   p r o -
t e i n s : Disc-electrophoresis was conducted according to Ornstein and Davis (1964) at pH 8,3 in 7,5% polyacrylamide gel, isoelectric focusing in a pH-gradient from pH 3,5 to 10 in 7,5% polyacrylamide gel according to Wrigley (1968).

E n z y m e   a s s a y s : Presence of peroxidase on malt-agar was determined after Lyr (drop-test, 1958) with a solution of 0.1% benzidine in extracts and cultural filtrates peroxidase was assayed for by a colorimetric test at 436 nm with guajacol and H_2O_2 according to Pütter (1974a,b).

T y r o s i n a s e on malt-agar was determined following Lyr (1958), using an aqueous solution of 0,01 M p-cresol containing 0,05 glycine. Tyrosinase in cultural filtrates and mycelial extracts was tested colorimetrically at 436 mm with 0,02 M L-tyrosine (Esser, 1963).

L a c c a s e   on malt-agar was assayed for according to Lyr (1958) with an aqueous solution of 0,005% α-naphthol or 0,005% guaiacol. Presence of laccase in cultured filtrates or mycelial extracts was investigated colorimetrically at 436 nm with 0,65mM 2,6-dimethoxyphenol in 0,1M Na-citrate/NaOH buffer, pH 5,0 according to Prillinger (1976). Laccase in the electrophoretic experiments was determined by incubation of the gels in 0.02 M guaiacol in 0.05 M phosphate buffer, pH 6.0. The gels after isoelectric focusing were equilibrated in buffer for 30 min before addition of the substrate.

1 unit of laccase-activity (1E) corresponds with a difference in absorption of 0.2/min at 436 nm and 1 cm light path.

The specifity of the enzyme assays was tested with pure enzymes, single and in mixtures. Mycelial extracts and cultural filtrates were prepared and concentrated at $4^{\circ}C$, electrophoreses and enzyme assays were conducted at $25^{\circ}C$.

All determinations on petri-dishes were done in triplicate, for the assays in liquid culture always 10 parallel flasks were used.

H e r b a r i u m   s p e c i m e n s   of species and strains treated in this paper are deposited in the Botanische Staatssammlung München (M) together with notes on the fruit bodies.

A c k n o w l e d g m e n t s :   We gratefully acknowledge the receipt of the following strains from: Prof. J. Arita (Tottori mushroom Inst., Japan; No 122/ 1u), P. Costa (Pleos, Italy; No 311-2/2v), Dr. K. Mori (Mori mushroom research Inst., Japan; No 4612/ 1w), J. Schliemann (Mykofarm, Hamburg; No Pfl 1/ 4b; "Florida") and Dr. M. Semerdzieva, Prague, CSSR (4r).

We are also indebted to G. Fuhrmann, Rastatt and A. Runge, Münster for fresh specimens of Pleurotus.

References

Anderson, N.A., S.S. Wang & J.W. Schwandt 1973: The Pleurotus ostreatus-sapidus species complex. Mycologia 65: 28-35.
Beadle G.W. & E.L. Tatum 1941: Genetic control of biochemical reactions in Neurospora. Proc.Natl.Acad.Sci. USA. 27: 499-506.
Blaich, R. 1972: Intra- und extrazelluläre Esterasespektren von Basidiomyceten: Ein Beitrag zur Artdiagnose von Holzpilzmyzelien. Nova Hedwigia 23: 923-936.
Blaich R. & K. Esser 1975: Functions of enzymes in wood destroying fungi. II. Multiple forms of laccase in white rot fungi. Arch.Microbiol. 103: 271-277.
Blumer, S. 1967: Echte Mehltaupilze, Fischer Jena.
Bucher, J.G. 1974: Anwendung der diskontinuierlichen Polyacrylamidgel-Elektrophorese in der Taxonomie der Gattung Nodulosphaeria Zürich. Viertelj.Schr.Naturforsch.Ges. 119: 125-164.
Burnett, J.H & M. Partington 1957: Proc.R.phys.Soc.Edinburgh 26: 61-68. Citation after Burnett, J.H. (1968): Fundamentals of Mycology, London, Arnold publ.
Davis, P.H. & V.H. Heywood 1963: Principles of Angiosperm taxonomy. Nostrand; Princeton, New Jersey, New-York.
Eger, G., Eden, G. & E. Wissig 1976: Pleurotus ostreatus - Breeding potential of a new cultivated mushroom. Theoretical and applied Genetics 47: 155-163.
Esser, K. 1963: Die Phenoloxidasen des Ascomyceten Podospora anserina I. Die Identifizierung von Laccase und Tyrosinase beim Wildstamm. Arch.Microbiol. 46: 217-226.
Esser, K. 1974: Breeding systems and evolution. In: M.J. Carlile and J.J. Skehel (edts.): Evolution in the microbial world. Univ.Press Cambridge, 87-104.
Esser, K. & R. Kuenen 1965: Genetik der Pilze. Springer, Berlin, Heidelberg, New York.
Eugenio, C.P. & N.A. Anderson 1968: The genetics and cultivation of Pleurotus ostreatus. Mycologia 60: 627-634.
Gäumann, E. 1923: Beiträge zu einer Monographie der Gattung Peronospora Corda. Beitr.Kryptogamenflora Schweiz 5 (4).

Gustavsson, A. 1959: Studies on nordic Peronosporas. II General account. Opera Botanica 3 (2).
Hackl, M. 1975: Nachweis von Phenoloxidasen bei höheren Pilzen unter besonderer Berücksichtigung der Agaricales, Versuch einer ökologischen und systematischen Bewertung. Zulassungsarbeit Wiss. Prüfung f.d. Höhere Lehramt. Regensburg.
Hall, R. 1969: Molecular approaches in taxonomy of fungi. Bot.Rev. 35: 285-304.
Hall, R. 1973: I. Electrophoretic protein profiles as criteria in the taxonomy of fungi and algae. Bull. Torrey Bot.Club. 100: 253-259.
Harkin, J.M., M.J. Larsen & J.R. Obst 1974: Use of syringaldazine for detection of Laccase in sporophores of wood rotting fungi. Mycologia 66: 469-476.
Heywood, V.H. 1973a: The role of chemistry in plant systematics. Pure and appl. Chem. 34: 355-375.
Heywood, V.H. 1973b: Chemosystematics - and artificial discipline. Nobel 25: 41-54.
Jablonsky, L. 1975: Einfluss der Belichtungsintensität und anderer Faktoren des Milieus auf die Entwicklung der Fruchtkörper des Austernseitlings- Pleurotus ostreatus. Ceska Mycologie 29: (3), 140-152.
Jonsson, M., E. Pettersson and B. Reinhammer 1968: Isoelectric fractionation, analysis, and characterization of ampholytes in natural pH gradients. VII. The isoelectric spectra of fungal laccase A and B. Acta chem. Scand. 22: 2135-2140.
Käärik, A. 1965: The identification of the mycelia of wood-decay fungi by their oxidation reactions with phenolic compounds. Stud. Forest. Suec. 31: 1-80.
Kemp, R.F.O. 1975: Breeding biology of Coprinus species in the section Lanatuli. Trans.Brit.Mycol.Soc. 65: (3), 375-388.
Kornerup, A. & J.H. Wanscher 1967: Methuen Handbook of Colour. Methuen, London, 2.Aufl.
Lamoure, D. 1965: Clitocybe rivulosa (Pers.ex.Fr.) Kumm.var. dryadicola Favre et Clitocybe candicans (Pers.ex.Fr.) Kumm.Bull.Soc.Mycol.France 81: 487-508.
Lamoure, D. 1972: Agaricales de la zone alpine. Genre Clitocybe.Trav. Scient.Parc National de la Vanoise 2: 107-152.

Leonowicz A. & J. Trojanowski 1975a: Induction of a new laccase form
    in the fungus Pleurotus ostreatus by ferulic acid.
    Mycrobios 13: 167-174.
Leonowicz A. & J. Trojanowski 1975b: Induction of laccase by ferulic
    acid in basidiomycetes. Acta bioch.Polon. 22: 291-295.
Lyr, H. 1958: Über den Nachweis von Oxidasen und Peroxidasen bei
    höheren Pilzen und die Bedeutung dieser Enzyme für
    die Bavendamm-Reaktion. Planta. 50: 359-370.
Macaya-Lizano A.V. 1974-1975: Pleurotus ostreatus (Jacq.ex Fr.)
    Quélet, formes et espèces affines. Comportement cul-
    tural et systématique. Rev. de Mycol. 39: 3-42.
Molitoris, H.P. & K. Esser 1971: The phenoloxidases of the ascomycete
    Podospora anserina. VII. Quantitaive changes in the
    spectrum of phenoloxidases during growth in submerged
    culture. Arch.Mikrobiol. 77: 99-110.
Nealson, K.H. & E.D. Garber 1967: An electrophoretic survey of este-
    rases, phosphatases, and leucine aminopeptidases in
    mycelial extracts of species of Aspergillus.
    Mycologia 59: 330-336.
Nobles, M.K. 1965: Identification of cultures of wood-inhabiting hy-
    menomycetes. Can.J.Bot. 43: 1097-1139.
Ornstein, L. & B.J. Davis 1964: Disc electrophoresis I u. II. Ann.N.
    Y.Acad.Sci. 121: 321-349, 404-427.
Prillinger, H. 1976: Genetische Kontrolle der Phenoloxidase "Laccase"
    des Ascomyceten Podospora anserina. Bibliotheca myco-
    logia, 51: Cramer, Lehre.
Prud'Homme, H. 1965: In: "Incompatibility of Fungi" 48-52, ed. Esser
    u Raper. Citation after Burnett, J.H. (1968): Funda-
    mentals of Mycology, Arnold publ., London.
Pütter, J. 1974a: in Methoden der enzymatischen Analyse (Bergmeyer,
    H.U. ed.), 725-731, Verlag Chemie, Weinheim.
Pütter, J. 1974b: in Methods of Enzymatic Analysis (Bergmeyer, H.U.
    ed.) p. 685, Verlag Chemie, Weinheim & Academic Press,
    New-York, London.
Reddy, M.M. & S.F.H. Threlkeld 1971: Genetic studies of isozymes in
    Neurospora I.A. study of eight species. Can.J.Genet.
    Cytol. 13: 298-305.
Rösch, R. & W. Liese 1970: Ringschalentest mit holzzerstörenden Pil-
    zen. I. Prüfung von Substraten für den Nachweis von
    Phenoloxidasen. Arch.Mikrobiol. 73: 281-292.

Romagnesi, H. 1969: Sur les Pleurotus du groupe Ostreatus (Ostreomyces Pilat). Bull.Soc.Mycol.France 85: (3), 305-314.
Schweizer, J. 1919: Die kleinen Arten bei Bremia Lactucae Regel und ihre Abhängigkeit von Milieu-Einflüssen. Verh.thurgauisch.Naturf.Ges. 23: 1-61.
Stankovicova, L. 1973: Hyphal structure in some pleurotoid species of Agaricales. Nova Hedwigia 24: 61-120.
Swiezynski, K.M. 1963: Genet.Poloniae 4: 21-36. Cited after Burnett, J.H. (1968): Fundamentals of Mycology, Arnold publ., London.
Snyder, R.D. & C.L. Kramer 1974: Polyacrylamide gel electrophoresis and numerical taxonomy of Taphrina caerulescens and Taphrina deformans. Mycologia, 66: 743-753.
Tyrell, D. 1969: Biochemical systematics and fungi. Bot.Rev. 35: 305-316.
Vandendries, R. 1933: De la valeur du barrage sexuel, comme critérium dans l'analyse d'une sporée tétraplaire de basidiomycète: Pleurotus ostreatus. Genetica 13: 202-212.
Vaughan, J.G. 1975: Proteins and taxonomy in: Proc.Phytochem.Soc., Symposium, Ghent, Belgium 1973 (Harbone, J.B. and C.F. Van Sumere, eds.) Academic Press, New York.
Wang, S.S. & J.R. Raper 1969: Protein specificity and sexual morphogenesis in Schizophyllum commune. J.Bact. 99: 291-297.
Wang, S.S. & N.A. Anderson 1972: A genetic analysis of sporocarp production in Pleurotus sapidus. Mycologia 64: 521-528.
Wrigley, C. 1968: Gel electrofocusing - a technique for analyzing multiple protein samples by isoelectric focusing. Science Tools 15: 17-23.
Zadrazil, F. & M. Schneidereit 1973: Die Grundlagen für die Inkulturnahme einer bislang nicht kultivierten Pleurotus-Art. Der Champignon 135: 25-32.

1 *Pleurotus cornucopiae* Paul. ex Fr. ss. Romagn.: strain 4r. 2 *Pleurotus "cornucopiae"*: strain 2v. 3 *Pleurotus columbinus* Quél. apud Bres.: strain 1n. 4 *Pleurotus columbinus* Quél. apud Bres.: strain 1n.

5

6

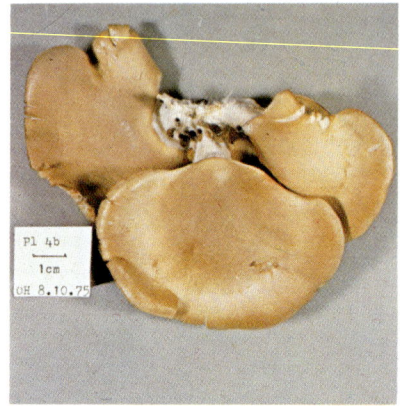

7

8

5 *Pleurotus pulmonarius* Fr.: strain 1r (f_1-fruit bodies). 6 *Pleurotus pulmonarius* Fr. ("Florida"): strain 4b. 7 *Pleurotus pulmonarius* Fr. ("Florida"): strain 4b. 8 *Pleurotus ostreatus* (Jacqu. ex Fr.) Kummer: strain 1w (left), strain 1s (right).

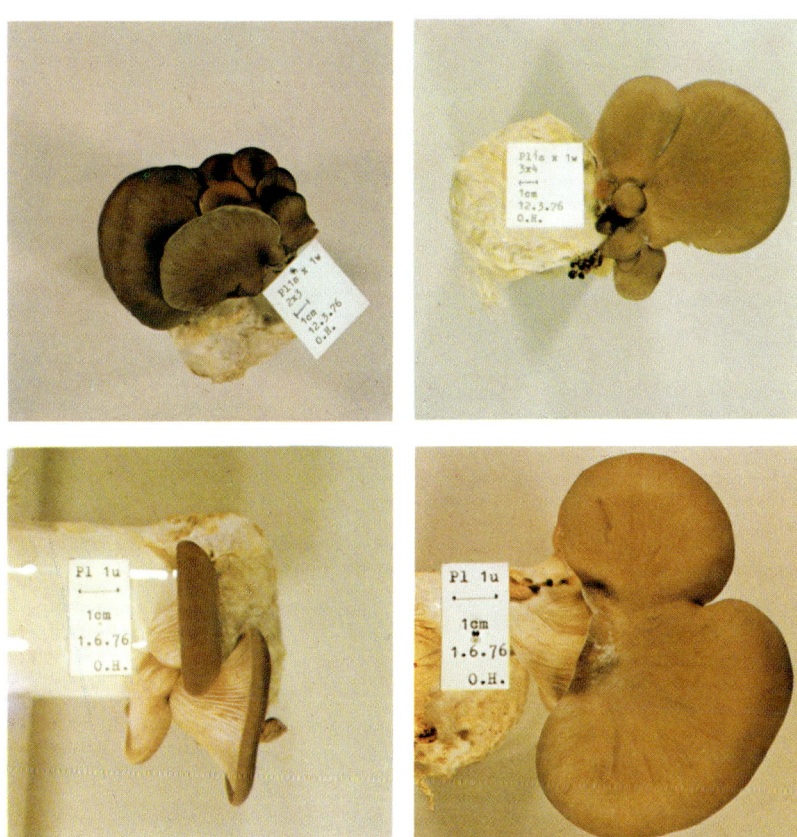

9 *Pleurotus ostreatus* (Jacqu. ex Fr.) Kummer: dark fruit bodies, f_1-generation from 1s × 1w. 10 *Pleurotus ostreatus* (Jacqu. ex Fr.) Kummer: light fruit bodies, f_1-generation from 1s × 1w. 11 *Pleurotus ostreatus* (Jacqu. ex Fr.) Kummer: strain 1u; young fruit body. 12 *Pleurotus ostreatus* (Jacqu. ex Fr.) Kummer: strain 1u; older fruit body.

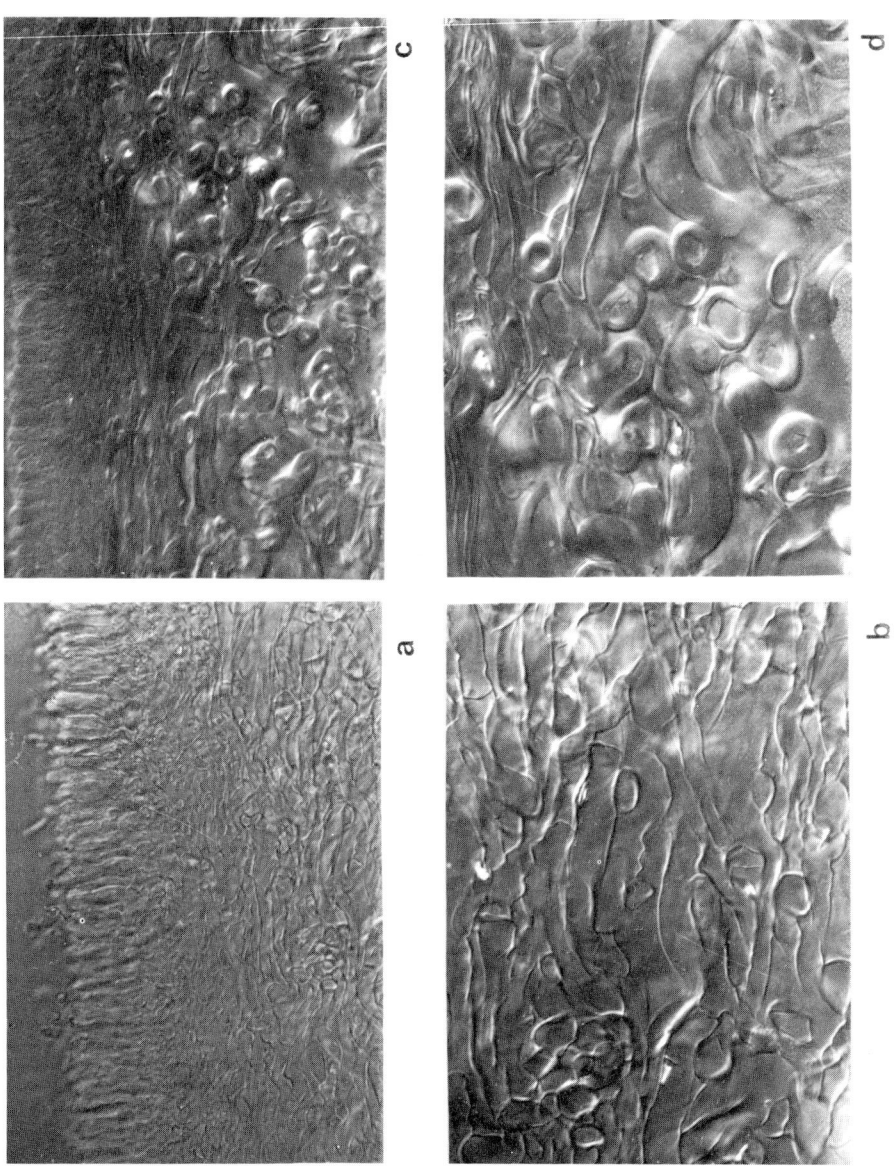

Fig. 2: P. ostreatus (lw): Trama of gills from fruit bodies cultivated on wheat grains (a-b) and on wheat straw (c-d). Note sclerified hyphae in c and d.-b and d magnified sections of a and c.

2. Survey of the Pleurotus-ostreatus-complex in Europe

Key

1 (2) Stipe dark brown in contrast with the cap which is much lighter. In the Mediterranean: Sicily
   P. nigripes Inzenga

2 (1) Stipe not dark brown; equalling in color to the cap or even lighter

3 (6) Gills decurrent on long, ± centric to excentric stipes and cap at the same time ± infundibiliform

4 (5) Spores elliptical at the inner side being slightly convex. - Cap infundibiliform, with ± light or deeper brownish colors. Gills running down the stipe reaching the base, being there repeatedly connected by anastomoses; first white then changing color to cream and becoming yellowish if dried. Stipe nearly centric or excentric, often fairly long, 3- 6.5 x 0.8 - 2.3 cm. Spore deposit pale and dingy greyish lilac. Odor specific, pleasant sweetish to nearly disagreeable, at the base of the stipe also farinaceous. Hyphae mostly with thin walls, partly also with increasingly thick walls (so called false skeletal hyphae). Fructification partly early in the season
   P. cornucopiae Pahl.ex Fr. ss. Romagn.

5 (4) Spores cylindric at the inner side being concave or straight. Cap ± umbilicate according to the age; light brownish-grey colored; covered by whitish scales, within the depressed center also with white felt. Gills running down the long stipe (5 - 6.5 x 0.5 - 1 - 2 cm) to the base, connected by crossveins and forked at least near the top of the stipe; of the same color like the cap, getting orange - ochraceous in older specimens. Mycelium with strong anislike odor.
   "P. cornucopiae ?"

6 (3) Gills not strongly decurrent and cap not at the same time infundibiliform. Stipes often laterally attached. Spores at the inner side being concave or straight

7 (8) Cap with light colors even when young: whitish, pale brownish or pale grey. Trama partly with skeletal hyphae, the walls of which are 1 - 3.5 um thick. Gills strongly decurrent, without anastomoses at the top of the stipe; first cream colored, later pale lemon yellowish tinged.

Stipe often very short, 1 - 1.5 x 0.3 - 0.8 cm, rarely longer. Spore deposit greyish white, sometimes getting lilac grey when dried. Odor comparable with that of 4 and 5, sweetish, however not ao strong. Fructification starts early in the season (May, July), partly also later (e.g. November)

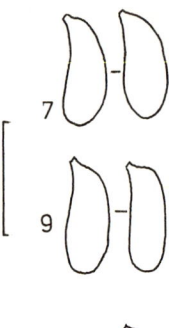

P. pulmonarius Fr.
On Opuntia in the Mediterranean: *P. opuntiae*

8 (7) Cap with deeper colors and trama (mostly?) lacking skeletal hyphae. Fructification late in the season

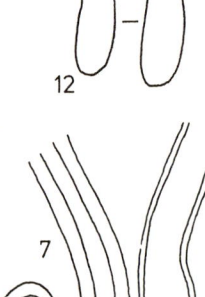

9 (10) Cap if young dark steelblue to light blue; when older more and more ochraceous starting from the center and leaving the margin of the cap greyish blue, blue greenish or dove-colored. Gills whitish with a weak lilac tinge, getting yellow patches and finaly somewhat ochraceous; partly connected by anastomoses near the top of the stipe. Stipe laterally attached, claviform to fusiform, lighter in color than the cap, however with comparable colors; strigose at the base, 1 - 2 x 0.5 cm. Spore deposit white.

P. columbinus Quél.ap.Bres.

10 (9) Cap not remaining blue at the margin while the center changes color to ochraceous; uniformly colored, mostly brownish to greybrown

11 (12) On Salix and Populus; separating features against P. ostreatus still unknown
"*P. salignus*"(Pers.ex Fr.)

12 (11) On different kinds of wood including that of conifers and frondose trees (incl. Salix). - Cap deep brown, greyish brown, black brown; partially also rather pale in older stages of development. Gills whitish to slightly brownish or grey, decurrent and covering the upper 1/3 of the stipe. Stipe 1.2 - 9 x 0.7 - 2.7 cm, whitish or with slight brownish tinge, sometimes strigose at the base, mostly distinct excentric or even lateral. Spore deposit whitish to cream colored whith slight incarnate tinge. Context with strong odor resembling that of some species of Polyporus.
P. ostreatus (Jacqu.ex Fr.) Kummer

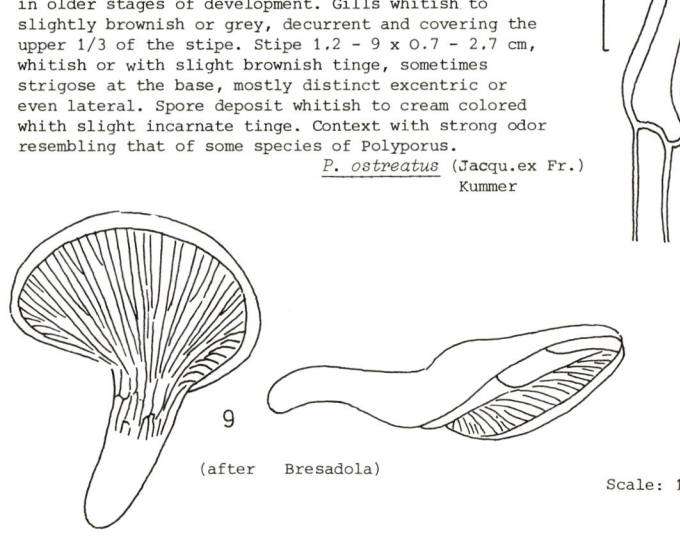

(after Bresadola)

Scale: 10 μm

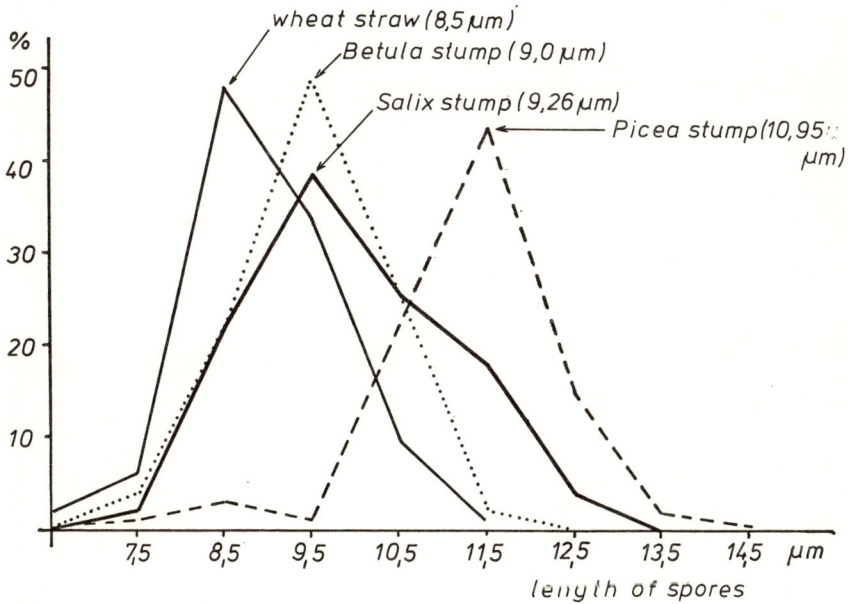

Fig. 3: Frequency of spore length in <u>Pleurotus</u> <u>ostreatus</u>, strain 2y, on various substrates. The values in brackets indicate average spore length.

Merkmalsdivergenz und Intersterilität bei Clitocybe
zusammengestellt nach Ergebnissen von D. Lamoure, 1965 und 1972

	Inter-fertili-tät	
1a C. pausiaca Herkunft: Schwedisch Lappland	+	1b C. pausiaca Herkunft: Französischer Jura
2a C. candicans Sporen: 5 x 3,5 µm Vorkommen: mit versch. Bäumen der Waldstufe und im subalp. Bereich, z.T. mit Helianthemum p-Kresol: mittlere Reakt.	+	2b C. candicans var. dryadicola Sporen 5-5,5 x 3,5 - 4 µm Vorkommen: in der alpinen Stufe mit Dryas p-Kresol: starke Reakt.
3a C. gracilipes Lamellen weniger gedrängt, bald beigefarben. Sporen: 4,5-5,5 x 3,5-4,5 µm	–	3b C. candicans var. dryadicola Lamellen sehr gedrängt, rein weiß, später gelblich Sporen: 4-5 x 3-3,5 µm Stämme aus Lappland und den Alpen interfertil
4a C. festiva Hut 1,2 - 3,5 cm Frk einzeln wachsend Stieloberfläche nicht faserig, hell Geruch: kampferartig Vorkommen: mit Dryas Sporen: 5-6 x 3-4 µm	–	4b C. festivoides Hut 1,1 - 5,1 cm Frk in Gruppen und Ringen Stieloberfläche faserig, dunkeler gefärbt Geruch: sehr unangenehm Vorkommen: mit Salix reticu- lata und retusa Sporen: 5,5 x 3-4 µm
5a C. gibba Sporen: 5,5-6,5 x 3,5-4,5 µm	–	5b C. catinus Sporen: 7-8 x 5-6 µm

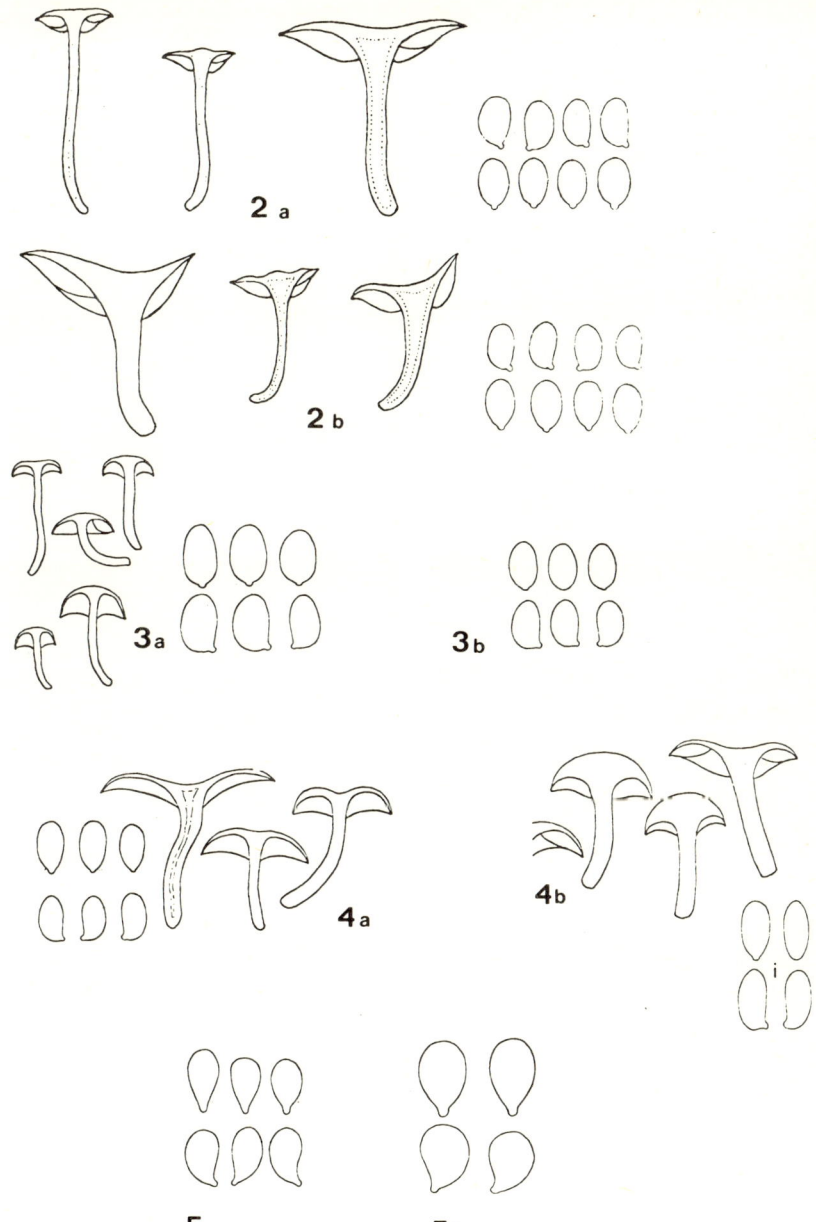

Table 1

a) P.pulmonarius (1r)
Confrontation of monocaryotic mycelia isolated from a single sporocarp

		A_1B_1					A_2B_2				A_1B_2			A_2B_1
		3	8	11	12	14	2	18	7	10	15	9	1	5
A_1B_1	3	-	-	-	-	-	+	+	+	+	-	-	-	-
	8	-	-	-	-	-	+	+	+	+	-	-	-	+
	11	-	-	-	-	-	+	+	+	+	-	+	+	-
	12	-	-	-	-	-	+	+	+	+	-	-	-	-
	14	-	-	-	-	-	+	+	+	+	-	-	-	-
A_2B_2	2	+	+	+	+	+	-	-	-	-	-	-	-	-
	18	+	+	+	+	+	-	-	-	-	-	-	-	-
	7	+	+	+	+	+	-	-	-	-	-	-	-	-
	10	+	+	+	+	+	-	-	-	-	-	-	-	+
A_1B_2	15	-	-	-	-	-	-	-	-	-	-	-	-	+
	9	-	-	+	-	-	-	-	-	-	-	-	-	+
	1	-	-	+	-	-	-	-	-	-	-	-	-	+
A_2B_1	5	-	+	-	-	-	-	-	-	+	+	+	+	-

b) P. columbinus (1n)
Confrontation of monocaryotic mycelia isolated from a single sporocarp

		A_1B_1			A_2B_2			A_1B_2			A_2B_1	
		4	7	11	3	6	10	5	9	12	1	8
A_1B_1	4	-	-	-	+	+	+	-	-	-	-	-
	7	-	-	-	+	+	+	-	-	-	-	-
	11	-	-	-	+	+	+	-	-	-	-	-
A_2B_2	3	+	+	+	-	-	-	-	-	-	-	-
	6	+	+	+	-	-	-	-	-	-	-	-
	10	+	+	+	-	-	-	-	-	-	-	-
A_1B_2	5	-	-	-	-	-	-	-	-	-	+	+
	9	-	-	-	-	-	-	-	-	-	-	+
	12	-	-	-	-	-	-	-	-	-	-	+
A_2B_1	1	-	-	-	-	-	-	+	-	-	-	-
	8	-	-	-	-	-	-	+	+	+	-	-

c) P. pulmonarius (1r) x P. columbinus (1n)   **d) P.pulmonarius x P.ostreatus:**

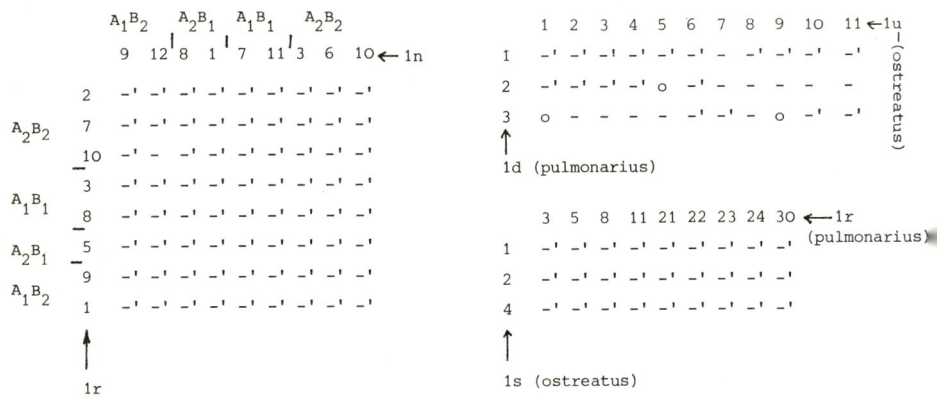

+ = clamp connections formed
- = no clamp connections formed
' = with barrage phenomenon (see fig.4)
o = outfall of confrontation

equivalence of mating factors in interspecies crosses could not be determined

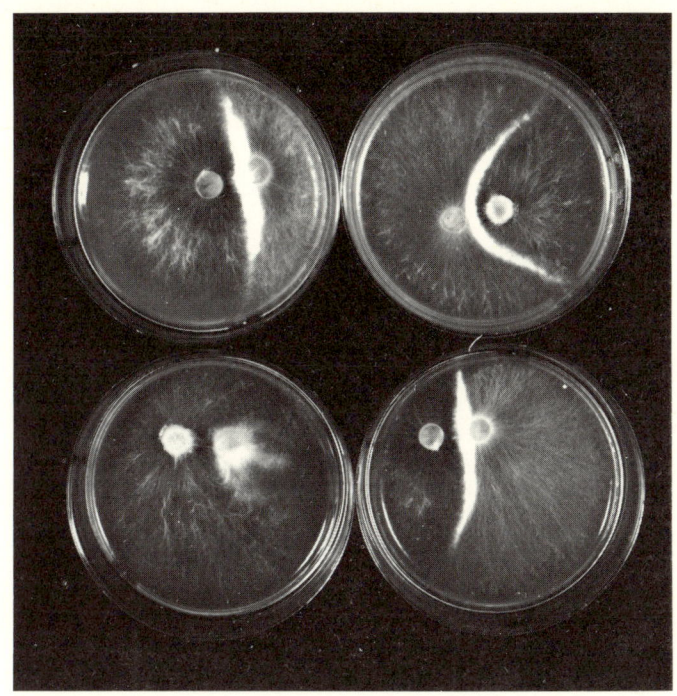

Fig. 4: *P.columbinus* (1n) x *P.pulmonarius* (1r)

Barrage phenomen in interspecies crosses:

Upper row from left to right: 11 x 13; 6 x 8

Row below from left to right: 12 x 10; 6 x 7

Table 2: Survey of non fertile crosses of Pleurotus:

cornucopiae [1]	x	*pulmonarius*	(4r x 4j)
	x	*columbinus*	(4r x 1n)
	x	*ostreatus*	(4r x 1u)
pulmonarius	x	*cornucopiae* [1]	(4j x 4r)
	x	*columbinus*	(1r x 2w)
	x	*ostreatus*	(1r x 1u)
pulmonarius "Stamm Florida"	x	*ostreatus*	(4b x 1a, 1u)
columbinus	x	*cornucopiae* [1]	(1n x 4r)
	x	*pulmonarius*	(1n x 1r)
	x	*ostreatus*	(1n x 1w)
ostreatus	x	*cornucopiae*	(1u x 4r)
	x	*pulmonarius*	(1u x 1r)
			(1s x 4x)
	x	*columbinus*	(1w x 1n)

[1] ss.Romagn.   non fertile = no clamp connections formed

Table 3

a) P. pulmonarius:

	1	2	3	4	5	6	7	8	9	10	11	12
4i → 1	o	+	+	o	-'	+	+	-	+	-	o	-
2	+	o	+	+	-'	+	+	-	+	-'	+	+
3	+	+	+	+	+	+	o	-	+	+	+	+
4	+	+	+	+	-'	+	+	-'	+	-	+	-

	1	2	3	4	5	6	7	8	9	10	11	12 ← 4h
1	+	+	+	+	o	+	+	+	+	+	+	+
2	o	o	+	+	+	+	+	+	+	+	+	o
3	+	+	+	+	+	+	+	+	+	+	+	+
4	+	+	+	+	-	+	+	-'	+	-'	+	+
5	+	+	+	+	-	+	+	-	+	-'	+	-'
↑4j												

Origin: Bayerischer Wald; Zwieseler Waldhaus; 4i und 4j growing 1 km appart from 4h

b) P. pulmonarius:

	3	8	7	1	18	19	20	24	29	30 ← 1r
1	+	+	+	+	+	+	+	+	(+)	+
3	+	+	+	+	+	+	+	+	+	+
4	+	+	+	+	+	+	+	+	+	+
5	+	+	+	+	+	+	+	+	+	+
6	+	+	+	(+)	+	+	+	+	+	+
7	+	+	+	+	+	+	+	+	+	+
8	+	+	+	+	+	+	(+)	+	+	+
9	+	+	+	+	+	+	+	+	+	+
11	-	+	+	+	+	+	+	+	+	+
12	+	+	+	+	+	+	+	+	+	+
↑4h										

Origin: 4h: Bayerischer Wald, Zwieseler Waldhaus;
1r: Österreich, Almsee near Scharnstein

c) P. pulmonarius:

	10	11	12	13	14	15	16	17	18	19 ← 4b
1	+	+	+'	+	+	+	+	+	+	+
2	+	+	+	+	+	+	+	+	+	+
4	+	+	+	+	+	+	+	+	+	+
6	+	+	+	+	+	+	+	+	+	+
7	+	+	+	+	+	+	+	+	+	+
8	+	+	+'	+	+	+	+	+	+	+
9	+	+	+	+	+	+	+	+	+	+
10	+	+	+'	+	+	+	+'	+	+	+
11	+	+	+	o	+	·+	+	+'	o	+
16	+	+	+	+	+	+	+'	+	+	+
↑1r										

Origin: 1r: Österreich, Almsee near Scharnstein;
4b: North America; "strain Florida"

d) P. ostreatus:

1w ↓

	1	2	3	4	5
1	+	+	+	+	+
2	+	+	+	+	+
3	+'	+'	+'	+'	+
4	+	+	+	+	+
5	+'	+'	+'	+'	+
7	+	+	+'	+	+'
8	+	+	+	+	+
9	+	+	+	+	+

	1	2	3	4	6 ← 1s
1	+	+	+	+	+
2	+	+	+	+	+
3	+	+	+	+	+
4	+	+	+	+	+
5	+	+	+	+	+
6	+	+	+	+	+
7	+	+	+	+	+
8	+	+	+	+	+
9	+	+	+	+	+
10	+	+	+	+	+
↑1u					

Origin:
1s: Westfalen, BRD
1w: Japan
1u: Japan

Pleurotus species Isolate[a]	pulmonarius 1 r (D)	columbinus 1 n (D)	ostreatus 1 w (D)	ostreatus 1 w 8 (M)	ostreatus x ostreatus 1 w x 1 s (D)	ostreatus 1 s 4 (M)	ostreatus 1 s (D)
Growth							
Malt-agar (mm/day)[b]	4.7	4.2	4.2	3.9	6.6	3.2	6.1
Liquid culture[c] (g wet weight/l medium)	17.0	9.7	12.2	19.5	4.0	14.1	13.7
Enzyme activity							
Peroxidase							
Drop test[d]	–	–	–	–	–	–	–
Mycelial extract (E/ml)[e]	0	0	0	0	0	0	0
Culture filtrate (E/ml)[e]	0	0	0	0	0	0	0
Tyrosinase							
Drop test[d]	–	(+)	(+)	–	–	–	(+)
Mycelial extract (E/ml)[e]	0	0	0	0	0	0	0
Culture filtrate (E/ml)[e]	0	0	0	0	0	0	0
Laccase							
Drop test[d]	+++	+++	++	+++	++	++	++
Mycelial extract (E/ml)[e]	1.6	0.2	8.4	1.4	4.6	1.2	7.2
total act. (E/l medium)[f]	79.5	7.7	436.8	152.8	64.6	76.5	390.8
Culture filtrate (E/ml)[e]	4.0	12.3	11.7	4.6	13.7	5.6	15.5
total act. (E/l medium)	3400	10886	10296	4094	12467	4928	13950

Table 4 Comparison of growth and enzyme production on solid and liquid medium between different dicaryotic (D) and monocaryotic (M) mycelia of Pleurotus and of a cross between two monocaryotic strains of P. ostreatus.
a) D = dicaryon, M = monocaryon. b) Malt-agar, petri-dishes. c) Malt-broth, surface culture, 14 days.
d) Malt-agar, petri-dishes, 21 days. Enzyme activity: – = none, (+) = trace, + = small, ++ = strong, +++ = very strong. e)Malt-broth, surface culture, 14 days. E/ml = units of activity/ml sample. f) Total activity in the mycelial extract from the mycelium of 10 flasks (= 1 l medium). All incubations at 27°C and diffuse light.

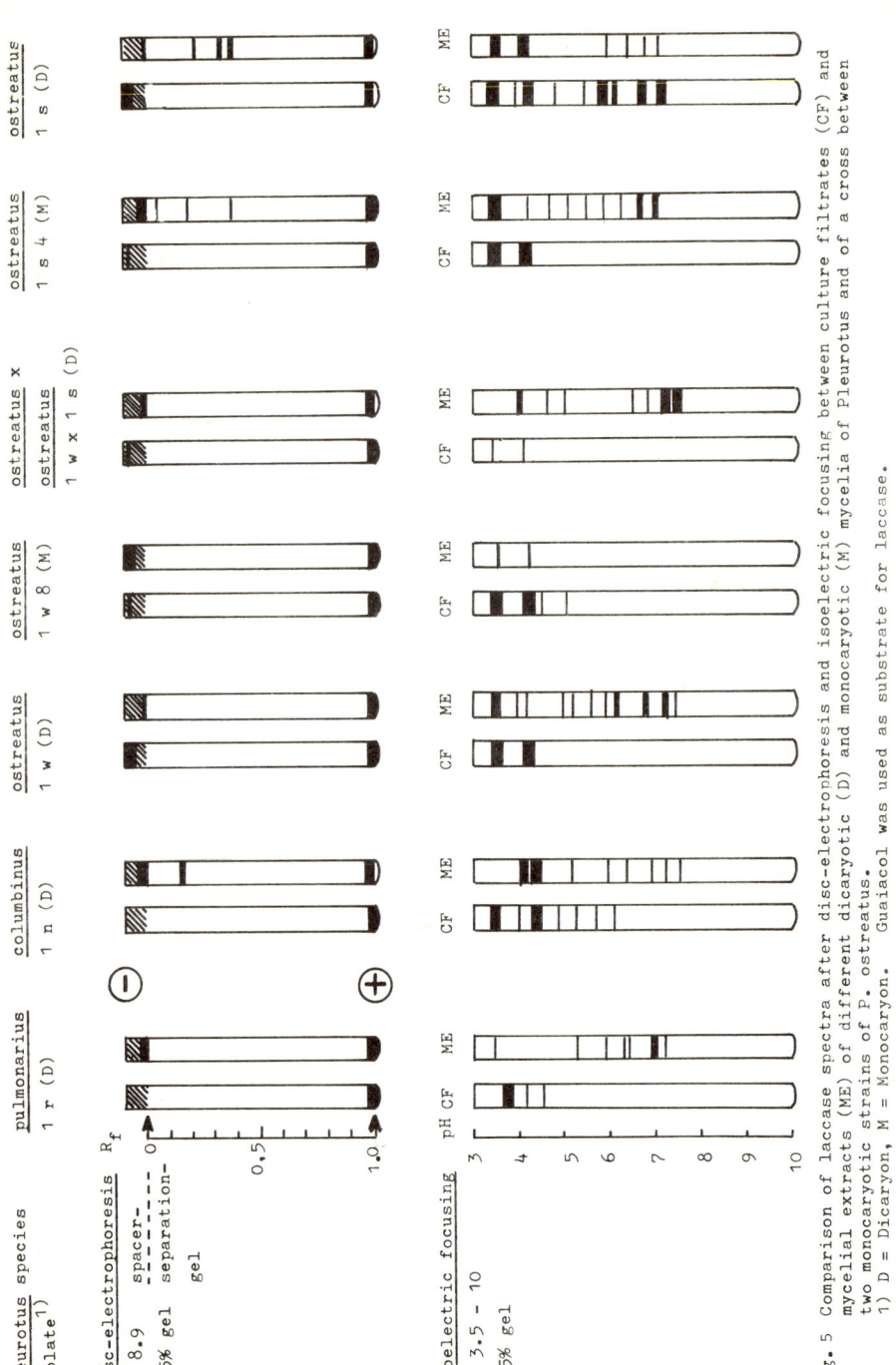

Fig. 5 Comparison of laccase spectra after disc-electrophoresis and isoelectric focusing between culture filtrates (CF) and mycelial extracts (ME) of different dicaryotic (D) and monocaryotic (M) mycelia of Pleurotus and of a cross between two monocaryotic strains of P. ostreatus.
1) D = Dicaryon, M = Monocaryon. Guaiacol was used as substrate for laccase.

DISCUSSION FOLLOWING DRS. BRESINSKY, HILBER AND MOLITORIS' PAPER

CLEMENCON: I just wanted to add a small observation of my own. You showed us that the substrate, Picea or Salix, influences the spore length. I was playing around with Hypholoma (or Naematoloma) fasciculare collected from nature, and I found consistently that the spores collected from fruiting bodies growing on conifers were longer than spores from fruiting bodies growing on Fagus. I was so much troubled that I was asking myself, are there two different species or not, but in view of what you are telling us I think we have not two different species. The differences were relatively small but statistically significant.

BAS: I have the impression that in the Pleurotus that I always used to call P. ostreatus, but I think now that it has to be called P. columbinus, at the end of the year, october, november, spores are getting longer. I collect these fungi along the coast, and I always thought the spores are getting longer because of higher humidity, but this is just an impression. Another thing I noticed: when you keep them too long in the refrigerator, the spores are getting longer also.

WATLING: I have measured literally thousands of spores in various members of the Bolbitiaceae to see the sort of effect one gets with different substrates etc.; I think in the Bolbitiaceae changes may be extreme, because the fruitbodies have very thin flesh and whatever happens to the environment is immediately expressed in the spores. In fact, in Bolbitiaceae we also see a slight change in shape. If you measure the Q-Value for the spore deposit you find there too you get not only a shortening of the spores or a lengthening of the spores, but also a change in shape; this change in shape may also be correlated with age of the fruiting body. These are not isolated caps, this is a fruiting body which is still growing.

CLEMENCON: I did this with many fungi, also with fleshy fungi, like Hebeloma leucosarx that is growing in my backyard, and I noticed one very interesting thing. Usually the length of the spore and the width, that means also the volume, are increasing; then decreasing. Once I had a strong north wind that somewhat dried the mushrooms. And even in the fleshy fruit bodies the spore volume, the length and width, dropped sharply and came back again when normal conditions were restored.

WATLING: Yes, you can alter the spores with the environmental conditions.

CLEMENCON: And even in fleshy fungi, not only in thin things.

WATLING: That is very interesting, because I always thought that in species like Tricholoma or other fleshy fruiting bodies this would not be so much the case.

BRESINSKY: To what extend do you have those differences? How many microns?

WATLING: Oh! 3 microns!

SMITH: What about the color of the spore deposit?

BRESINSKY: I did not include this character here. Of course we noted the color of spore print, but we did not include it in any experimental setup so far. Certainly it will be a very important feature, and I know that you in the States have different species, P. sapidus and P. ostreatus, that are pretty well distinguished by different colors of the spore print. This does apparently not work in Europe.

SMITH: Does your dark Pleurotus ostreatus have a distinctly colored spore deposit?

BRESINSKY: Mostly it starts with a color whitish and then we get more and more lilac colors.

SMITH: Yes, you have not worked on this whitish ostreatus-like fungus that we have from our Biological Station in Northern Michigan?

BRESINSKY: No, I have not. I would like to have spores.

BLAICH: You told us about segregation of the cap color. What type of segregation? One to one?

BRESINSKY: So far as we could make calculations, on a limited material only, it is one to one.

BLAICH: So here is certainly the possibility of a monogenic difference between the two strains. And this would be a difference which is clearly not a specific difference.

ESSER: I was very pleased to see that you have observed about the same phenomena in your Westfalen strain, where does it come from?

BRESINSKY: Münster, Westfalen, leg.A.Runge 13.12.1973; isolated by O. Hilber 17.12.1973.

ESSER: You use the word "intersterility". And I have a tendency to look carefully at intersterility, because intersterility might be true sterility, but it might be due to other mechanisms which have nothing to do with sterility, like heterogenic incompatibility. At the moment I would say failure of interfertility, I would not say intersterility.

WATLING: I do not know whether our last speaker suffered from the restless nights trying to put together his experimental work and his field work as I have because in the Bolbitiaceae we have similar variabilities of spore size on different substrates. When you go into the field you do not find the same degree of variability. In the field I really never reached the extremes so that in fact nature seems to put a buffering effect on variation. You did mention selection. There is a strain of _Psilocybe merdaria_ which was found in 1968; at the base of the fruiting body small gastroid fruiting bodies were observed. They did not have all the normal characteristics, and by selection of these fruiting bodies, and not the agaricoid fruiting bodies, after three generations a culture was produced which was basically gastroid.

BRESINSKY: A published paper?

WATLING: Yes, New Phytologist 70: 307-326, 1971. The problem, of course, is whether it is cytoplasmically inherited, and at the moment we are investigating this possibility; is it a virus particle which in fact we are selcting for each time we select these "aberrant" fruiting bodies? It so we finish up with a culture which gets more and more of this viral "compound"; it is a hypothesis, but we have yet to find the virus particles.

SINGER: I understand that McKnight, who first produced such gastroid fruiting bodies in _Psilocybe_, considers this now as a mutation triggered by the conditions.

WATLING: From comparing McKnight's experimental work it certainly does not look to be the same thing. My material has been passed over to a geneticist who is much more competent than I to work out the problem. Professor Burnett now has cultures of all the different strains which we have been able to select out. Hopefully in the future we will have some evidence in the Basidiomycetes of selection of this type.

ROMAGNESI: Monsieur Bresinsky, vous avez pu faire ce que je n'ai pas pu faire à mon grand regret. Je voudrais surtout vous demander quelques précisions en me placant surtout du point de vue du systématicien. Je dois vous donner mon accord complet en ce qui concerne la présence d'hyphes à paroi épaissie dans la trame du _Pleurotus_ _pulmonarius_. J'ai toujours été préoccupé de savoir s'il s'agissait d'un caractère constant. Parce que je récoltais souvent dans la nature des carpophores, ou des basidiomes, si vous préférez, qui avaient tous les caractères extérieurs de _pulmonarius_ sans en avoir les hyphes épaissies. Vous savez peut-être, qu'au Laboratoire de Cryptogamie du Muséum de Paris, Monsieur Roger Cailleux cultive artificiellement tous les _Pleurotus_ du groupe _ostreatus_, ainsi que ceux du groupe eryngii. Je lui ai donc confié une sporée d'un _pulmonarius_ à hyphes épaissies qu'il a cultivée sur compost. Et, de fait, les hyphes à parois épaissies qui se trouvaient dans le carpophore trouvé dans la nature ne se trouvaient plus, ou tout au moins ce caractère était extrêmement atténué. Il est donc clair, comme vous le pensez, que ce caractère est en partie lié à l'environnement. Cependant, j'ai fait personnellement dans une forêt des environs de Paris la récolte simultanée, sur un même tronc de _Fagus_ abattu, la récolte d'une grande série de carpophores ayant des hyphes à paroi mince et, au bout, un petit groupe de trois ou quatre carpophores qui avaient des hyphes à paroi épaissie. Dans ce cas il n'est pas possible d'évoquer l'envi-

Pleurotus

ronnement ou des conditions extérieures puisqu'elles étaient les mêmes. Deuxièmement je voudrais vous demander quelques précisions sur la façon dont vous concevez le Pleurotus columbinus. Est-ce que vous avez récolté cette espèce en hiver?

BRESINSKY: Pl. columbinus should be a species in accordance with Bresadola, Iconographia, tab. 291. Our Pl. columbinus has been collected in November (1n; Karlsruhe, leg.Fuhrmann) and in October (4t; Poland; leg.M.Moser). I don't think that sclerified hyphae are an absolutely reliable character for Pl. pulmonarius.

ROMAGNESI: Je veux vous donner la raison de la question. Je ne sais si vous connaissez une thèse sur le Pleurotus ostreatus qui a été soutenue, il y a un ou deux ans, au Laboratoire de Cryptogamie de Monsieur Heim à Paris. Il s'agissait d'une étudiante Sud-américaine, dont j'ai malheureusement oublié le nom compliqué, mais qui avait étudié surtout le comportement du groupe ostreatus et eryngii en culture portant principalement sur l'incidence des caractères de température et d'autres facteurs physiologiques qui accompagnaient ces cultures. Il a été ainsi trouvé d'une façon extrêmement claire que dans les récoltes de Pleurotus ostreatus il y avait deux groupes d'espèces qui se différenciaient très nettement par la température optimale du développement. Et ce que les mycologues français ont toujours considéré comme Pleurotus ostreatus est une espèce d'hiver dont l'optimum de développement est relativement bas, tandis que pulmonarius, cornucopiae et les autres espèces qui poussent au printemps ou en automne ont un optimum de température beaucoup plus élevé. Or, des formes d'hiver que nous avons toujours considérées comme étant le type Pleurotus ostreatus sont des formes qui présentent des colorations bleues puisque les carpophores virent au brun dans la vieillesse. Je me demande si vos Pleurotus columbinus ne seraient pas nos Pleurotus ostreatus au sens des mycologues français.

BRESINSKY: I am sorry that I cannot give a final answer, since this question should be solved on the base of confrontation experiments. What we certainly know is that there are at least two different dark-colored species which are fruiting in the late season and which we are calling P. ostreatus and P. columbinus.

ROMAGNESI: Sur l'interprétation de Pleurotus columbinus il y a des difficultés. En effet, je crois vous rappelez que la planche d'Icones

de Bresadola, dans les fungi tridentini vous avez Pleurotus ostreatus, représante un champignon qui n'est pas bleu, ni bleu d'acier,
mais qui est vert fuligineux.

BRESINSKY: Ce n'est pas ostreatus, c'est columbinus.

ROMAGNESI: Or, il y a peut-être quarante années qu'au jardin des
plantes de Paris j'ai récolté sur un Cercis siliquastrum ce que je
pense être le columbinus de Bresadola. Eh bien, ce champignon, lorsqu'il est jeune a un chapeau fauve ou ocre, et quand il se développe
il prend exactement la couleur gris-olivâtre verdâtre de la Planche
de Bresadola. Et sur la planche même de Bresadola la couleur rougâtre
à laquelle vous faites allusion est très proche, en effet, de celle
que j'observe. D'autre part, vous savez qu'il y a dans la littérature
des données sur la toxicité de Pleurotus columbinus. Or, je puis vous
affirmer, que le Pleurotus bleu qui pousse en hiver est parfaitement
comestible. Je me demande donc, si en fait, votre columbinus n'est
pas le véritable ostreatus et si votre ostreatus n'est pas le champignon qu'à tort ou à raison j'ai donné dans mon travail sur ce groupe sous le nom de salignus, lequel est en effet un champignon brun,
et qui pousse aussi bien sur feuillu que sur conifères. Il aurait la
peut-être un problème de systématique à résoudre.

BRESINSKY: Some of the species, like Pleurotus columbinus, are very
ambiguous. This shows that breeding experiments are necessary here in
order to find the border lines between the species and in order to
decide what is conspecific or not. We should have the opportunity to
investigate your Pl. salignus.

BLAICH: Are there breeding experiments columbinus-ostreatus?

BRESINSKY: Columbinus x ostreatus ..... are in preparation.

SMITH: I am wondering if your situation here in Europe is not confused somewhat by invading American variants. Now we have columbinus in
the western United States and it is blue and it is on conifers. I have never seen this blue thing on anything but conifer. I take it you
get it on hardwoods here. In the spring we have what we call ochraceous to white Pleurotus and I tried to identify it with a number of
European species. It has a spore print that is white yellowish to

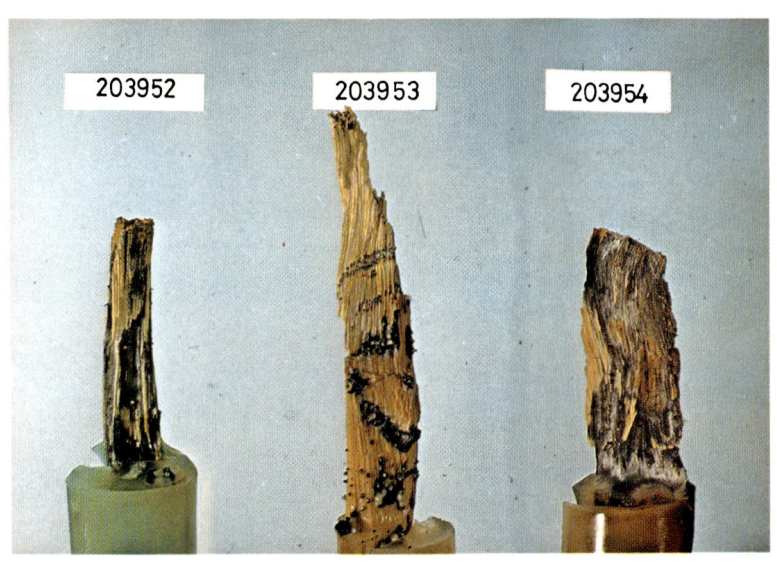

sometimes pale lilac. This grows on aspen. This is the one we are
going to send you. And then we have the big dark ostreatus with a
rather bitter taste, it is almost black and it grows in the spring
after the snow melts on the ski slopes. It is very distinct from the
ordinary Pleurotus that we get on the elm and other hardwood.

BRESINSKY: I think that it is very interesting that we can observe
so much variation in Pleurotus within one species as well as between
the species.

ESSER: I think in future general regulations have to become efficient. Some years ago commercial firms have sold cultures of Pleurotus ostreatus to everybody who wanted to breed this fungus in his
basement. After harvesting the fruit bodies the people has thrown the
cultures substrate on the trash pile. These races of Pleurotus ostreatus became a chance to interbreed with each other and with many
other fungi.
There is another question: I have heard that Pleurotus ostreatus may
cause allergies. Is this a rumor? Or is this possible due to the fact
that the spores of Pleurotus ostreatus are able to germinate at temperatures up to $40^{o}C$ and therefore may cause after inhaling inflamma-
tions of the lung? Are these allergic reactions caused by all ostreatus strains or only by certain races? Or even by other species like
columbinus?

BRESINSKY: We have put it on our program.

ESSER: What is known?

BRESINSKY: It is known that Pleurotus causes allergies. We don't
know wether this is true for all species.

BAS: This is possibly a matter of the amount of spores. Pleurotus
ostreatus starts to sporulate very early. They have been growing
Pleurotus ostreatus in mushroom houses in the Netherlands. The workers refused to go in because there was a mist of spores. They had to
wear a kind of mask because they could not stand the spores. I think
if you inhale clouds of spores, there must be some effect.

SINGER: The only information I have is that in Ohio there was an allergic reaction.

Herbette Symposium on Species Concept in Hymenomycetes 1976

OIDIAL HOMING AND THE TAXONOMY AND SPECIATION OF BASIDIOMYCETES
WITH SPECIAL REFERENCE TO THE GENUS COPRINUS

R . F . O . K e m p

S u m m a r y
Breeding studies on species of Coprinus (Kemp 1975) and Psathyrella
(Jurand 1975) suggest that sympatric speciation at the cellular level
may be common in basidiomycetes. Loss of hyphal recognition and fusion occur after the species have evolved and do not cause speciation. The widespread occurrence of suitable habitats for dung species
suggests that it is unlikely that allopatric speciation needs to be
at all common. Basidiomycetes therefore offer particularly suitable
material for the study of internal speciation in contrast to the external speciation which occurs in many angiosperms and in animals.

By testing for crossability between strains, and for the homing and
lethal reactions which occur between oidia and hyphae of related species, it is possible to define the species in breeding terms, in addition to the morphological characters normally used. The reactions
of hyphae to oidia provide a breeding test which may indicate the relationship between taxa above the level of the species. Once the different species have been detected on a breeding basis it is then possible to correlate morphological characters with them and to decide
which characters are valid. The angiosperm flower is the site of
plasmogamy, karyogamy and meiosis which together are the key events
of sexual reproduction. The delimitation of angiosperm species on the
basis of floral structure owes much of its success to this. A similar
system of sexual classification in basidiomycetes must necessarily
involve hyphal fusion as well as the structure of the fruit body. The
fruit body is the site of karyogamy and meiosis but not plasmogamy so
the integration of breeding and classical studies might help to give
a truly Linnaean system of classification.

The rapid vacuolation or lethal reaction which occurs after oidial
homing tests between closely related species suggests that the hybrid
cell dies because there is a rapid blocking and breakdown of the metabolism. There appear to be similarities between gene-for-gene resistance to disease (Flor 1942) and speciation in certain basidiomy-

cetes. Speciation might be due to the polymorphism of membranes and the presence of metabolic blockers, for example, so that hybrid fusions would be lethal when the mycelia which fused each had a membrane which was sensitive to a metabolic blocker formed by the other. Gene-for-gene speciation would thus be a form of heterogenic incompatibility (Esser and Blaich 1973).

Gene-for-gene speciation at the cellular level, or symplasmic speciation, could result in two populations being completely intersterile in the absence of morphological differences. In contrast two widely separated populations of a species forming mycorrhiza on an angiosperm species having a disjunct distribution might accumulate morphological differences while being completely or partially interfertile in culture. The species concept of the taxonomist ought to recognise intersterility as a valid species difference and that of the geneticist should recognise that speciation can occur in the absence of complete intersterility.

Introduction

Before considering some of the more detailed work on basidiomycete breeding systems it might be useful to examine what sexual reproduction in this group involves. Sexual reproduction in all groups involves repeating cycles of plasmogamy, karyogamy and meiosis. In angiosperms all three events occur in the flower and this probably explains why the Linnaean system basically works very well in this group. In the basidiomycetes only karyogamy and meiosis occur in the fruit body and plasmogamy normally occurs between ordinairy vegetative hyphae. Angiosperm species consist of distinct individuals each with a genotype fixed at fertilisation and they have evolved a very complex reproductive apparatus which is necessary if the haploid gametes are to combine together to form a diploid zygote. A basidiomycete does not consist of distinct individuals with a fixed genotype as hyphal fusion and plasmogamy can give new combinations of haploid nuclei when vegetative hyphae fuse. One group therefore has problems in fusing genetic information and the other in preventing its fusion. Whereas changes in floral morphology may be important in speciation in insect pollinated angiosperms changes in the morphology of a basidiomycete fruit body are probably the result of genetic isolation and not the cause of it. As plasmogamy occurs in the hyphal tip then it is this process and this organ which are involved in symplasmic speciation.

In higher organisms speciation is referred to as being allopatric or sympatric but in basidiomycetes it would seem in most cases that it might have to be symplasmic. This term is used to indicate speciation which involves the evolution of populations between which the recognition of self and non-self takes place in the hybrid cell formed by the fusion of two hyphae or a hypha with an oidium. It would appear that genetical speciation in basidiomycetes involves the formation and selection of genotypes which when combined together in the same cell are metabolically lethal. If this does not occur then hybrid dikaryotic mycelia could be ubiquitous. If the fusion cell is the site of speciation it follows that inter-specific hybrids are impossible and there should be no problems with vegetative hybridisation.

It would be surprising however, if the fungi with their great diversity did not show similar diversity in their mechanisms of speciation. Grant (1963) suggested that isolating mechanisms had three main components namely spatial distance, the nature of the environment and the reproductive characteristics of the organism. Although speciation at the cellular level would seem to be necessary in species with a global distribution such as many of those in Coprinus and Psathyrella this need not be so in species which are pathogenic on or form mycorrhiza with an angiosperm species having a very disjunct distribution. At one extreme genetical isolation could be complete in the absence of any morphological differences as in Psathyrella candolleana, P. gracilis, P. coprobia and P. coprophila (Galland 1975, Jurand 1975). Further studies might reveal morphological differences which would enable each species to be recognised by the taxonomist but this need not be so. If symplasmic speciation had occurred a long time ago without associated changes in morphology then morphological differences could have accumulated during the period up to the present. This would seem to be the most likely mechanism in most basidiomycetes. At the other extreme the spatial isolation of the mycorrhizal host could result in fungal populations having morphological differences acceptable to the taxonomist at the species level without there being complete inter-sterility. This is the situation found in Auricularia (Duncan and MacDonald 1969). The type concept in basidiomycetes does not permit the application of a binomial on the basis of crossability

tests alone. Although the geneticist might like this it is probably
not desirable for all those who have to collect and to refer to specimens by a name. However there should be some way of indicating that
a given binomial is really a macrospecies consisting of several microspecies. Microspecies have done all that is necessary to be completely isolated at the cellular level and a lack of recognition of this
in the specific epithet might be considered some what unjust. Jurand
(1975) suggested for species of Psathyrella that a prefix (a,b,c,d or
the Greek equivalent) should be added to the specific name so that
the macrospecies P. candolleana would contain the four microspecies
P. a-candolleana, P. b-candolleana and so on. The prefix would only
be valid if crossability tests were made against type cultures or
strains tested against them. The use of a prefix for each microspecies could act as a reminder that the binomial applied to a macrospecies and thus focus attention on the need to look for new morphological characters which might enable each microspecies to have a binomial.

The structure and function of oidia
Many species of basidiomycetes produce oidia of one type or another
and in Coprinus about 75% of the species do so. In general they are
formed only on monokaryotic mycelia but C. comatus is an exception.
They can be classified into the broad categories wet and dry depending on whether the oidia coalesce into droplets or remain in chains.
Wet oidia are formed on various types of oidiophores and E.M. studies
on the wet oidia of Psathyrella coprobia (Jurand and Kemp 1972) have
shown that they have a capsule and oidial mounts in Indian ink indicate that capsules are probably present on all wet oidia. Dry oidia
or arthrospores are formed in chains on oidiophores which differ only
slightly between species. In general dry oidia of most species germinate well whereas wet oidia do so less frequently. Wet oidia usually
cause homing and an absence of homing within a species is more often
found in those with dry oidia. They are more often present in heterothallic than homothallic species. Perhaps wet oidia which cause homing should be called spermatia.

To test for oidial reactions it is only necessary to smear the oidia
with a wet or dry sterilised loop from the surface of one mycelium to
a position just in front of the hyphal tips of the mycelium to be

tested. If many tests have to be made it is better to make a concentrated oidial suspension and to use a loopfull as required. If the mycelium being tested forms oidia very close to the edge of the colony care should be taken in placing the oidia in front of the leading hyphal tips and not over the colony edge. Care should also be taken not to include too many hyphal fragments as these may be viable and when overrun by the advancing hyphal tips could give a localised positive result in error. It is usually best to test for oidial homing against monokaryotic mycelia as those of a dikaryon usually show less homing. In C. cinereus the main tips of a dikaryotic mycelium very rarely show homing but it is shown by lateral branches some of which are monokaryotic. In Flammulina velutipes dikaryotic and monokaryotic tips both respond with equal vigour to the oidial stimulus. Although some species usually show homing from a range of 20-30 microns others form very short peg-like branches which appear to knot themselves round the oidia.

It must be emphasised that the oidial reaction works at the level of the species and above and is not dependent on the mating types of the strains. The hyphal tips of a monokaryotic mycelium respond equally well to the oidia from the centre of the same mycelium (in a tetrapolar species A = B =) as they do to tests involving A $\neq$ B =, A = B $\neq$ and A $\neq$ B $\neq$ combinations. If the oidia of each species only stimulated its own hyphae then little more information would be obtained from oidial tests than from tests for intersterility, apart from having a quick means of identifying mycelia from the same species. The homing response can however occur between different species and the results of the tests fall into three general categories namely good homing followed by a lethal reaction; a homing response ranging from good to poor without a lethal reaction; and finally no homing response at all. The responses are often not reciprocal.

H o m i n g   f o l l o w e d   b y   a   l e t h a l   r e a c - t i o n

The most useful homing test is that which results in a lethal reaction in the hyphal tip after it has homed onto and fused with the oidium. When this happens the initial homing response is indistinguishable from that involving oidia of the same species. In many species very few of the leading hyphal tips fail to home onto an oidium.

Lethal reactions seem to occur when two closely related species are
tested. After 2-3 hours the tip cell begins to vacuolate and this effect may spread to one or two cells below the tip. When vacuolation
has occurred the cell is dead as it makes no more growth and does not
respond to an oidial stimulus. That the lethal reaction occurs between closely related species is shown by the very good homing, by taxonomic similarities and by the fact that it often occurs between
microspecies. For example the oidia of C. congregatus are lethal on
the hyphae of C. bisporus although the reciprocal tests show no homing. The two species are morphologically similar in that they both
lack clamp connections on their dikaryotic mycelia and both are bipolar. They are the only two species in the section Setulosi which have
this type of incompatibility. Their main difference is that one is
4-spored and the other 2-spored. Although the two species are morphologically distinct the hyphae of C. bisporus are unable to distinguish the oidia of C. congregatus from its own. After homing it seems
likely that the nucleus from the oidium of C. congregatus migrates
into the hyphal tip of C. bisporus but their combined metabolic activities are lethal. The two species may have become genetically isolated at the cellular level and only later diverged morphologically.
They clearly did not become genetically isolated by the loss of the
homing response or the loss of hyphal fusion. If speciation in this
case occurred first at the symplasmic level it might be expected that
oidial and hyphal recognition would be lost next to be followed eventually by the accumulation of morphological differences. There may be
a good reason why this latter sequence of events has to occur. If speciation occurs at the cellular level then two nuclei which carry between
them a potentially lethal combination of genes can only express this
lethality if they are brought together into a common cytoplasm. Hyphal fusion is therefore essential for speciation at this level and
it seems to be lost only very slowly after speciation.

Symplasmic speciation has many similarities with gene for gene resistance of the type described by Flor (1942) in his studies on the inheritance of the resistance of flax varieties to rust. The systems of
heterogenic incompatibility reviewed in detail by Esser and Blaich
(1973) are probably similar to those which occur in basidiomycetes.
A possible mechanism of symplasmic speciation will be considered below. The most important similarity between symplasmic speciation and
gene for gene resistance is that it is not possible to test if a flax

plant is resistant without inoculating it with the rust: likewise it
is not possible to test if the nuclei of different genotypes will
form a lethal combination except by testing their activities in a
single cell. The Flax by developing resistance,is preventing the rust
from sharing its metabolism and this also applies to symplasmic speciation as two diverging populations are forced apart by being pre-
vented from sharing a common metabolism. Much of the speciation in
basidiomycetes and other fungi may only be possible at the cellular
level which is an extreme form of sympatric speciation. It could be
more rapid because of the lethal reaction and this may be why fusion
and lethality still occur when two species have diverged morphologically. If symplasmic speciation occurs it is not surprising that C.
congregatus and C. bisporus both grow on mixtures of dung and straw.
The hyphae of C. bisporus could avoid the lethal weapons of C. congregatus if its spores germinated after the monokaryotic mycelia of
C. congregatus had become dikaryotised and were no longer producing
oidia. C. bisporus in forming dikaryotic spores does not produce monokaryotic mycelia which respond much more readily to the oidial stimulus than dikaryotic mycelia do. The retention of oidial homing and
lethality after genetical isolation is complete could be important in
determining the sequence of species in an ecological succession.

The lethal response which follows homing seems to provide a reasonably objective means of showing that two distinct species are closely
related because the phenotypic effect is caused by the expression of
the genes which control the mechanism of speciation. If symplasmic
speciation has occurred then the failure to form a dikaryon between
two intrafertile groups of strains indicates that two species are involved and conversely if a dikaryon is formed then the strains must
belong to the same species even if fruit bodies cannot be produced in
culture. Although many different characters are used by the taxonomist to distinguish between species intersterility is not a character in the same sense. Plasmogamy, karyogamy and meiosis are the cardinal events which provide a route for the transmission of characters
from one generation to another and intersterility at the cellular level is therefore not simply a character - it is the result of a mechanism being present which completely prevents characters from being
inherited at all. The recognition of self from non-self at the cellular level means that hybrid dikaryons cannot be formed at least between closely related species. Test to see whether the sexual cycle

can be completed and viable spores formed are probably only necessary
when allopatric speciation is suspected.

It is clear from the pattern of lethal reactions found between nine
species in the section Lanatuli of the genus Coprinus (Kemp 1975,
Fig.2) that it is the exception for reciprocal tests between two species to be lethal. Only two out of 15 pairs of species tested showed
a reciprocal lethal reaction. Although it may be reciprocal when speciation actually occurs it would not be surprising if the oidial stimulus and the response to the stimulus by the hyphal tip diverged independently. The system is similar to a radio transmitter and receiver in which a message may fail to be transmitted because of a change
in the wavelength of the transmitter or of the receiver. These independent changes could apply at the cellular level for the lethal reactions or to the homing reactions which occur between hyphae and oidia. It is not yet possible to delimit accurately the different levels of relationship above the species level due to the complex net-
work of oidial reactions. The following series indicating a decrease
in similarity might be expected: reciprocal lethality, unilateral
lethality, reciprocal homing, unilateral homing and finally no response at all.

Homing responses not followed by lethality

When lethality follows the homing response between two species there
is no difference in the extent or rate of homing between self and
non-self tests. But when homing is not followed by a lethal response
the amount of homing can range from being frequent to very rare. Only
two or three hyphal tips of C. radiatus respond to a smear containing
about 10^4 oidia of Flammulina velutipes. The hyphae of C. lagopus
(from woodland soils) home onto oidia of C. congregatus sufficiently
frequently to be quickly detected although the number of tips or side
branches which home may be only 5-10%. Homing responses between species in different sections of a genus may also be useful in indicating which species are closely related. For example the oidia of both
C. bisporus and C. congregatus in the section Setulosi stimulate homing of hyphae belonging to C. lagopus, C. bilanatus nom.prov., isolate no.160 and no.174 all of which belong to the section Lanatuli.
The four species react identically to the oidial stimulus of both

C. bisporus and C. congregatus and cannot distinguish between them, thus providing further evidence that the two species are closely related. Although homing reactions can occur between different sections of a genus and also between species in different families lethal reactions have only been found between species in the same section. As C. disseminatus has lethal reactions with species in the section Setulosi it would seem that this species is correctly placed in the genus Coprinus. The absence of a lethal reaction after a good homing response may be due to a failure of nuclear migration from the oidium into the hyphal tip or to the oidial nucleus being killed by a lethal reaction.

The identification of mycelia by their characteristic patterns of oidial responses

Homing studies on the species in the section Lanatuli indicate that each species has a characteristic pattern of oidial reactions with other species. Table 1 shows the patterns of five of the species in this section. Note that only C. macrocephalus has the same pattern in reciprocal tests. For example the oidia of C. lagopides cause a homing response on hyphae of the four other species (the pattern H,H,H,H.). The reciporcal test give the pattern H,H,L,L. No other species has this pattern and all the species differ from each other.

Table 1: Oidial reactions between 5 species of Coprinus in the section Lanatuli.

Oidia	Hyphae					
	cin	mac	pid	pus	rad	
cinereus	/	O	H	O	H	H = Homing
macrocephalus *)	O	/	H	O	O	L = Lethal
lagopides	H	H	/	H	H	O = No response
lagopus	O	O	L	/	O	/ = Same species
radiatus	O	O	L	O	/	

*) See notes on this species in Kemp (1975).

It would therefore seem possible in the absence of a set of stock cultures to determine from oidial reactions the identity of five intersterile groups of strains which had been previously studied. It need not always be necessary, when microspecies have been identified on the basis of crossability tests, to retain a permanent set of cultures as a new set giving the same pattern of reactions could be collected. In an emergency oidial reactions could be used to identify a set of stock cultures in the event of all the labels being lost.

The identification of mycelia of homothallic species

Mycelia of heterothallic species are easily identified by the formation of dikaryotic mycelia in compatible matings. Matings are impossible in homothallic species. Preliminary studies on species in the section Setulosi (C. pellucidus and C. stellatus) and in the section Micacei (C. domesticus gp.) indicate that it should be possible to identify the mycelia of homothallic species by studying their patterns of oidial reactions with related species. Each distinct pattern in a set of tests is presumably characteristic of each species and from this it may be possible to find correlated morphological differences.

Chimaeras

During early studies on dung Coprini it sometimes proved difficult to distinguish between young specimens of C. miser and C. pellucidus. Some seemed to be intermediate between the two sections Nudi and Setulosi. Mycelia growing from sections of stipes sometimes had an outer zone of 5-7 mm which lacked clamps (= C. pellucidus) and an inner zone which was denser and had clamps (= C. miser). A detailed examination of the gills of some of the fruit bodies showed that they had tetrads of C. miser with lentiform spores and also tetrads of C. pellucidus with smaller narrow spores. The mycelia of these two species are clearly not antagonistic and they are able to share in the production of a fruit body. These fruit bodies were found on horse dung which had been collected in the wild and incubated in the laboratory. Efforts to produce fruit bodies of this type in culture were unsuccessful. Chimaera fruit bodies although not hybrids in the usual sense can show characters which are intermediate between two species. It is hoped that they are not too frequent as they are a taxonomists and geneticists nightmare.

Gene-for-gene resistance to disease and gene for gene speciation

Gene-for-gene resistance was first described by Flor (1942) for the situation he found in flax and its fungal parasite, the rust <u>Melampsora lini</u>. Working with several varieties of flax and several strains of rust he found that the flax was resistant when it carried a specific dominant allele and the rust also carried a specific dominant allele.

Table 2: The genotypes of flax and rust studied by Flor (1942) and alternative symbols used in a discussion of symplasmic speciation.

	Flax					
Rust race	Ottawa LLnn	Bomby 11NN			blocker 1^+ blocker 2^-	blocker 1^- blocker 2^+
22 (vLvL VNVN)	S	R	site 1^-	site 2^+	S	R
24 (VLVL vNvN)	R	S	site 1^+	site 2^-	R	S

One of the handicaps in considering related problems in different species is that each system has its own set of symbols. In order to attempt to overcome this I will use a terminology based on the concept of blockers and sites as a mechanism of this type might possibly be the basis for several unrelated systems. The right side of Table 1 introduces the symbols. A flax variety which produces blocker 1 (1^+) is resistant to the race of rust having site 1^+ but is susceptible to a rust strain which carries site 2^+. A mechanism of this type would involve each metabolic blocker pairing with a highly specific site which may be a membrane. The two loci in flax and also the two in the rust are inherited independently. It could presumably be possible for a species to be polymorphic and to have multiple alleles at a single locus controlling site specificity. As mentioned above the flax has to be inoculated with the rust to test for resistance.

Gene-for-gene speciation

It is now necessary to see the significance of host-parasite relationships to speciation in basidiomycetes. From studies on the homing of hyphae towards oidia, and the events which follow their fusion, in species of <u>Coprinus</u> and <u>Psathyrella</u> it is clear that the recognition of self and non-self between closely related species occurs in the hybrid tip cell when the two nuclei begin to function together. It follows that speciation is controlled by the genes which determine this metabolic lethality. For speciation of a symplasmic type to occur it would only be necessary for both of the diverging taxa to produce a blocker substance to which the other was sensitive. In other words it might be similar to the gene-for-gene relationship found by Flor in flax and its rust. Two intersterile populations could evolve on the basis of two pairs of alleles at each of four loci although many more may be involved in nature. The four dominant alleles could be blocker-1^+, site-1^+, blocker-2^+ and site-2^+. One intersterile strain would have the genotype blocker-1^- site-1^+ blocker-2^+ size-2^- and the other blocker-1^+ site-1^- blocker-2^- site-2^+. Table 3 shows the nine possible viable genotypes and the arrows indicate the mutations which would be necessary to form two intersterile strains from an original strain having no blockers or sites. The inviable gene combinations contain a matching blocker and site within the same nucleus. It is likely that all strains have active sites which can potentially be blocked.

Table 3: The 9 viable genotypes for a metabolic blocking system controlled by two blocker loci and two site loci.

blo	site	blo	site	blo	site
1^- 2^-	1^+ 2^+	1^+ 2^-	1^- 2^+	1^- 2^-	1^- 2^+
1^- 2^+	1^+ 2^-	1^+ 2^+	1^- 2^-	1^- 2^+	1^- 2^-
1^- 2^-	1^+ 2^-	1^+ 2^-	1^- 2^-	1^- 2^-	1^- 2^-

In *Podospora anserina* Esser and others (see Esser and Blaich 1973) have studied the formation of perithecia in crosses involving strains from different races. Perithecia can be formed on both sides of the junction-line, on one side or on neither side. The four pairs of alleles involved are a/al, b/bl, c/cl, v/vl. Although there are 16 combinations of alleles those containing al with b or cl with v are not viable. When a strain having the allele b makes contact with a strain having the allele al perithecia are only formed on the b side of the mating. The al allele seems to block the migration and the survival of the b nuclei which enter its trichogynes. In Table 3 there are only nine viable genotypes because a nucleus cannot contain the alleles blocker-1^+ with site-1^+ or blocker-2^+ with site-2^+ because these strains would block their own metabolism. Using Esser's terminology al and b cannot exist together and neither can cl with v. The pairing of the strain alblcv with abclvl is sterile and no perithecia are formed. As the system in flax and also in Podospora can be interpreted in terms of blockers and sites it is possible that a similar system might determine symplasmic speciation in basidiomycetes. All of these systems involve heterogenic incompatibility which have alleles of certain loci reacting with alleles of different loci. There are clearly many sites in the cell which could be blocked ranging from those which cause complete lethality to those which affect only nuclear migration or possibly later stages of development and fruiting. Species in most groups of organisms appear to have genes controlling homogenic incompatibility which are responsible for promoting outbreeding within a species and also genes controlling heterogenic incompatibility which give a species the potential to diverge into two or more intersterile populations and eventually species. If many basidiomycetes have symplasmic speciation it would not be surprising if they all showed great diversity in the success of plasmogamy, hyphal interactions and nuclear migration as these are essential if speciation is to occur.

Species of ascomycetes with 4 spores instead of 8 and basidiomycetes with 2 spores instead of 4 are not uncommon and it is possible that they may be involved in speciation. In both groups meiosis occurs normally but the position of the spindles is usually altered. The effect is to form dikaryotic spores without there being any plasmogamy. This extreme form of self-fertilisation might at first sight suggest that the two nuclei in each spore would rapidly become homozygous for all loci. In Neurospora tetrasperma compatible nuclei are included in most spores because the mating type locus nearly always segregates at the first meiotic division. In Podospora anserina the two nuclei in a spore are heterozygous for mating type because the locus nearly always segregates at the second meiotic division. In N. tetrasperma all loci on all chromosomes will be permanently heterozygous in each spore if they always segregate at the first meiotic division. The same would apply to all second division segregant loci in Podospora.

There is however very little variation in the growth rates of mycelia
of the two types derived from each spore in Podospora. This is not so
in C. bisporus and is has always proved possible to identify the two
monokaryotic types of mycelia when making macerate platings of dikaryotic mycelia. In some strains the growth of one of the components
is so thin that it can hardly be detected without use of the microscope. A two-spored basidiomycete is able to remain permanently heterozygous at the mating type locus which controls homogenic incompatibility at the same time as being able to become homozygous for the
heterogenic incompatibility loci which is necessary if each population is to be fertile. In this way two populations diverging from a
common ancestor could remain heterozygous and have the potential to
adapt while at the same time achieving rapid homozygosity of the speciation genes which make the populations intersterile. The formation
of dikaryotic spores in basidiomycetes thus achieves the same genetical effect as chromosomal ring-complexes in Oenothera and Periplane-
ta. Whereas it is not possible for a species of Oenothera with a ring
of 14 chromosomes at meiosis to return to having 7 bivalents there
would seem nothing to prevent the original spindle orientations being
restored in fungi. The formation of two-spored species may be connected with the process of speciation. Would a two-spored species be
able to evolve on more than one occasion? In a two-spored species related to C. patouillardii I have strains from Sweden, Norway and
England and all strains contain the same two mating type alleles.
This species has bipolar incompatibility without clamp connections.
In contrast I have 5 strains of the new 2-spored tetrapolar species
C. bilanatus nom.prov. from Holland and they only have one allele in
common. Although 2-spored populations might arise in separate localities at the same time it would seem likely that the alleles controlling lethality in the hyphal tips would be different.

Heterogenic incompatibility in Podospora is associated with the formation of barrage lines and crosses between races of the 2-spored
species C. sassii also show weak lines of this type. However the pigmented border lines formed between different species of Polyporus
(Esser and Hoffman, this symposium) are not common in Coprinus. They
may be associated with allopatric but not sympatric speciation in basidiomycetes. Similarly the A-B-C pattern of interfertility between
species may be found where allopatric speciation has occurred but not
with symplasmic speciation. There are no examples of this in Coprinus

species growing on dung.

Concluding remarks on the Species Concept

The species concept must be flexible enough to deal adequately with both symplasmic and allopatric speciation. The taxonomist must recognise that in many cases morphological differences only accumulate after genetical isolation. Similarly the geneticist should recognise that two species need not be completely intersterile to belong to different species.

The diversity of form and methods of speciation which may occur in the basidiomycetes suggest that the species concept should be flexible enough to recognise this diversity.

References

Esser K. & R. Blaich 1973: Heterogenic incompatibility in plants and
         animals. Adv. Genet. 17: 107-152.
Flor H.H. 1942: Inheritance of pathogenicity in Melampsora lini.
         Phytopathology 32: 653-669.
Galland M.C. 1973: Contribution à l'étude du genre Psathyrella (Fr.)
         Quel. (Agaricales) Thèse pour Docteur des Sciences
         Naturelles, Lyon
Jurand M.K. 1975: Breeding biology of the genus Psathyrella.
         Ph.D. Thesis, University of Edinburgh.
Jurand M.K & R.F.O. Kemp 1972: Surface ultrastructure of oidia in the
         basidiomycete Psathyrella coprophila
         J.Gen.Microbiol. 72: 575-579.
Kemp R.F.O. 1975: Breeding biology of Coprinus species in the section
         Lanatuli. Trans.Brit.mycol.Soc. 65: 375-388.

DISCUSSION FOLLOWING DR. KEMP'S PAPER

CLEMENCON: Does your method work on oidium-like cells, or with cell fragments from broken, slightly homogenized mycelia? Can you use these single cells instead of oidia?

KEMP: The oidia of some species of Coprinus do not even cause homing of their own hyphae, (generally homothallic species). Single celled mycelial fragments do not function like oidia.

ESSER: I am very fascinated by your results. We have, some 20 years ago, found similar phenomena in Podospora. Evidently Basidiomycetes have developed in their evolution a very similar system. Basically, there is a reaction of aversion, whatever the mechanism is, the mechanism might be different, but this aversion between the hyphae, that they cannot fuse, and if they fuse, they kill each other, is a basic phenomenon which I have called heterogenic incompatibility, that means that genetic different material is not able to co-exist. I am fascinated to see that in the Basidiomycetes more and more the same systems are present that in the Ascomycetes. I think it should help taxonomists to distinguish on a quantitative unequivocal manner between different species.
There is one thing I probably did not understand. You said that the bisporic species are heterogenous. In the fifties we have published a little paper about Podospora strains that, in contrast to any other Ascomycetes we took out of nature, were very homogenous. They are very homogenous because they act like true inbreeders.

KEMP: Yes, it does seem that 2-spored species of Coprinus can be quite heterozygous. This can be detected by macerating the dicaryons and isolating the monocaryotic components.

ESSER: The Podospora races isolated from nature are different, since they are heterokaryotic for the mating type alleles (more and less). Each race is pseudocompatible and submitted to a permanent inbreeding. This explains that all races we isolated from nature were homogeneous except for the mating type alleles. I wonder whether this is also true in Coprinus.

KEMP: Perhaps it would be useful if I described the technique. I isolate diads with a micro-manipulator, grow up the dicaryotic mycelia and then macerate them in a test-tube grinder. After the plated fragments have been incubated isolates are made onto fresh plates so that the two component monocaryons can be identified. In all species it is usually possible to distinguish the two monocaryons by slight differences in growth form. In C. bisporus it is often much easier as it is not unusual for one of the components to have very weak mycelial growth. I do not think these strains would survive in the absence of permanently heterozygous dicaryotic spores.

ESSER: The question is, if you isolate from nature a fruit body of bisporus, and isolate the spores, then I would suggest that you have not many other genes present. But if you take a four-spored, tetrapolar bisporus then you may isolate many genes different, because they have to come together.

KEMP: I do not think that this is so. For studies from the field I think it is essential to take hyphae growing from stipe material to find out exactly the genotype of the dicaryon which made the fruit body: So you get this dicaryon, fruit it in culture and then isolate diads. Maceration of the diads then shows that some of the monocaryons have a very poor growth. The growth is often better, by cross-feeding, when the two component monocaryons grow close together on the plate. The genetic variability of the monocaryons is very high.

ESSER: Thank you. I may come back later to this problem.

CLEMENCON: I have just one more question: is it possible to isolate, with a micromanipulator, the nucleus and inject it into a hyphae and to see what happens?

KEMP: It is possible, but it is tricky. In Coprinus the hyphae are very narrow. In Ascomycetes it is easier.

INTERET DES CULTURES DANS LA DELIMITATION DES ESPECES
CHEZ LES APHYLLOPHORALES ET LES AURICULARIALES

J. Boidin

Les champignons lignicoles sont fréquemment de culture facile, et peuvent servir d'exemple pour illustrer l'aide qu'apporte leur culture pure dans la difficile délimitation des espèces.

On objectera peut-être que l'éventail arbitrairement choisi (Aphyllophorales et Auriculariales) pourrait enlever aux conclusions le caractère général souhaitable. Je ne le crois pas, et d'ailleurs quelques exemples complémentaires pourront être pris chez les Agaricales. Par contre, il m'apparaît que qu'elle-que soit la technique mise en oeuvre - l'interprétation exige impérieusement une connaissance suffisante du groupe systématique en investigation. C'est la deuxième justification des limites du cadre choisi.

Les points suivants seront abordés
  (I)   Les caractères observables en culture
  (II) Les critères d'interfertilité-interstérilité
  (III) Les anastomoses végétatives
  (IV) Séparation protéique et immunologie
mais seul le point II sera largement détaillé.

(I) LES CARACTERES OBSERVABLES EN CULTURE

L'étude macroscopique et microscopique tout comme l'étude biochimique des cultures peuvent apporter des critères diagnostics complémentaires au morphologiste qui, souvent, se contente de l'étude du carpophore.

Bien des articles cités plus loin à propos des interfertilités et interstérilités font état de différences entre cultures d'espèces affines. Les spécialistes peuvent reconnaître bien des espèces à leurs seuls caractères culturaux (voir notamment Nobles 1965). J'illustrerai par un seul exemple chacun des trois points suivants: aspect macroscopique, morphologie microscopique, caryologie mycélienne.

Trois <u>Podoscypha</u>, dont la microscopie du basidiome est très semblable, et pour cela longtemps confondus, peuvent aisément se distinguer par leurs couleurs notamment celle de leurs cultures: mycélium orangé de <u>P</u>. <u>involuta</u>, jaune soufre de <u>P</u>. <u>vespillonea</u>, blanc de <u>P</u>. <u>gillesii</u> (Boidin et Lanquetin 1973).

Après Biggs (1937), les généticiens qui se sont intéressés au <u>Sistotrema</u> <u>brinkmanii</u> (Lemke 1969, Ullrich 1973) différencient facilement les groupes ou sous-espèces homothalle, bipolaire et tétrapolaire par la présence de bulbilles non pigmentées en chaine (sous-espèce homothalle), de bulbilles pigmentées en grappe (sous-espèce tétrapolaire, claires dans le sous-groupe 1, brun-sombre dans le sous-groupe 2) ou leur absence (sous-espèce bipolaire).

La spécification dans le genre <u>Leucoporus</u> est le sujet de longues controverses. David et Romagnesi (1972) ont montré que les comportements nucléaires étaient différents pour <u>L</u>. <u>lepideus</u> du type "normal", <u>L</u>. <u>arcularius</u> hétérocytique et <u>L</u>. <u>brumalis</u> et <u>meridionalis</u> astatocénocytiques. *)

Si la morphologie des basidiomes peut être influencée par la diversité et les variations des conditions ambiantes naturelles, celle des cultures est observée dans des conditions fixées qui facilitent les comparaisons. Mais là ne se limite pas, loin de là, l'intérêt des cultures.

---

*) pour les définitions de ces termes voir Boidin 1964 ou 1971 et tableau I.

(II) LES CRITERES D'INTERFERTILITE ET D'INTERSTERILITE *)

Si ces critères, utilisés sporadiquement, ont toujours apporté à celui qui les emploie un éclairage fort utile et souvent décisif, de grosses réticences sont généralement exprimées par les autres mycologues qui craignent l'énorme travail à fournir, l'infirmation de leurs vues de systématicien ou plus souvent la multiplication d'espèces difficiles à distinguer par les moyens classiques de la morphologie.

Je vais essayer de faire le point de l'expérience acquise par les systématiciens au cours d'un demi-siècle, c'est-à-dire après les travaux fondamentaux de Bensaude (1918) et surtout de Kniep (1920-1923-1928) et la publication du principe de Vandendries (1923) **), en montrant les côtés positifs, les problèmes en suspens, et en comparant les interprétations. Cette mise au point permettra, je l'espère, de demander ensuite aux généticiens leur point de vue, leurs critiques et leurs conseils.

Les motivations qui ont incité des mycologues à tenter des interfertilités sont diverses, mais toutes recouvrent le même désir, celui d'une meilleure approche de l'espèce.

Citons les principales:
1. Savoir si des variations d'aspect ou de microscopie sont dues à l'âge ou au développement, sont assujetties au milieu, et alors d'intérêt limité pour le spécificateur.
2. Savoir si des différences d'origine génotypique - dont il sait que l'observation et la mesure dépendent trop de ses sens (vue, odorat ...) et des techniques qui en améliorent l'usage - sont importantes ou secondaires.
3. Savoir si deux champignons semblables ou proches morphologiquement mais à habitats différents: l'un lié aux conifères, l'autre aux

---

*) Dans ce qui suit, il faut comprendre le terme "interstérilité" dans le sens d'absence de dicaryotisation.
**) Vandendries écrit p. 15: ... Si les haplontes de deux carpophores sauvages sont toujours et indéfiniment fertiles entre eux, ces deux carpophores appartiennent à la même espèce" et ajoute "En prenant ce principe pour critère, il sera intéressant d'étudier une foule de formes voisines dont la spécificité est encore un objet de controverse pour les mycologues".

feuillus par exemple, ou récoltés dans 2 ou plusieurs zones floristiques doivent ou non porter le même nom.

4. Confirmer ou infirmer la détermination de champignons étudiés en des laboratoires éloignés, en fonction de traditions orales ou même écrites pas toujours concordantes.

5. Connaître (cf. 3 et 4) la répartition géographique précise d'une espèce à large extension.

6. Assurer avec le maximum de sécurité l'identification de formes imparfaites ou de mycéliums obtenus à partir de bois attaqué, de galeries d'insecte, de sclérotes, en l'absence donc des basidiomes.

Bien souvent les auteurs on poursuivi simultanément plusieurs de ces buts.

On peut ajouter, bien que rares soient les recherches de ce type,
- Juger de la microévolution d'une espèce à large répartition en confrontant un grand nombre de populations allopatriques.

En outre une Aphyllophorale signalée à l'attention des généticiens par les mycologues, Sistotrema brinkmanii, a fait l'objet d'études approfondies d'un grand intérêt (Lemke 1969), Ullrich (1973), Ullrich et Raper (1974, 1975).

Nous allons passer en revue les données de la littérature en classant les résultats sous les rubriques suivantes:
A. Interfertilité totale ou presque totale
B. Interstérilité totale
C. Interfertilités ou interstérilités partielles

Rappelons que, de manière habituelle, les appariements de monocaryons issus de carpophores différents donnant lieu à l'obtention d'un mycélium dicaryotique sont dits positifs. Dans la pratique, on se contente donc de noter l'apparition de boucles, un trop faible nombre de champignons pouvant être menés, rapidement et à coup sûr, à la fructification dans les conditions usuelles de culture au laboratoire. En l'absence de boucles, la formation de dicaryons peut être mise en évidence par des colorations nucléaires. On peut aussi obtenir la dicaryotisation d'un monocaryon par un dicaryon (phénomène de Buller, ou confrontations "di-mon").

A. Interfertilité totale ou presque totale

Lorsque les auteurs disposent de cultures de plusieurs souches d'une même espèce, il est fréquent qu'ils s'assurent de leur identité par appariements de leurs monocaryons (voir entre autres Mounce et Macrae 1938, Nobles 1943, Eriksson 1950, White 1951, Maxwell 1954, Boidin 1958, Weresub et Gibson 1960, Davidson et all. 1960, Lanquetin 1973 a et b...) et les résultats positifs confirment très généralement leurs déterminations, et permettent, avant des appariements entre entités plus ou moins dissemblables, de s'assurer du bon fonctionnement des cultures monocaryotiques.

1. Confrontations positives malgré des différences d'aspect.

Eriksson (1950), p.36) écrit: "P.(eniophora) cinctula... differing in having smoother fructifications with a narrow, white, sterile margin, ... The combinaison tests show that P. violaceo-livida and P. cinctula contrary to current opinion can not be considered as separate species".
Peniophora versiformis montre un hyménium violacé à sombre sur le frais et gris rosâtre en herbier; il diffère nettement de P. carbonicola dont la surface est brune à l'état frais comme à l'état sec; au vu de leurs caractères micrographiques semblables, les auteurs ont placé ce dernier comme sous-espèce, forme, ou même synonyme du premier; ils sont interfertiles (Boidin 1958).

Mc Kay (1962) écrit: "Some basidiocarps from hardwoods may differ so greatly from the usual concept of P. (Polyporus) palustris on conifers as to defy identification except by use of interfertility studies of monocaryotic cultures".

Des récoltes camerounaises de Cymatoderma dendriticum qui: "montrent une grande variété d'aspect, d'épaisseur et de couleurs", sont cependant interfertiles en confrontation di-mon (Berthet et Boidin 1966). La forme rose du Podoscypha bolleana est interfertile avec la forme type (Boidin 1966). Des Lopharia mirabilis peuvent avoir des allures très différentes: plusieurs récoltes de Côte d'Ivoire et de la Réunion sont charnues, fortement irpicoïdes, beige rosâtre ou plus colorées, tandis qu'une autre récolte de Côte d'Ivoire et une de Singapour sont minces, à pores polygonaux bas, blancs sur le frais. Les confrontations de leurs monocaryons sont cependant positives et produisent aisément des fructifications (inédit).

2. Confrontations positives malgré des différences d'ordre microscopique.

Robak (1942) nous apprend que si, selon Overholts, le Stereum rugosiusculum américain diffère du Stereum purpureum "mainly in possessing cystidia", opinion cependant non retenue par divers systématiciens européens, les deux sont interfertiles. Malgré leur richesse très diverse en dendrophyses et en gloeocystides, les diverses récoltes françaises (Boidin 1958 , p.454) ou suédoises (Hallenberg et Hallingbäck 1974) du Peniophora lycii sont régulièrement interfertiles. Ces derniers font en outre remarquer qu'Amylostereum laevigatum existe chez eux sous deux formes, l'une, liée au Juniperus communis a des spores longues de 7-9 microns, l'autre, liée au Taxus baccata a des spores plus grandes, 9-12 microns; ils sont cependent compatibles.

Mais le plus bel exemple est celui donné par Duncan (1972) qui constate que des récoltes d'Auricularia polytricha de l'Assam et d'Australie, totalement interfertiles, ont des spores et des basides de tailles nettement différentes:

	longueur des spores en microns		longueur des basides en microns	
	valeurs absolues	moyennes	valeurs absolues	moyennes
Australie	13,4 - 20,4	17,0 $\pm$ 0,3	64,9 - 94,6	80,9 $\pm$ 1,2
Assam	9,8 - 14	11,8 $\pm$ 0,1	34,1 - 57,6	45,7 $\pm$ 1,0

3. Confrontations positives de champignons récoltés sur des supports différents.

Pour les champignons lignicoles, le végétal ligneux support est un des principaux composants du milieu, et représente à lui seul l'ensemble des facteurs édaphiques. Il peut jouer un rôle important dans la spéciation modifiant les chances de rencontre des monocaryons dans un même secteur géographique. Aussi constate-t-on souvent qu'une espèce saprophyte est inféodée à un type végétal (conifères, feuillus) ou même de manière plus précise à une famille, un genre, ou même

une seule espèce ligneuse. S'assurer de l'identité de champignons récoltés sur des supports différents est une fréquente nécessité.
On peut reprendre ici l'exemple cité plus haut (II - A.2) de l'_Amylostereum_ _laevigatum_. Citons en quelques autres: Mounce et Macrae (1936) montrent que les _Trametes_ _saepiaria_ du _Pinus_ et du _Picea_ sont compatibles, et (1938) que _Fomes_ _pinicola_ croît sur conifères et feuillus. Eriksson (1950) montre l'interfertilité de _Peniophora_ _pithya_ récoltés sur _Picea_ _Abies_ et du _Corticium_ _plumbeum_ du _Pinus_ _silvestris_, mais aussi d'une récolte d'abord identifiée _P._ _cinerea_ trouvée sur _Sorbus_ _aucuparia_; _P._ _versiformis_ existe sur feuillus et sur _Abies_ (Boidin 1958 b). Mc Kay écrit en 1962: "Because dicaryotic cultures from basidiocarps identified as _P._ (_olyporus_) _palustris_ could not be distinguished from those identified as _P._ _durescens_ except by determination of the host and distributions, it seemed of value to apply the technique of pairing single spore cultures. ... The evidence show in table 2 demonstrate complete compatibility".

4. Confrontations positives entre récoltes de zones géographiques
   éloignées ou de zones climatiques différentes.

L'isolement géographique est un facteur important de spéciation et ce n'est pas sans quelque appréhension que l'on se décide à identifier des récoltes issues de points très éloignés. L'objectivité des tests d'interfertilité peut alors être mise à profit.
Mounce (1929) et Mounce & Macrae (1938) ont confronté des _Fomes_ _pinicola_ américains, japonais et européens; Robak (1942) a apparié avec succès divers polypores canadiens et norvégiens tel _Lenzites_ _sepiaria_, Harmsen (1960), des _Merulius_ _himantioides_ et _lacrymans_ danois, américains et indiens, et Macrae (1967) des _Hirschioporus_ _fusco-violaceus_ canadiens d'une part, norvégiens et esthoniens d'autre part (Tableau III). Nobles et Frew (1962) ont fait de même entre _Pycnoporus_ _cinnabarinus_ canadiens, européens et japonais, entre _P._ _coccineus_ du domaine indo-pacifique et entre _P._ _sanguineus_ américains du Nord et du Sud, africains, asiatiques et océaniens. Boidin et des Pomeys (1961) ont montré l'interfertilité entre des _Peniophora_ canadiens et français, français et marocains, et Boidin et all. (1968) entre des _Aleurodiscus_ _livido-coeruleus_ de Colombie britannique et des Alpes françaises, Maas Gesteranus et Lanquetin (1975) entre récoltes ivoiriennes et guadeloupéennes de _Stecchericium_ _seriatum_ et entre récoltes centrafricaines, ivoiriennes et américaines (Louisiane) de _Gloeodontia_ _discolor_.

La littérature n'est pas très riche en exemples de cette sorte, sans doute parce que l'échange de cultures monospermes ou au moins de sporées vivantes n'est pas encore très courant. Aussi ajouterai-je quelques résultats inédits tels que l'interfertilité entre des _Punctularia_ _tuberculosa_ de France méridionale, du Maroc et de Centrafrique, des _Phaephlebia_ _strigoso-zonata_ de Sibérie et de Centrafrique, entre des _Duportella_ _tristicula_ du Pakistan, du Gabon et de République Centrafricaine, entre des _Scytinostroma_ _galactinum_ d'Europe (France), d'Asie (Kamtschatka), d'Amérique (Canada, Guyane) et d'Afrique (Côte d'Ivoire), entre des _Scytinostroma_ _duriusculum_ d'Asie (Ceylan), des Amériques (Texas, Martinique, Argentine) et d'Afrique (Rép. centrafricaine, Côte d'Ivoire).

5. Confrontations positives permettant d'éliminer des erreurs de détermination ou de dénomination.

La littérature mycologique de chaque région est encombrée de dénominations superflues. D'autre part, en des régions éloignées, des traditions mènent à l'emploi de noms différents pour une même espèce, et l'on ne peut pas toujours au vu de spécimens anciens assurer la synonymie sans hésitation. Là encore le critère d'interfertilité a rendu de grands services.

On peut reprendre l'exemple des Stereum purpureum et rugosiusculum (Robak 1942). M.K. Nobles (1943) écrit: "I may be seen that of the isolate of this species used in the interfertility tests ... only F 2045 and 10.618 were received under the name P. palustris, the remainder having been labellé T. serialis. Interfertility tests have allowed for positive identification of these cultures". J. Eriksson (1950) montre que Peniophora maculaeformis sensu Hoehn. et Litsch. et P. syringae (Karst.) sont des P. nuda, Mc Keen (1952) que "Peniophora" allescheri est "P." mutata, Weresub et Gibson (1960) que si le P. pini européen est interstérile avec le P. pini canadien, il est partiellement fertile avec le P. duplex américain (voir II C c 1), Harmsen (1960) que M. (Merulius) americanus est M. himantioides, Boidin et des Pomeys (1961) que le Phlebia radiata européen et le Phlebia merismoides canadien sont un même champignon, de même le Peniophora piceae français et le P. separans nord-américain. Mac Kay (1962) prouve que Polyporus palustris est identique à un champignon japonais déterminé Tyromyces squalens, puis Polyporus versiporus Yasuda. Roed (1969) identifie le Typhula incarnata du Japon à la fois au T. itoana et au T. graminum du Canada. Par des colorations nucléaires, vu l'absence de boucles, P. Lanquetin (1973 a) démontre que deux cultures américaines dénommées Scytinostroma portentosum sont l'une interfertile avec Scytinostroma hemidichophyticum non encore signalé en Amérique, l'autre avec Scytinostroma duriusculum.

Il est bien clair que c'est grâce à l'accumulation de telles données, que nous pourrons un jour connaître avec le maximum de précision et de sécurité, la répartition des champignons sur notre planète.

6. Identification de formes imparfaites et de mycéliums.

Même lorsqu'il est possible d'obtenir en conditions artificielles la fructification de cultures non identifieées, la détermination sûre peut nécessiter des essais d'interfertilité. Ces essais deviennent indispensables si, ce qui est fréquent, la fructification n'est pas obtenue. Lorsque l'on ne veut s'astreindre à l'obtention de néohaplontes par traitements chimiques (notamment sels biliaires) pour l'ensemble des cultures en étude, la confrontation dicaryon à déterminer, monocaryons connus peut être employée.

C'est ainsi que Nobles et Frew (1962) identifient 103 stocks, 57 par interfertilité, 46 par phénomène de Buller - Weresub et Gibson (1960) montrent que le "fungus T" de Nobles (1956) est Peniophora pini subsp. duplex. Si Talbot (1964) avait pu, par fructification in vitro,

rapporter au genre Amylostereum le champignon commensal des galeries
de Sirex noctilio en Australie, Gaut (1969) le détermine par interfertilité comme A. areolatum. Bruehl et all. (1975) ont par appariements di-mon cherché à identifier à des Typhula connues, 85 dicaryons
isolés à partir de sclérotes trouvés sur blé et sur des graminées ma-
lades. Sur les 19 cultures isolées de pêchers vivants par Petersen,
Mc Kay (1962) en identifie 16 à Polyporus palustris (Tableau II).
Une récolte niçoise du champignon imparfait Ceriomyces venulosus
(Berk. et Curt.) Torrend s'est montré interfertile avec Punctularia
tuberculosa (Pat.) Pat., non encore signalé en France, en conformité
avec l'hypothèse de P.H. Talbot (1958) (inédit).

B. I n t e r s t é r i l i t é s   t o t a l e s

On peut résumer la position des systématiciens en empruntant cette
phrase à J. Eriksson (1950, p.24): "Negative results have been taken
as a sign of non identity which has been considered as certain if
both forms also show different morphological characters".

En caricaturant quelque peu, on peut considérer plusieurs cas

1. Le Systématicien s'assure que des différences morphologiques constatées vont de pair avec l'interstérilité, et ce double
constat lui permet d'affirmer l'existence d'hologamodèmes considérés
comme autant d'espèces taxonomiques distinctes.

2. Le Systématicien attend le verdict de tests tentés systématiquement à l'intérieur d'un ensemble difficile, à propos duquel
distinctions et regroupements ont été également proposés par des mycologues. Il constate alors très généralement que les mêmes monocaryons issus de souches plus ou moins nombreuses sont, dans le même
temps, interfertiles avec les uns, interstériles avec les autres, dé-
finissant ainsi les groupes de souches interfertiles entre elles et
interstériles de groupe à groupe. Les groupes sont alors considérés
comme les unités élémentaires et sont pour beaucoup, si des caractères distinctifs peuvent être détectés, de bonnes espèces.

3. Des interstérilités "inattendues" sont constatées.
Après réétude détaillée des caractères microscopiques des carpophores
et des cultures, de l'habitat, ...

a) le systématicien réussit à définir des unités qui lui
avaient échappé auparavant.

b) il ne trouve aucune explication.

On pourrait ironiser en disant que deux types de résultats sont à distinguer: ceux qui corroborent les vues ou prévisions du systématicien et ceux qui les démentent. Dans ce dernier cas, celui-ci a parfois tendance à les refuser et à retirer toute signification à des interstérilités qui l'obligeraient à l'éclatement de maintes espèces linnéennes en entités plus petites qu'il ne cerne pas.

Il serait difficile, l'opérateur ne précisant pas toujours ses hypothèses de travail, de classer les données de la littérature selon les 3 rubriques distinguées ci-dessus. Aussi je distinguerai successivement les interstérilités associées

    1) à des différences morphologiques,

    2) à des différences de thallies,

    3) non liées à des critères morphologiques ou de thallie

1. Interstérilités acceptées par le systématicien parce que corroborées par des critères morphologiques.

Il faut tout d'abord éliminer les interstérilités dues à des erreurs de déterminations ultérieurement corrigées.
On se contentera de deux exemples. Robak (1942) s'étonne que le Trametes serialis norvégien soit interstérile avec un Tr. serialis canadien déterminé par Overholts. Nobles (1943, note page 214) nous apprend qu'Overholts a corrigé cette détermination en Polyporus palustris. Boidin (1958) s'étonne que les mycéliums d'une souche suédoise et d'une souche française appelées par le même déterminateur Peniophora pithya diffèrent, et soient de ce fait interstériles. Il les décrit séparément. La souche française se rapportait en fait au P. piceae reconnaissable à ses spores plus grandes, l'absence de gloeocystides et les boucles inconstantes, mais ces caractères distinctifs n'avaient pas été relevés à cette époque.

De tels cas sont intéressants, en ce qu'ils soulignent l'objectivité du test, objectivité qu'ont mise à profit de nombreux auteurs.

Mounce et Macrae (1937) constatent l'interstérilité des Fomes roseus et subroseus et écrivent " ... the failure to form clamp-connections in any of the 502 pairings of a monosporous mycelium from any on the five different F. roseus cultures with a monosporous mycelium from any of 15 different F. subroseus cultures is in accordance with the conclusion of Weir, Snell, Overholts, and others, than these two fungi ... are distinct".

Nobles (1943) distingue 4 espèces dans le "Trametes serialis complex". Eriksson (1950), dans son excellente mise au point sur le genre Peniophora en Suède, s'assure de l'interstérilité des espèces qu'il croit distinguer morphologiquement dans le difficile "groupe cinerea". Mc Keen (1952) différencie 3 "Peniophora" à conidies dans la stirpe mutata. Nobles (1956) prouve que 6 espèces de Peniophora

Cultures des Auriculariales   287

des conifères, interstériles entre elles, cohabitent au Canada.
Boidin (1958) s'assure, avant publication, de l'autonomie d'une nouvelle espèce, P. meridionalis, vis-à-vis de P. lycii et versiformis.
Weresub et Gibson (1960) distinguent après interstérilité P. pini et
et P. pseudopini. Harmsen (1960) et la même année Lentz et Mc Kay
ainsi que Davidson et all., s'assurent de même de l'indépendance le
premier de Serpula himantioides d'avec S. lacrimans, les autres de
Laurilia taxodii d'avec L. sulcata. Boidin et des Pomeys (1961) concluent que dans le groupe de P. nuda, une espèce distincte existe au
Canada, qui interstérile ne peut être assimilée ni à P. nuda ni à
P. violaceo-livida. Après Mc Kay (1959) qui prouve l'incompatibilité
de monocaryons de Polyporus sanguineus et cinnabarinus, Nobles et
Frew (1962) élargissent beaucoup cette étude en y incorporant Pycnoporus coccineus, toutefois 4 souches incomplètement interstériles
leur posent quelque problème (cf. II-C-3b). Mc Kay (1962) reconnait 3
entités spécifiques affines, Polyporus palustris, spraguei et un P.
sp. à dénommer (TableauII). Macrae (1967) montre que dans l'"Hirschioporus abietinus aggregate" existent en Amérique et en Europe des espèces distinctes par l'ornementation de l'hyménium (lamellé chez H.
laricinus, irpicoïde chez H. fusco-violaceus et poré chez H. abietinus (Tableau III). Boidin et Lanquetin (1973) ont prouvé que 3 espèces, interstériles entre elles, étaient confondues en Afrique sous le
nom de Podoscypha involuta. P. Lanquetin (1973 a) montre dans le
groupe des Scytinostroma sans boucles et à spores sphériques, l'autonomie des Sc. duriusculum, hemidichophyticum et portentosum (sensu
auct. europ.), puis (1973 b) clarifie la spécification des Vararia
subg. Dichostereum en reconnaissant au moins 6 espèces interstériles
(Tableau IV) et supputant l'autonomie de V. peniophoroides; celui-ci
récolté depuis lors s'est bien révélé interstérile avec ses voisins.
A. David (1974) s'assure de la non dicaryotisation entre Tyromyces
caesius et une espèce nouvelle très proche qu'elle dénomme T. subcaesius. Les résultats négatifs de confronations effectuées entre
Trametes rubescens et Lenzites tricolor souvent considérés comme des
formes de Daedaleopsis confragosa montrent qu'il s'agit de deux espèces distinctes (David, inédit).

Dans tous le cas, le systématicien conclut à la distinction d'espèces
affines et précise les caractères distinctifs. Il considère que l'interstérilité a confirmé la valeur des différences, parfois minimes,
observées par lui. A l'inverse, un résultat positif l'aurait amené
(cf. II-A- 1 et 2) à considérer que les divergences notées n'étaient
que phénotypiques, ou que, génotypiques, elles étaient de trop faible
importance pour interdire l'appariement nucléaire et la conjuguaison
ultérieure des mitoses. Ne seraient en cause que des formes ou variétés.

   2. Interstérilités liées à des différences de thallie.
   L'expérience a montré que des différences, qui pouvaient apparaître mineures aux yeux de beaucoup, comme des différences de thallie, empêchent toute interfertilité.

On peut citer le cas du Corticium coronilla qui comporte, selon Biggs
(1937) au moins 4 types interstériles plus ou moins reconnaissables,
une fois prévenu, à de petits détails morphologiques: des types bipolaire. (gr.I), homothalle (gr. II), tétrapolaires (gr.IIIa et b) en
partie retrouvés en France (Boidin 1958, sous le nom de Trechispora
brinkmanii). Ce Sistotrema brinkmanii a été l'objet d'études génétiques très approfondies de Lemke (1969), Ullrich (1973) et Ullrich et
Raper (1974, 1975), qui sont pour notre sujet du plus haut intérêt.
Sur 96 représentants, 16 sont homothalles, 74 bipolaires et 6 tétrapolaires. Dans leur répartition sur le globe, aucune liaison n'est
apparente avec les substrats ou la géographie. En culture, on peut
noter des différences morphologiques: les homothalles montrent des
bulbilles incolores en chaînes, les tétrapolaires des bulbilles plus
ou moins colorés en grappe; les bipolaires n'en possèdent pas. L'homothallie est primaire, la spore uninucléée et il y a fusion, dans la
baside, du dicaryon homocaryotique puis méïose. Aucun mutant homothalle ne réinvente la bi- ou la tétrapolarité. Par utilisation de
mutants auxotrophes à déficiences complémentaires (forçage nutritionnel) des types prototrophes apparaissent montrant que, dans ces conditions, des échanges sont possibles entre toutes les souches homothalles, qui forment donc un seul groupe *). Par forçages répétés,
aucune hétérocaryose n'a pu être obtenue entre bipolaires et tétrapolaires, homothalles et tétrapolaires, par contre des hybrides de
bipolaire et d'homothalle sont parfois obtenus (réaction allothalli-
que) montrant que l'isolement n'est pas ici total malgré une évolution divergente déjà marquée. Il n'est peut-être pas sans à propos de
rappeler que par des voies bien différentes, nous avions suggéré
(1971) que: "bipolarity seems to be, apparently, the access to homothallism".

Corticium (maintenant Sistotremastrum) niveo-cremeum contient aussi
des souches hétérothalles et des souches homothalles (Boidin 1958).
Les confrontations di-mon sont toujours négatives (entre dicaryon homothalle et monocaryons de l'hétérothalle). On peut assurer aujour-
d'hui qu'il en est de même entre types bipolaires et homothalles de
Gloeocystidium (maintenant Hyphoderma) tenue (Boidin 1950), d'Hyphoderma setigerum (Boidin et Lanquetin 1965), de Typhula micans
(Berthier 1974), de Coriolellus malicola (David, inédit).

L'amphithallie (homothallie secondaire, ou pseudocompatibilité) souvent liée à la bisporie est peu signalée chez les Aphyllophorales.

A part le cas classique d'Aleurodiscus canadensis (Skolko 1944) où
n'est connu qu'un type amphithalle, on peut citer le cas du Vararia
firma subsp. amphithallica (Boidin et Lanquetin 1975); les représentants amphithalles, récoltés au Gabon et en Côte d'Ivoire, qui ne
diffèrent du type tétrasporique décrit de République Centrafricaine
que par la bisporie et la thallie, ne peuvent dicaryotiser en phénomène de Buller les monocaryons de V. firma. Il en est de même entre
type hétérothalle et type amphithalle tétrasporique de Typhula culmi-
gena (Berthier 1974).

Pour élargir nos vues, il faut ici faire appel aux Agaricales chez
lesquelles sont connus de nombreux cas de parthénogénèse fixée

---

*) Whitney et Parmeter ainsi que Garza-Chapa et Anderson mettent de
même en évidence les échanges nucléaires chez Rhizoctonia solani, espèce homothalle sans boucles.

(haploid parthénogenesis, Bauch 1926). Lorsque, dans une même entité
morphologique, existent dans la nature, des carpophores tétrasporiques bouclés et des carpophores bisporiques sans boucles et monocaryotiques, par ex. Mycena galericulata, les mycéliums producteurs de
fructifications monocaryotiques ont perdu toute aptitude à copuler
entre eux (Lamoure 1960) et avec des monocaryons de l'hétérothalle.
(Lamoure, inédit). Tout laisse penser que leur origine est dans la
fructification de monocaryons du type hétérothalle correspondant
puisque cette fructification d'homocaryons est assez fréquente en laboratoire et souvent associée à une diminution du nombre des stérigmates (Laxitextum bicolor, Stecchericium seriatum, Sistotrema brinkmanii par ex.). Toutefois les spores issues de ces fructifications
homocaryotiques d'espèces hétérothalles obtenues en laboratoire sont
capables de donner des mycéliums copulants, tous du même pôle, celui
de leur générateur, tandis que les carpophores parthénogénétiques de
la nature ont perdu toute capacité de dicaryotiser. Aucun essai toutefois n'a semble-t-il, été tenté par forçage nutritionnel.

Chez les Aphyllophorales pourraient être parthénogénétiques: le Corticium rhodoleucum (Boidin 1958), non cultivé, et qu'il faut sans
doute rattacher à l'Amylocorticium cebennense bouclé et tétrasporique, et les deux cultures sans boucles de Sistotrema brinkmanii dont
parlent Ullrich et Raper (1974, 1975) qui refusent de copuler avec
toutes les souches hétérothalles testées. Peut-être faudrait-il faire
appel pour expliquer l'isolement des parthénogénétiques fixés vis-à-
vis de leur collègue hétérothalle à une incompatibilité hétérogénique
(voir II-3 ci-après), mais cette dernière suffit-elle à expliquer que
les 20 souches parthénogénétiques de Mycena galericulata soient interstériles?

Donc, toute différence de thallie mène dans la pratique à un isolement génique, qu'il y ait une différence morphologique (nombre de
stérigmates) ou non.

3. Interstérilités apparemment sans lien avec des différences
   morphologiques ou de thallie.

Il n'est pas exceptionnel cependant que des résultats négatifs ne
soient pas interprètables c'est-à-dire ne semblent associés à aucune
différence morphologique stable. Bien des résultats isolés pourraient
être signalés. Il a semblé préférable de n'exposer ici que des cas
ayant fait l'objet d'une investigation poussée.

Mounce et Macrae (1938) distinguent deux groupes A et B pratiquement
interstériles parmi les Fomes pinicola américains; toutefois le

groupe C, formé arbitrairement de récoltes étrangères, européennes et
japonaises, totalement interfertiles entre elles, est pratiquement
compatible avec le groupe américain A et partiellement avec le groupe
B (Tableau IV). De même (voir Tableau III), Macrae (1967) constate
l'existence de deux groupes interstériles parmi les Hirschioporus poroïdes américains (H. abietinus), toutefois ils sont tous deux partiellement compatibles avec des souches norvégiennes et totalement
interfertiles avec une récolte allemande (voir II-C-3). Duncan (1972)
montre qu'une récolte américaine (Louisiane) d'Auricularia polytricha
est interstérile avec toutes les autres récoltes (non américaines)
bien qu'elle se situe par la taille des basides et des spores entre
des récoltes indiennes et australiennes interfertiles. Les études ré-
pétées du Sistotrema brinkmanii (loc. cit.) nous apprennent que les
74 représentants bipolaires forment 5 groupes interstériles et les 6
représentants tétrapolaires se répartissent en 2 groupes interstériles. Le genre Laxitextum, bien caractérisé, est actuellement considéré comme monospécifique. L. bicolor est largement répandu sur feuillus divers; de nombreuses récoltes ont pu être confrontées: 5 groupes
interstériles se sont déjà dégagés; le groupe D comprend les 4 récoltes africaines (Gabon, R.C.A.) cultivées à ce jour; le groupe A, formé de 13 récoltes existe sur 3 continents (Europe: Allemagne, France;
Asie: Indes; Amérique: Canada, Louisiane); le groupe B comprend 3 récoltes européennes (France et Allemagne) et une américaine (Louisiane), le groupe C deux spécimens originaires de France et d'Allemagne.
A l'heure actuelle le systématicien ne peut prévoir dans quel groupe
se placera sa récolte, sauf peut-être si elle provient d'Afrique tropicale. Enfin, le groupe E est défini par son homothallie, il comprend une récolte de Singapour et une de la Réunion; son interstérilité en appariements di-mon avec les autres groupes était attendue
(cf. II-B-2).

On doit remarquer que dans ces 5 études un nombre élevé de récoltes
est en cause. Un tel résultat interviendrait-il dans tout travail
d'une certaine envergure? A part l'homothallie du Laxitextum bicolor
gr. E, aucun des sous-ensembles interstériles n'a encore montré à
Macrae, Duncan, ni à moi-même de divergences distinctives, notamment
de nature morphologique.

On ne peut expliquer ces résultats totalement négatifs obtenus de manière répétée lors de multiples confrontations intersouches par des
erreurs expérimentales, des mycéliums souffreteux, ... etc. ... vu
leur bon comportement intragroupe. Une barrière existe dont il faudrait comprendre la nature.

Elle est parfois matérialisée par des phénomènes d'interactions avec
ou sans coloration de la ligne de contact.

Duncan et MacDonald, puis Duncan (loc.cit.) parlent de "monokaryon
interactions" nettes entre "microevolutionary units" à l'intérieur
des Auricularia auricula et polytricha, mais aussi entre croisements
interspécifiques de A. auricula, fusco-succinea et polytricha, et les
comparent aux interactions notées par Vandendries (1927a et b) entre
souches partiellement interstériles de Coprinus micaceus, avec ligne

de démarcation "rouge comme du sang", et à celles notées par Mounce
(1929) entre sous-unités de Fomes pinicola.

Les cas d'interactions cités ne sont pas fréquents *) et il apparaît
nécessaire d'attirer l'attention des chercheurs sur ces phénomènes
qu'ils ont peut-être omis de noter ou de signaler. Ces interactions
seraient à rechercher entre souches interstériles qu'elles soient
considerées comme appartenant à des espèces très affines ou à de
sous-unités d'une même espèce. Duncan et MacDonald précisent que "No
intermingly of the monocaryons appears to occur, and on the surface
the interaction between them is generally evident as a ridge of hyphae".

Ce "ridge of hyphae" peut-il être assimilé aux barrages provoqués par
un "phénomène de répulsion" comme l'écrit Vandendries (1933), et apparaîssant comme une zone où les hyphes arrivées en contact ont dégénéré laissant un "no man's land" étroit entre les deux cultures? On
sait que ces barrages peuvent apparaître chez certaines espèces tétrapolaires lors de l'appariements de monocaryons présentants des
facteurs B communs (A $\neq$ B =) ce qui n'est pas le cas ici. Des termes
nouveaux devraient être employés mais auparavant une description précise de ces interactions serait nécessaire.

Vandendries (1933), Mounce (1929), Duncan (loc.cit.) ont-ils observé
'es barrages au sens strict (c'est-à-dire interruption entre les 2
cultures), ou un bourrelet (c'est-à-dire une ligne de contact en relief) avec ou non excrétion de nécropigments? Ces réactions ont-elles
des significations différentes? Ce sont des bourrelets avec excrétion
brune qu'à observé P. Lanquetin entre souches interstériles de divers
Vararia subg. Dichostereum lors qu'elle écrit (1973b, p.171): "dans
les 350 confrontations négatives de souches interstériles une ligne
très brune se forme à la rencontre des mycéliums, tandis que cette
ligne n'apparaît pas dans les confrontations d'haplontes incompatibles d'une même récolte". Il serait nécessaire que les termes de
barrage, de "border line", de "monokaryon interactions", de "ridge of

---
*) On peut cependant citer les études de Verrall (1937) qui signalent des interactions entre 3 types écologiques de Fomes igniarius,
espèces sans boucles et les études de Barrett et Uscuplic (1971)
entre souches de Polyporus schweinitzii, autre espèce sans boucles. De
telles réactions sont aussi notées chez Fomes carjanderi, entre monocaryons à facteurs de polarité identiques, entre mono- et dicaryons
sexuellement compatibles et lors de la confrontation de dicaryons
différents (Adams et Roth 1967, Neuhauser et Gilbertson 1971).

hypnae" soient bien définis, et que les auteurs leur portent à l'avenir une attention toute particulière.

La seule étude génétique détaillée a été faite sur l'Ascomycète <u>Podospora anserina</u>, chez lequel Rizet avait en 1952 *) observé un barrage entre races géographiques. Esser (1962, voir aussi Esser et Blaich 1973) appelle incompatibilité hétérogénique ce phénomène qui n'a lieu qu'entre races différentes d'une même espèce, ce refus dépendant de gènes autres que les gènes de l'incompatibilité classique due, chez les Basidiomycètes, aux facteurs de polarité (incompatibilité homogénique). Il signale aussi le cas du <u>Sordaria fimicola</u> analysé par Olive (1956) où des races A1 et C1 ne peuvent se croiser par suite d'incompatibilité hétérogénique; toutefois A1 et C1 sont fertiles avec la race C4; nous rencontrerons de tels résultats "triangulaires" ci-après.

Des résultats négatifs imprévisibles associés à un phénomène de barrage pourraient donc être interprétés chez les Basidiomycètes comme résultant de la confrontation d'unités désormais isolées génétiquement par cette incompatibilité hétérogénique.

Il faut reconnaître que pour le Taxonomiste il y a ici une grosse difficulté. Savoir que des interstérilités inattendues peuvent s'expliquer par l'existence de facteurs d'incompatibilité particuliers ne résout pas son problème. Il ne peut parler d'espèces, car il n'est pas en mesure de les reconnaître morphologiquement. Il est d'ailleurs prouvé pour <u>Fomes pinicola</u> comme pour <u>Hirschioporus abietinus</u> que les 2 sous-groupes américains interstériles sont l'un et l'autre plus ou moins compatibles avec des souches européennes.

De même pour les <u>Auricularia auricula</u> (Duncan et MacDonald 1967) européens et américains des feuillus (non cités dans cette rubrique II B3 parce que des différences significatives de taille des basides et basidiospores les distinguent) interstériles deux à deux sont l'un et l'autre plus ou moins interfertiles avec les souches américaines sur conifères (voir II C 3).

Avant de parler de ces interfertilités incomplètes, on peut déjà con-

---

*) K. Esser et G. Rizet, C.R. Acad. Sciences (Paris) 1954.

clure que des interstérilités totales, non corroborées par des caractères morphologiques distinctifs ne devraient pas permettre d'ajouter un nom spécifique tant que l'on n'aura pas la certitude (et comment l'avoir?) que ces entités ne seront jamais toutes deux interfertiles au moins partiellement avec une troisième.

Si l'isolement génique est un facteur de spéciation, il ne prouve pas que la spéciation soit réalisée.

C. Interfertilités ou Interstérilités partielles

A côté de nombreux résultats sans ambiguité, quelques publications font état de l'apparition de boucles dans une partie des appariements, ou sur certaines hyphes.

Il faut essayer maintenant de classer ces données, parfois difficilement comparables, les auteurs les ayant obtenues dans des conditions différentes. Leur mise en ordre et leur éventuelle explication, même si elle reste hypothétique, doivent être tentées.

1. Interfertilités ou interstérilités partielles duoc à des allèles identiques de polarité.

Des résultats partiellement négatifs sont à attendre, lorsque des allèles identiques de polarité sont en présence. Ce fait, relativement rare lorsque les récoltes sont distantes est cependant d'autant plus probable que l'origine géographique des souches utilisées est restreinte.

La littérature nous fournit quelques exemples chez les Aphyllophorales: Tout d'abord citons les résultats de Fries et Jonasson (1941) entre <u>Polyporus abietinus</u> tous suédois. Les souches 10230 et 10234 de <u>Trametes serialis</u> (Nobles 1943) dont les "fruit bodies ... collected by two collectors working over a small area in a mill yard ... from the same plant" ont les mêmes gènes d'incompatibilité. J. Eriksson (1950) a de même rencontré des allèles identiques dans deux <u>Peniophora cinerea</u> du même lieu sur arbres différents, mais aussi dans deux <u>P. fraxinea</u> suédois recueillis l'un sur <u>Fraxinus</u>, l'autre sur <u>Syringa</u> dans deux provinces différentes. Les résultats des appariements de monocaryons de 5 "Fungus T" (devenu <u>P. pini</u> subsp. duplex) par Nobles (1956) qui présentés de manière très succinte dans l'angle inférieur droit de la figure 8 par des <u>+</u>, le sont de manière plus explicite dans les figures 87 et 88 de ce même travail et dans la figure 1 de

Weresub et Gibson (1960). Il est clair que l'obtention de 47 résultats positifs sur les 160 attendus s'expliquent par l'origine très confinée des 5 souches et la présence des mêmes allèles de gènes de polarité. La démonstration eut été plus belle encore si les 4 pôles de DAOM 31232 avaient figuré dans le Tableau (fig. 88) de Nobles - mais la tétrapolarité de cette souche n'a peut-être été établie que postérieurement. Antérieurement, Mounce et Macrae (1937) avaient clairement montré l'existence d'un même allèle A dans deux récoltes de Fomes roseus du Québec et de l'Ontario, et en 1971, Neuhauser et Gilbertson apportent des preuves semblables pour un autre Fomes bipolaire: F. cajanderi.

Un autre exemple est à citer, celui du commensal du Sirex noctilio de Tasmanie déterminé Amylostereum areolatum par confrontation avec des monocaryons français de 4 pôles. A_1B_1, A_2B_2, A_1B_2 et A_2B_1 (Gaut 1969, tabl.2). L'interfertilité partielle observée est: "explicable if one assumes an incompatibility factor genotype of A_1B_1 + A_2B_3 for the fungus isolate P2, and if one assumes that all four homocaryons derived from it have the A_1B_1 nucleus". Ici on ne peut accuser la proximité géographique - les partenaires proviennent pratiquement des antipodes - mais peut-être faire appel à la dissémination récente par l'homme dans la zone australienne de cet Amylostereum en provenance d'Europe (?).

2. Interfertilités incomplètes dues au vieillissement des cultures.

Robak (1942) constate l'interfertilité de 2 souches canadiennes de Lenzites sepiaria avec 5 souches norvégiennes, mais ajoute: "The interfertility has not been absolute ... It must be taken into consideration that the Canadian strains were 3-4 years old at the time of pairing in question ... In one case, ... the interfertility was almost total. L 90 was the youngest and most vigorous among the Norvegian strains. Consequently, the age of the cultures must in all probability be held responsible for the greater percentage of negatively resulting pairings". Weresub et Gibson (1960) font remarquer que les monocaryons du Stereum pini américain (= Peniophora pseudo-pini) qui "had paired completely in 1953; in 1958, ... some of the pairings were negative and some with only rare clamps". Boidin et des Pomeys (1961) exposent des observations qui peuvent s'expliquer notamment dans le genre Peniophora, par une diminution avec l'âge des possibilités de copulation.

Toutefois les investigations sont encore fort insuffisantes en ce domaine, et une étude rigoureuse reste à faire.

3. Interfertilités partielles non expliquées par une communauté d'allèles ou le vieillissement.

Lorsque l'on ne peut faire appel ni à l'identité d'allèles de polarité ni au dépérissement d'une souche, il faut rechercher ailleurs l'explication des interfertilités partielles.

Très souvent les expérimentateurs soulignent que les monocaryons qui avec certains partenaires montrent une interstérilité partielle sont, dans le même temps, parfaitement interfertiles avec d'autres.

On peut essayer de regrouper les données de la littérature de manière schématique.

a) Faible pourcentage de résultats négatifs.

Les auteurs les signalent et ne s'y arrêtent pas ne leur donnant aucune signification. L'explication de ces manques n'est généralement pas recherchée.

On peut citer l'exemple des confrontations Peniophora duplex x P. duplex ou P. pini x P. pini (fig. 3 de Weresub et Gibson 1960); les auteurs concluent: "But again, sufficient mating capacity was retained to demonstrate clearly the intimate relationship within each group of tested collections ...". Il en est de même entre souches de Merulius lacrymans (Harmsen 1960 fig. 26). Macrae (1967) signale que les tests entre formes irpicoïdes (=Hirschioporus fusco-vidaceus) sont positifs avec les seules exceptions de 23 croisements sur 783 c'est-à-dire dans 5 séries d'appariements sur 52.

Par contre Duncan et MacDonald (1967) écrivent: "attention is drawn to the degree of interfertility within the N. American "coniferous" unit, namely 94% ... Therefore, the figure of 94%, although only a little short of the expected 100%, is taken to indicate that additional, but indefinable, units of micro-evolution are present". (Voir ci-après c, "Cas intermédiaires").

b) Faible pourcentage de résultats positifs.

Il est difficile d'interpréter aujourd'hui l'apparition de boucles dans un très faible pourcentage de confrontations, de l'ordre de 2 à 3%, et les systématiciens qui les signalent, les négligent. De tels résultats, s'il ne sont pas la conséquence d'erreurs expérimentales, sont bien embarrassants.

C'est ainsi que Nobles et Frew (1962) qui notent une fois sur 41 croisements entre Pycnoporus cinnabarinus et coccineus de rares boucles sur quelques hyphes, et Macrae (1967, fig. 16) qui constate que dans les confrontations entre la forme lamellée (=Hirschioporus laricinus) américaine et la forme porée norvégienne (=H. abietinus) sont apparues des boucles dans 4 appariements sur 153, n'en reconnaissent pas moins des espèces distinctes: "these four positive results may be ignored as probable errors".

Duncan et MacDonald (1967) qui n'obtiennent que 12 résultats positifs sur 624 croisements entre des Auricularia auricula européens, sur Sambucus et des A. auricula américains des conifères, concluent (p.808): "The European unit was virtually intersterile with the N.-American "coniferous unit", mais n'en font pas une espèce distincte. Il est vrai que, vu les techniques utilisées, les résultats de Duncan ne sont pas tout à fait comparables à ceux de Nobles et Frew (voir ci-dessous II C 3c).

c) Cas intermédiaires.

Doivent se situer ici tous les cas où un pourcentage significatif de résultats à la fois positifs et négatifs ont été constatés. Si l'on peut affirmer un résultat positif après une seule observation indiscutable, par exemple lorsque, à la première observation, ligne de contact comme marges opposées des deux monocaryons appariés montrent des hyphes régulièrement bouclées, il n'en est pas de même des résultats négatifs.

C'est ainsi que Duncan et MacDonald, puis Duncan (loc.cit.) sur un matériel difficile, aux boucles ou médaillons souvent délicats à voir sur des hyphes plus ou moins gélifiées, opèrent en observant des subcultures de la ligne de contact effectuées le 28ème jour. De ce fait, ils ne pourront sans doute voir les boucles rares ou fausses de Nobles et Frew, ni de Bruehl et all. qui ne se maintiennent généralement pas en subcultures. On peut aussi se demander si des prélèvements faits par exemple au 42ème jour, lorsque l'observation des subcultures n'a pas montré de boucles, n'auraient pas permis de noter un beaucoup plus grand nombre de confrontations positives. Les résultats des auteurs écossais sont donc essentiellement comparables entre eux, et même si les pourcentages variés d'interfertilité observés dans des conditions standard peuvent être considérés comme un moyen d'estimer les affinités, ce ne sera qu'entre Auriculaires étudiés de la même manière.

On n'est pas toujours aussi bien renseigné sur les techniques d'études des autres auteurs, ce qui serait cependant indispensable lors de résultats intermédiaires comme ceux qui nous retiennent actuellement.

Dans un travail tout récent, J. Ginns (1976) croise les deux pôles de trois Phlebia rufa des U.S.A. avec ceux de deux récoltes françaises et note: "... the resultant mycelium, in most cases, produced at least a few clamp connection. The absence of clamps in a few crosses (2 sur 45), the scarcity of them in other cases, as well as the presence of pseudoclamps in pairings which produced few clamps (25 sur 45) illustrate that the matings were only partially compatible and suggest that this species is fragmenting by means of a sterility barrier".

Cultures des Auriculariales   297

Pour juger ce résultat, il faut savoir que, comme tous les vrais
Phlebia, Ph. rufa fait partie du groupe 54 de Nobles (1958) c'est-à-
dire est astatocénocytique (Boidin 1964 et 1971); il est donc à boucles variables. Selon les conditions d'aération, les boucles sont
constantes, rares et limitées aux hyphes aériennes les plus grêles,
ou absentes. Le mycélium reste cependant hétérocaryotique et les boucles peuvent réapparaître, d'abord fausses, c'est-à-dire non anastomosées et pouvant renfermer un noyau en surnombre, puis vraies et de
plus en plus nombreuses. La remarque de Ginns: "few clamps on narrow
aerial hyphae, some pseudoclamps" est très significative. Il faut
donc conclure que les résultats qui nous sont présentés sont parfaitement explicables par l'astatocénocytie et qu'il n'est pas nécessaire d'imaginer une barrière de stérilité.

    c1) Boucles vraies, abondantes et durables dans certaines con-
        frontations, pas de fausses boucles.

On peut rappeler tout d'abord les importantes contributions de
Duncan (loc.cit.). Chez Auricularia polytricha, si l'on excepte la
souche américaine (Louisiane), interstérile peut être par incompatibilité hétérogénique (voir II B3), on constate de grandes variations
dans les pourcentages d'interfertilité (25 à 100%) selon l'origine
géographique des partenaires.

C'est ainsi que l'on approche ou atteint 100% entre Hawai et les autres contrées, entre Australie et Nouvelle-Zélande, Bengal et Assam,
Assam et Australie bien que ces derniers aient des spores de deux
tailles extrêmes (cf. II Al). A l'inverse, les subcultures qui sont
bouclées 46 fois sur 64 entre Assam et Kenya ne le sont que 16 fois
sur 64 entre Bengal et Kenya ...

On ne peut parler ici d'interstérilité. Les "microevolutionary units"
do Duncan portent bien leur nom et ne peuvent être assimilées à des
taxons même sous-spécifiques du morphologiste. Peut-être en sont-elles de lointains précurseurs.

Parmi les Auricularia auricula nord-américains, dans la ligne des résultats antérieurs de Barnett (1937), Duncan et MacDonald distinguent
ceux des conifères, et ceux des feuillus: "While their overall degree
of interfertility is 26%, the degree of interfertility between monokaryons from individual basidiocarps varies from 0 to 75%. Intersterility can neither be correlated with particular mating type factors
nor with particular substrates".

A la suite des remarques techniques faites plus haut, on doit considérer que les résultats positifs correspondent à l'existence et à la
permanence d'un mycélium bouclé et que ces deux unités sont encore
étroitement apparentées. L'évolution semble se dessiner cependant
plus nettement que chez A. polytricha vers la disjonction de 3 sous-

unités: celles des feuillus à petites basides (moyenne 57,5 microns), celles des conifères à petites basides (moyenne 57,4 microns) et celles des conifères à grandes basides (moyenne 77,9 microns). On aimerait cependant connaître le résultat d'appariements des monocaryons de ces deux dernières sous-unités, et, si possible la taille des basides et spores de l'éventuel hybride et de sa descendance.

Mc Keen (1952) dégage en Amérique du Nord 2 groupes A et B chez Peniophora (Hyphoderma) mutata; l'interfertilité est totale à l'intérieur de chaque groupe; les croisements intergroupes sont une fois tout à fait positifs, trois fois tout à fait négatifs (selon text. fig. 13), 32 fois partiellement positifs (dont 13 fois à plus de 50% et 19 fois à moins de 50%). Si ces deux groupes ne sont pas isolés géographiquement (ils cohabitent dans l'Ontario) et s'il n'y a pas de différences morphologiques suffisantes pour distinguer une variété, le groupe A n'a été récolté que sur Populus, le groupe B sur des feuillus autres que des Populus. (Tableau V)

Weresub et Gibson (1960) observent entre P. pini d'Europe et P. duplex américain des réactions très rarement toutes positives (mais non renouvelables), très généralement une compatibilité partielle avec des résultats variables, alors que les monocaryons suédois et français de P. pini sont entre eux interfertiles à 100% mais ceux de P. duplex cependant seulement à 87%.

Très voisins par leur caractères morphologiques et leurs cultures, mais à aires géographiques distinctes (allopatrides), ces deux entités sont considérées par ces auteurs comme sous-espèces: P. pini subsp. pini et subsp. duplex.

On peut très bien voir dans ces deux exemples des "microevolutionary units" de Duncan.

Par contre, les cas suivants ne peuvent être expliqués par le rôle du support ni de l'isolement géographique. Mounce et Macrae (1938) puis Macrae (1967) distinguent clairement par croisements deux groupes interstériles parmi les récoltes américaines d'une part de Fomes pinicola (Tabl. VI), d'autre part d'Hirschioporus abietinus, dénommés A et B; les européens, pratiquement interfertiles entre eux, le sont plus ou moins avec les deux groupes américains. On a ici un type de réaction triangulaire. Les auteurs ne reconnaissent entre A et B ni différences morphologiques ni corrélations avec les hôtes ou la répartition géographique et excluent le rôle d'allèles de polarité identiques. La seule explication pourrait être dans l'incompatibilité hétérogénique du type de celle qu'a rencontré Olive avec Sordaria fimicola qui montre lui aussi des réactions triangulaires (cf. II B 3).

Dans tous les cas signalés dans ce paragraphe, les unités lorsqu'elles sont plus ou moins distinguables, ont été considérées par les auteurs comme de niveau infraspécifique. On peut interpréter les divergences plus ou moins sensibles au niveau morphologique comme des marques de la spéciation en marche - d'où découlerait une diminution de l'interfertilité. Balbutiante chez Auricularia polytricha, un peu plus dessinée chez A. auricula, elle s'aide de l'isolement géographique chez Peniophora pini et de la nature du substrat (isolement écologique) chez P. mutata; mais dans ce dernier cas par exemple, on peut se demander si l'élargissement de l'étude de Mc Keen, grâce à des récoltes d'Europe et d'ailleurs n'amènerait pas à compliquer l'image actuelle, mettant par ex. en évidence des réactions triangulaires - Boidin et des Pomeys (1961) ont cependant montré qu'un Hyphoderma mutatum français sur Fagus silvatica était interfertile avec le DAOM 17.550 de Mac Keen appartenant au groupe B (feuillus autres que Populus), mais rien n'a été tenté avec un membre du groupe A.

On peut aussi, l'incompatibilité nucléaire semblant précéder parfois les divergences morphologiques ou écologiques penser que l'incompatibilité hétérogénique pourrait être dans ces cas, le facteur premier - efficace, mais non suffisant - de la spéciation.

c2) Hyphes bouclées éparses, boucles rares et fausses boucles.

Certains signalent l'apparition de boucles très espacées, parfois incomplètes, et non anastomosées (fausses boucles) parfois uniquement à la ligne de contact. Celles-ci ne sont pas en mesure d'envahir les territoires haploïdes et même bien souvent ne peuvent être décelées régulièrement lors de prélèvements successifs. Elles ne se maintiennent pas (ou pas durablement) par subculture. Il est apparent ici que ces résultats ne peuvent être dits positifs et que la communauté nucléaire qui a pu s'installer est vouée à l'échec; les divisions conjuguées s'instaurent mal et la caryogamie sera pratiquement interdite. Rappelons qu'il ne faut pas ici prendre en considération la présence de fausses boucles due à la communauté du facteur B ou à l'astatocénocytie.

On peut sans doute classer ici les observations de Nobles et Frew (1962) dans les confrontations entre divers Pycnoporus sanguineus et 3 souches de P. coccineus comme entre un P. sanguineus (52195) et divers P. coccineus où: "a few hyphae with rare clamp connections, often incomplete, were found", celles de Bruehl et all. (1975) entre Typhula idahoensis et T. ishikariensis qui écrivent: "these products of these crosses were abnormal. Hyphae possessed a few scattered clamp connections, and growth by the mating product was slower than that of the parent monokaryons. We refer to the matings ... as "spurious". We believe the two species are closely related, but that significant genetic separation exists between them". David et all. (1974) distinguent Auriporia aurulenta d'A. aurea et écrivent: "Sur vingt confrontations, quatorze se sont révélées négatives: dans les six autres de très rares boucles ont été observées, toujours localisées dans la zone de contact des cultures confrontées. Des subcultures ont été faites à partir de ces zones bouclées: dans aucun cas les boucles ne se sont maintenues".

Dans ces exemples, les résultats sont obtenus entre des unités considérées par les auteurs comme des espèces proches mais distinctes.

On pourrait sans doute parler des croisements de P. pini et duplex, mais à côté des boucles limitées à la ligne de contact, comme dans des confrontations illégitimes, pouvant disparaître par transferts, il y a aussi des dicaryotisations unilatérales et des dicaryotisations complètes qui ressortent du paragraphe précédent.

Un cas assez semblable concerne Peniophora malençonii (Boidin et Lanquetin 1977) qui est un champignon méditerranéen pratiquement indistinguable de certains Stereum heterosporum californiens (type exclus). Confrontés, seuls certains monocaryons montrent des boucles rares, incomplètes; parfois cependant des anses d'anastomose se forment presque à chacune des cloisons de certaines hyphes, et se maintiennent, ou même peuvent envahir le territoire d'un des deux monocaryons. Ici encore, l'isolement géographique semble se doubler d'un isolement génique assez poussé.

Cultures des Auriculariales   301

Conclusion

De cette analyse détaillée des résultats en ma possession, il apparaît

1. que, contrairement aux craintes de certains systématiciens, l'emploi intensif des test d'interfertilité ne mène pas à une pulvérisation sans limite de l'espèce. Certes, la notion d'espèce qui en découle est relativement étroite, ou jugée telle avec nos moyens d'investigations actuels, et ces tests donnent plus souvent raison à ceux qui distinguent qu'à ceux qui regroupent; par exemple toutes les espèces de <u>Vararia</u> subg. <u>Dichostereum</u> gardées par les systématiciens modernes sont "bonnes", et les synonymies proposées par d'autres sont toutes injustifiées.

Mais ces mêmes tests conduisent aussi à des regroupements, des synonymies sûres, donc à une indiscutable clarification. Ils sont un outil indispensable à tous ceux qui voudront connaître la véritable répartition des espèces d'Aphyllophorales: les zones climatiques qu'elles supportent, les zones géographiques qu'elles habitent, les supports qu'elles colonisent.

2. qu'un accord assez général se fait sur les conclusions que tirent les auteurs de ces tests

a) Si les résultats sont totalement ou en grande majorité positifs, tous considèrent comme secondaires ou comme seulement phénotypiques les différences morphologiques éventuellement remarquées, et font référence à une seule espèce.

b) Si les résultats, tous négatifs sont associés à des différences morphologiques, tous considèrent qu'ils ont affaire à des espèces distinctes. Il en est de même s'il y a un très petit nombre de résultats positifs ou partiellement positifs.

c) Si les résultats sont imparfaits (confrontations rarement et imparfaitement positives, fausses boucles souvent caduques ...) deux positions sont adoptées

1) Quand il y a différences morphologiques, Nobles et Frew, Bruehl et all., David et all. parlent d'espèces distinctes mais très affines.

2) Quand il n'y a pas de différences morphologiques suffisantes, sont alors décelées soit une différence de support (<u>Peniophora</u> <u>mutata</u> A et B de Mc Keen) soit une séparation géographique

(P. pini subsp. pini et subsp. duplex de Weresub et Gibson, l'Auricularia auricula européen de Duncan et MacDonald, P. malençonii circumméditerranéen et P. "heterospora" californien). Les auteurs, sauf Weresub et Gibson qui parlent de sous-espèces (il est vrai qu'elles soulignent quelques minimes critères morphologiques), ne donnent pas de rang à ces sous-ensembles ou parlent comme Duncan d'unité de microévolution.

d) Si les résultats sont négatifs, sans différences morphologiques, deux cas sont distingués.

1) Il y a diffèrence de thallie (et parfois du nombre des stérigmates): Lemke parle de sous-espèces chez Sistotrema brinkmanii ainsi que Boidin et Lanquetin chez Vararia firma.

2) En l'absence de telles différences: qu'il s'agissent des Fomes pinicola de Mounce et Macrae, des Hirschioporus abietinus américains de Macrae, de l'Auricula polytricha de Louisiane de Duncan, des 5 groupes interstériles de Sistotrema brinkmanii bipolaires, et des 2 groupes du même Sistotrema tétrapolaires de Lemke, des 5 groupes décelés dans Laxitextum bicolor, les auteurs ne tirent aucune conclusion. L'explication actuellement la plus plausible est qu'il s'agit de "races" séparées par l'incompatibilité hétérogénique. Peut-être faudrait-il trouver un nom pour ces sous-unités génétiques et non taxonomiques.

Si l'évolution procède grâce à la mutation, la recombinaison et la sélection, on peut écrire en accord avec K. Esser (1971) que la recombinaison est ici sujette à diverses contraintes:

Les échanges de gènes sont rendus obligatoires par l'incompatibilité homogénique, et dans ce cas la spéciation doit s'aider de facteurs d'isolement (:isolement par le substrat à l'intérieur d'une même aire, isolement géographique) et d'une longue période. La spéciation est progressive et il ne faut pas s'étonner que la tâche du systématicien soit parfois compliquée par des interfertilités très imparfaites; cas rares encore, mais fort intéressants, ils sont la preuve tangible de cette évolution en marche.

A l'inverse les échanges de gènes sont limités par l'amphithallie et l'homothallie et pratiquement interdits par l'incompatibilité hétérogénique. Les sous-espèces à thallie différentes, les "races" incompatibles deviennent les unités évolutives, unités qu'il n'est généralement pas permis au morphologiste de repérer.

*   *
*

En appendice à ce chapitre, il peut être utile d'émettre quelques recommandations qui pourront paraître sévères mais qui sont indispensables chaque fois que des difficultés apparaissent: interfertilité partielles, fausses boucles etc...

    1. Utiliser des monocaryons de pôles connus et si possible plusieurs de chacun des pôles. Par cette précaution on s'assure de l'hétérothallie et de la bonne tenue des monocaryons, on exclut les cultures bispermes incompatibles, les homothallies lentes, et l'on pourra découvrir d'éventuelles identités d'allèles de polarité.

    2. Confronter en boite de Pétri, et pour la recherche des boucles prélever non seulement à la ligne de contact, mais encore à mi-rayon et à la marge opposée de chaque partenaire.

    3. Si les boucles sont rares ou fausses (non anastomosées en retour), faire des subcultures et en suivre le devenir = disparition ou généralisation des boucles, maintien ou non des boucles imparfaites et inconstantes. Ne pas perdre de vue l'existence d'espèces astatocénocytiques aux boucles variables en fonction de l'aération (ce sont généralement des espèces bipolaires).

    4. Si nécessaire répéter ultérieurement les prélèvements.

    5. En cas de résultats positifs inexplicables, vérifier que les cultures utilisées sont restées monocaryotiques.

(III) ANASTOMOSES VEGETATIVES

Robak (1942, p.44-45) et Léger (1968a, p.333) résument les travaux et opinions anciennes sur ce sujet. J'emprunte à Léger les lignes suivantes: "Les anastomoses sont extrêmement nombreuses à l'intérieur d'une bouture unique ou entre deux inoculums provenant de la même espèce. Dès 1892, Reinhardt montre que ces anastomoses ne se produisent pas entre mycéliums d'espèces différentes et qu'il y a alors attaque réciproque des filaments qui se tuent l'un l'autre. En 1919, Laibach observe chez _Septoria apii_ que les tubes germinatifs d'une même race s'anastomosent facilement tandis qu'entre races différentes il n'y a pas réalisation certaine d'anastomoses. De nombreux auteurs pensent que de la fusion des hyphes autorise l'assimilation des boutures anastomosées à une même espèce. Vandendries (1933) reprend cette idée et affirme que pour trancher le diagnostic entre espèces affines il suffit de confronter les diplontes obtenus par bouturage: "si des anastomoses sont visibles les deux carpophores appartiennent à la même espèce, sinon ils sont spécifiquement différents". Robak (1942) aboutit aux mêmes conclusions.

Une phrase de Vandendries (1933) doit être gardée en mémoire "Il reste entendu que l'observation est délicate, qu'elle exige un examen très attentif, que la croissance des hyphes doit être suivie de près pour ne pas confondre des fusions entre rameaux étrangers avec des anastomoses, toujours nombreuses entre hyphes de la même bouture".

Dans ce but bien des travaux proposent des techniques un peu différentes (Davidson et all. (1932), Van Uden (1951), Jorgensen (1954), Carmichael (1956), Macrae (1967) et Léger (1968a), tandis que d'autres travaux recherchent les conditions, milieu, pH, température ... optimales (Bourchier 1957), Ahmad et Miles (1970a et b).

La tendance a souvent été d'utiliser ces techniques lorsque les critères d'interfertilité-interstérilité ne peuvent être employés comme par exemple chez les _Dermatophytes_, les _mycelia sterilia_, les cultures isolées de bois, mais aussi pour les Aphyllophorales homothalles.

Robak (1942) note des anastomoses entre _Stereum_ rugosum et des _St. sanguinolentum_ tous deux rougissants à la blessure (sect. _Cruentata_) mais bien distincts pour le systématicien et inféodés le premier aux feuillus, le second aux conifères. Léger (1968a) confirme les résultats de Robak et montre que des anastomoses sont possibles entre beaucoup d'autres _Stereum_; soit de la section _Cruentata_ entre eux, soit de la section _Luteola_ entre eux, mais aussi dans un cas entre un

représentant de la section Luteola (St. insignitum) et un de la sect.
Cruentata (St. rugosum). Ceci peut inciter le systématicien à bannir
la distinction Luteola-Cruentata, et le genre Haematostereum Pouzar
proposé pour cette dernière section, car "basée sur un caractère
quantitatif: elle est plus commode que valable" (Boidin 1958), et à
retenir la présence ou l'absence de pseudoacanthophyses comme un caractère plus significatif.

On peut cependant objecter, la délimitation des espèces dans le genre
Stereum ne pouvant être précisée par des tests d'interfertilité-interstérilité, que ces exemples ne suffisent pas pour conclure que
l'anastomose végétative ne signifie pas identité spécifique.

Les Aphyllophorales nous apportent heureusement d'autres exemples
d'espèces bouclées hétérothalles chez lesquelles pouvaient être tentés parallèlement des appariements de monocaryons et des anastomoses
végétatives.

Cabral (1951) qui n'a pas pris cette précaution, conclut après étude
de 21 espèces de polypores (p.355): "between distinct species (dip-
lont) incompatibility was always observed". Il nous faut regarder de
plus près les résultats de Robak (1942), Macrae (1967), Harmsen
(1960), Gaut (1969) et Léger(1968b).

Sections Espèces	Luteola	Cruentata
sans pseudo- acanthophyses	complicatum sulphuratum ⇕ ↭ ⇕ subtomentosum striatum	
avec pseudo- acanthophyses	insignitum ⟵⟶	rugosum ⟷ gausapatum C ⇕ sanguinolentum

Tableau VII

Si Robak (1942) montre bien souvent l'existence d'anastomoses entre
champignons d'une même espèce, il ne peut en observer entre Stereum
purpureum d'une part, St. rugosum et sanguinolentum d'autre part,
mais le premier cité n'a guère de parenté avec les 2 autres et est
unanimement classé dans un autre genre (Chondrostereum) sinon dans
une autre famille. Par contre il observe certaines fusions hyphales
entre formes poroïdes et irpicoïdes, poroïdes et lamellées du complexe P. abietinus reconnues par Macrae (1967) d'après les constats
d'interstérilité comme des espèces autonomes (voir plus haut II B1),
mais ces résultats sont contestés par celle-ci qui, deux cas douteux
exceptés, n'a vu aucune anastomose entre les trois espèces poroïdes,
irpicoïdes et lamellées.

Harmsen (1960) qui montre l'interstérilité entre Merulius lacrymans
et M. himantioides, obtient entre eux et certains de leurs synonymes
des anastomoses, ce qu'il ne peut observer envers M. tignicola et
écrit "formation of anastomoses cannot be used for species separation

within the large spored species of Merulius investigated ... However the presence of anastomoses seems to indicate that the large spored species in question must be closely related".

Gaut (1969) qui, vu la difficulté d'obtention de monocaryons du champignon du Sirex, tente des anastomoses végétatives écrit "Homokaryons and heterokaryons of A. (Amylostereum) chailletii, A. areolatum, A. laevigatum and the S. noctilio fungus were paired in all possible combination; each pairing was repeated at least three times. Results were very clear: anastomosis was observed at least once in every possible combination a the S. noctilio fungus with A. areolatum; anastomosis was not seen in any of the other pairings". Il confirmait plus tard par interfertilité cette identification.

Enfin Léger (1968b) a choisi comme matériel le groupe cinerea du genre Peniophora, dans lequel un intense travail préalable d'interfertilité avait permis de bien préciser les unités (c'est-a-dire les espèces).

Il a pu montrer que, tant entre monocaryons qu'entre dicaryons, des anastomoses certaines (avec soudures et fusions des hyphes contrôlées après élimination du cytoplasme dans les meilleurs conditions d'observations à fort grossissement) ont lieu entre espèces interstériles. Des anastomoses sont décelées bien que parfois en très petit nombre par rapport au nombre de lames observées entre P. cinerea et P. fraxinea (=P. limitata), P. junipericola, P. rufomarginata, P. quercina, P. violaceo-livida, de même entre P. nuda et P. pini (ex Sterellum pini) entre P. quercina et P. malençonii *) (=Lopharia heterospora ss. auct. europ.) A l'inverse après d'autres, Léger insiste sur la fréquence, même entre mycéliums d'une même souche, de lames qui ne montrent pas d'anastomoses certaines. Avec presque tous les auteurs on peut considérer (et c'est une grave limite de cette technique) que de multiples résultats négatifs sont sans aucune signification. On ne peut par contre ignorer les résultats de Harmsen et de Léger qui, entre espèces interstériles ont pu constater d'indiscutables fusions d'hyphes. Ces résultats sont en accord avec ceux obtenus sur des Agaricales par Kemp (1975) après la mise en présence ingénieuse d'oïdies et d'hyphes d'espèces différentes.

Il apparaît à la lumière de ces observations, que les anastomoses peuvent avoir lieu non seulement entre monocaryons compatibles mais encore entre monocaryons incompatibles d'une même souche, entre mono- ou di-caryons d'une même espèce, et enfin, au moins dans certains cas, entre mycéliums d'espèces interstériles et que le critère des fusions hyphales ne recouvre pas celui de l'interfertilité mais le déborde.

Il convient enfin d'attirer l'attention sur de nombreuses remarques des auteurs cités plus haut concernant, l'attirance, ou au contraire des phénomènes d'interactions, de répulsion, épaississement des pa-

---

*) Nom proposé pour le P. (Duportella) heterospora (Burt) non sensu Burt (cf. Boidin et Lanquetin, 1977).

rois qui peuvent aller de pair avec l'échec, comme avec la réussite au moins momentanée de l'anastomose.
En effet, certains insistent, comme Harmsen, sur le devenir incertain des anastomoses effectives, et il écrit "Yet anastomoses between hyphae of different isolates within the same species as well as between isolates from different species always resulted in degeneration of the anastomosing cells. Degeneration was apparent after a few hours, but sometimes not befor 1 - 2 days after the anastomosing". Peuvent encore jouer ici, au niveau cellulaire, des mécanismes d'incompatibilité hétérogénique.

On ne peut que souhaiter des études plus poussées des phénomènes avant, pendant et après l'anastomose, et avec Gaut exprimer le désir de connaître "the nature and extent of genetic control over the phenomenon ...". Il sera peut-être alors possible de saisir la signification des interactions et des anastomoses elles-mêmes.

Malgré les difficultés pratiques et les résultats controversés, les anastomoses parfaites et les anastomoses condamnées à terme n'en expriment pas moins des degrés de parenté des constituants protoplasmiques, sans doute des constituants protéiques, parenté que l'on peut rechercher par des méthodes électrophorétiques et immunologiques.

(IV) COMPARAISON DES PROFILS PROTEIQUES ET IMMUNOLOGIE

A. P r o f i l s   p r o t é i q u e s
Depuis que Chang et all. (1962) ont montré que différentes espèces de Neurospora ont des profils protéiques distincts et que les différences interspécifiques sont beaucoup plus importantes que les différences intraspécifiques, un certain nombre d'études ont été faites sur les champignons, le plus souvent par électrophorèse en gel d'amidon ou de polyacrylamide. Elles s'intéressent à l'ensemble des protéines solubles ou seulement à des isoenzymes. Comme pour les anastomoses végétatives, la plupart des auteurs se sont adressés à des groupes de champignons à identification difficile notamment à des Fungi imperfecti. Rares et très fragmentaires sont les études portant sur les champignons supérieurs qui nous occupent. On peut citer à nouveau les

travaux de Gaut (1969) et la thèse de Léger (1974). Gaut a comparé les profils protéiques (une vingtaine de bandes) du commensal du Sirex, à ceux des Amylostereum chailletii et areolatum et trouvé une beaucoup plus grande ressemblance avec ce dernier (14 bandes communes) qu'avec A. chailletii (7 bandes communes), en accord avec les données de l'anastomose et de l'interfertilité. Léger a présenté les profils de 22 espèces de Peniophora (Tableau VIII) après s'être assuré que 7 récoltes de P. quercina donnaient des résultats similaires. Il n'observe aucun profil semblable. Vu le nombre de bandes et le nombres d'espèces, il fait appel à une méthode d'analyse factorielle et calcule les distances d'espèce à espèce dans un espace multidimensionnel; elles varient de 0,70 à 5,00. Ceci permet de mesurer les affinités et de partager le genre en sous-unités qui sont en assez bon accord avec les sous-genres du Systématicien. L'étude - dans les mêmes conditions de culture (milieu, âge ...), d'extraction, d'électrophorèse - du profil protéique de tout autre récolte de Peniophora doit permettre de la rapprocher plus précisément de l'une d'entre elles (identification) ou de plusieurs (affinités), mais la décision restera sans doute subjective. En l'absence d'une délimitation antérieure des espèces de Peniophora utilisés, par la morphologie classique, les caractères culturaux, et les interfertilités, nul ne sait le nombre d'espèces qui auraient été distinguées.

On peut aussi objecter que, malgré les perfectionnements techniques, des bandes homologues, c'est-à-dire de migration identique peuvent ne pas être formées de la même protéine, plusieurs facteurs (charge, encombrement moléculaire ...) réglant cette migration. On pouvait alors songer à utiliser l'extrême spécificité de la réponse immunitaire.

B. I m m u n o l o g i e

A de rares exceptions près, les techniques immunologiques utilisées d'un point de vue taxonomique sont soit l'agglutination, en présence d'antigènes figurés, soit la précipitation en milieu gélifié par les antigènes solubles. La plupart des travaux s'intéressent à des champignons pathogènes des animaux ou des végétaux, très peu à des Basidiomycètes. Ce sont alors soit les techniques d'immuno-diffusion qui sont employées (sur quelques polypores par Madhosingh 1964, 1969, Madhosingh et Ginns 1974, 1975), soit les techniques d'immunoélectrophorèse (Drosdova 1973 sur des Coriolus, Léger 1974 sur des Peniophora). Si la première technique ne permet de mettre en évidence que 5 à 8 bandes de précipitation, la seconde qui permet de distinguer dans les exemples cités, de 10 à 17 arcs est d'une finesse analytique supérieure. Les réponses n'étant pas égales suivant le sens de la réaction, il apparaît nécessaire d'obtenir des sérums de chaque espèce ou souche en étude, et d'effectuer toutes les réactions croisées.

Le pouvoir antigénique des extraits de champignons supérieurs s'est montré jusqu'ici faible par comparaison avec ce que peuvent donner des Candida ou certains Aspergillus. Il est cependant toujours apparu des différences entre les espèces proches comparées. Les auteurs en déduisent le plus souvent des degrés de parenté et il n'est guère question, contrairement à ce qui a été tenté avec d'autres champignons, de diagnostic ou de délimitation spécifique.

CONCLUSIONS GENERALES

La systématique doit présenter la synthèse de toutes les données d'ordre les plus divers réunies sur les êtres en étude.

Lorsqu'il dispose du plus grand nombre de caractères des basidiomes, morphologie externe, structure, morphologie microscopique photonique et électronique, réactions cytochimiques, analyses des constituants (pigments ...), du maximum de données sur les cultures: morphologie macro- et microscopique, conditions optimales de croissance (température, pH, besoins nutritifs ...), cycles, comportements nucléaires, chimisme, arsenal enzymatique, profils protéïques ... enfin des précisions sur l'habitat (supports, répartition géographique et climatique ...) le mycologue expérimenté utilise cette somme de renseignements accessibles pour définir les unités systématiques et les ordonner en un système qu'il veut naturel, ou parfois seulement pratique. De toute façon l'espèce sera le taxon essentiel de toute cette construction systématique.

Mais il y a dans le travail ardu du systématicien, en fonction de son expérience, des domaines particuliers d'observation où il s'est attardé, une inévitable subjectivité. La preuve en est que les accords sont rarement unanimes sur la délimitation des espèces et sur leurs affinités réciproques.

Des tests objectifs, de valeur établie, sont éminemment souhaitables et pourront seuls aider le systématicien dans ses prises de positions, régler les différents, les interprétations divergentes. J'ai ci-dessus particulièrement insisté sur les critères proposés dans cette optique.
- Comparaison des profils protéiques et immunologiques
- Anastomoses végétatives
- Interfertilités

Tous ces critères découlent de la spécificité des substances primordiales de la matière vivante: protéines et nucléoprotéines.

Dans l'état actuel de nos connaissances, dénombrer les protéines solubles semblables et différentes, les fractions communes ou non des tests immunologiques, la possibilité d'anastomoses végétatives permet d'estimer des affinités, mais ne répond pas de manière objective, automatique à la question: ces deux récoltes appartiennent-elles ou non à la même espèce, au même hologamodème de Gilmour et Heslop-Harrison?

Seuls, aujourd'hui, les critères d'interfertilité-interstérilité prouvent indépendamment de l'opérateur, que l'association des cytoplasmes est viable, que celle des noyaux va jusqu'à la répétition sans limite de la mitose conjuguée des noyaux des deux partenaires, et sans doute - mais cela n'a été suivi que de temps à autre, parce que nous ne maitrisons pas suffisamment les conditions de fructification - jusqu'à la caryogamie et la méïose.
Quelques difficultés subsistent cependant:

1) Si la très grande majorité des résultats sont parfaitement nets, on a pu constater quelques cas d'interfertilités imparfaites. Assurées dans des conditions rigoureuses - ce qui élimine certaines données de la littérature - elles sont la preuve de l'évolution en cours. Ce n'est sans doute pas par hasard que les quelques cas indiscutables vont de pair avec un isolement écologique (arbres-supports des <u>Hyphoderma mutatum</u>), ou le plus souvent géographique.

2) On a pu aussi constater des cas d'interstérilité totale inexplicables par le systématicien que ne peut mettre en évidence des différences morphologiques ou autres entre récoltes interstériles qu'il considère comme appartenant à une même espèce. La seule explication de ce refus a été proposée par Esser: l'incompatibilité hétérogénique. Il faudrait cependant savoir si le barrage inter-récolte est la preuve nécessaire et suffisante de ce type d'incompatibilité. S'il en est bien ainsi, dans la trilogie des facteurs de l'évolution: mutation - recombinaison - sélection, la recombinaison se trouverait donc rendue impossible selon deux processus différents

a) par spéciation graduée aidée par l'isolement spatial ou écologique. Il arrive un moment où l'interfertilité est partielle, puis nulle, l'accumulation de mutations divergentes crée l'espèce taxonomique en même temps que l'espèce biologique (hologamodème).

b) par blocage brusque, conséquence de mutations portant sur la thallie, ou créant l'incompatibilité hétérogénique. Les échanges de noyaux sont très difficiles ou impossibles, l'espèce biologique (hologamodème) précède dans le temps l'espèce taxonomique.

P. Lanquetin a montré que, chez les hétérothalles dicaryotiques sans boucles, les mêmes études peuvent être faites par le biais de colorations nucléaires. Pour les espèces homothalles vraies (homothallie primaire) avec ou sans boucles, Lemke a prouvé, par l'emploi de mutants auxotrophes la possibilité d'échanges nucléaires, ce qui autorise le systématicien à considérer les espèces homothalles comme des hologamodèmes et non - ce qui aurait rendu ses efforts pratiquement vains - comme un faisceau d'unités évolutives séparées. L'emploi commode des tests d'interfertilités ne peut cependant être envisagé puisque la formation de dicaryons ou de boucles, ou de fructification, lorsqu'ils ou elles existent, ne nécessitent qu'un seul partenaire. Il faudrait utiliser des mutants auxotrophes différents, et bénéficier de l'avantage du "nutritional forcing" qui en découle, pour essayer de résoudre objectivement les problèmes de délimitation chez des espèces homothalles.

Bibliographie

Adams D.H. & Roth L.F. 1967: Demarcation lines in paired cultures of
Fomes cajanderi as a basis for detecting genetically
distinct mycelia. Can.J.Bot., 45: 1583-1589.
Ahmad S.S. & Miles Ph.G. 1970a: Hyphal fusions in the wood-rotting
fungus Schizophyllum commune. 1. The effects of incompatibility factors. Genet. Res., 15: 19-28.
Ahmad S.S. & Miles Ph.G. 1970b: Hyphal fusions in the wood-rooting
fungus Schizophyllum commune. 2. Effects of environ-
mental and chemical factors. Mycologia, 62: 1008-
1017.
Barrett D.K. & Uscuplic M. 1971: The field distribution of interar-
ting strains of Polyporus schweinitzii and their ori-
gin. New.Phytol., 70: 581-598.
Bauch R. 1926: Untersuchungen über zweisporige Hymenomyzeten. I.
Haploid parthenogenesis bei Camarophyllus virgineus.
Z. Botan., 18: 337-387.
Barnett H.L. 1937: Studies on the sexuality of the heterobasidiae.
Mycologia, 29: 626-649.
Bensaude M. 1918: Recherches sur le cycle évolutif et la sexualité
chez les Basidiomycètes. Thèse, Paris, 156 p.
Berthier J. 1974: Reproduction et comportements nucléaires chez les
Typhula Fr. (Clavariacées) et genres affines. C.R.
Acad. Sc. Paris, sér. D, 278: 3307-3310.
Biggs R. 1937: The species concept in Corticium coronilla. Mycologia,
29: 686-706.
Boidin J. 1950: Sur l'existence de races interstériles chez Gloeo-
cystidium tenue (Pat.); étude morphologieque et com-
portement nucléaire de leur cultures. Bull. Soc.
Mycol. France, 66: 205-221.
Boidin J. 1958: Essai biotaxinomique sur les Hydnés résupinés et les
Corticiés; étude spéciale du comportement nucléaire
et des mycéliums. Mém. h. sér. No 6, revue de Myco-
logie, Paris 390 p. 103 fig. 10 pl.
Boidin J. 1958: Hétérobasidiopycètes saprophytes et Homobasidiomyètes
résupinés. V - Essai sur le genre Stereum Pers. ex
S.F. Gray. Rev. mycol. Paris, 23: 318-346.

Boidin J. 1964: Valeurs des caractères culturaux et cytologiques pour la taxinomie des Thelephoraceae résupinés et étalés-réfléchis (Basidiomycètes). Bull. soc. bot. Fr., 111: 309-315.

Boidin J. 1971: Nuclear behavior in the mycelium and the evolution of the Basidiomycètes. P. 124-148 in "Evolution of the Higher Basidiomycetes" an International Symposium. Par R.H. Petersen, University of Tennessee Press. Knoxville, U.S.A.

Boidin J. & des Pomeys M. 1961: Hétérobasidiomycètes saprophytes et Homobasidiomycètes résupinés. IX - De l'utilisation des critères d'interfertilité et de polarité pour la reconnaissance objective des limites spécifiques et des affinités. Bull. soc. Mycol. France, 77: 237-261.

Boidin J. & Lanquetin P. 1965: Hétérobasidiomycètes saprophytes et Homobasidiomycètes résupinés X-Nouvelles données sur la polarité dite sexuelle. Rev. Mycol., Paris, 30: 3-16.

Boidin J. & Lanquetin P. 1973: Podoscypha involuta (Klotzsch) Imaz. est une espèce composite. (Basidiomycète Podoscyphaceae). Persoonia, 7: 141-150.

Boidin J. & Lanquetin P. 1975: Vararia subgenus Vararia (Basidiomycètes Lachnocladiaceae): Etude spéciale des espèces d'Afrique intertropicale. Bull. Soc. Mycol. France, 91: 457-513.

Boidin J. & Lanquetin P. 1977: Peniophora (subg. Duportella) malençonii nov. sp. (Basidiomycète Corticiaceae), espèce méditerranéenne partiellement interstérile avec son vicariant californien. Rev. mycol., Paris, 41: 119-128.

Boidin J., Terra P. & Lanquetin P. 1968: Contribution à la connaissance des caractères mycéliens et sexuels des genres Aleurodiscus, Dendrothele, Laeticorticium et Vuilleminia (Basidiomycètes Corticiaceae). Hétérobasidiomycètes saprophytes et Homobasidiomycètes résupinés. 11ème Contribution. Bull. soc. mycol. France, 84: 53-84.

Bourchier R.J. 1957: Variation in cultural conditions and its effect on hyphal fusion in Corticium vellereum. Mycologia, 49: 20-28.

Bruehl G.W., Machtmes R. & Kiyomoto R. 1975: Taxonomic relationships among Typhula species as revealed by mating experiments. Phytopath. 65: 1108-1114.

Cabral R.V. de G. 1951: Anastomoses miceliais sen valor nodiagnostico das poliporoses. Bol. Soc. Broteriana, 25: 291-362.

Carmichael J.W. 1956: The cellophane technique for studying morphology and hyphal fusions in fungi. Mycologia, 48: 450-453.

Chang L.O., Srb A.M. & Stewart F.C. 1962: Electrophoretic separations of the soluble proteins of Neurospora. Nature, 198: 756-759.

David A. 1974: Une nouvelle espèce de Polyporaceae: Tyromyces subcaesius. Bull. Soc. Linn. Lyon, No spec. : 119-126.

David A. & Romagnesi H. 1972: Contribution à l'étude Leucopores français et description d'une espèce nouvelle: Leucoporus meridionalis nov. sp. Bull. Soc. Mycol. France, 88: 293-303.

David A., Tortic M. & Jelic M. 1974: Etudes comparatives de deux espèces d'Auriporia: A. aurea (Peck.) Ryv. espèces américaine et A. aurulenta nouvelle espèce européenne. Compatibilité partielle de leur mycélium. Bull. Soc. mycol. France, 90: 359-370.

Davidson A.M., Dowding C.S. & Buller A.H.R. 1932: Hyphal fusions in Dermatophytes. Canad. J. Res., 6: 1-20.

Davidson R.W., Lentz P.L. & McKay H.H. 1960: The Fungus Causing pecky cypress. Mycologia, 52: 260-279.

Drosdova T.N. 1973: Immunoelectrophoretic studies of proteins of fungi belonging to family Polyporaceae. Mikol. Fitopatol., 7: 502-507.

Duncan E.G. 1972: Microevolution in Auricularia polytricha. Mycologia, 64: 394-406.

Duncan E.C. & MacDonald J.A. 1967: Micro-evolution in Auricularia auricula. Mycologia, 59: 808-813.

Eriksson J. 1950: Peniophora Cke sect. Coloratae Bourd. et Galz. A taxonomical study with special reference to the swedish species. Symbolae Botan. Upsalienses, 10 (5) 76 pages.

Esser K. 1962: Die genetik der sexuellen Fortpflanzung bei den Pilzen. Biol. Zentr., 81: 161-172.

Esser K. 1966: Incompatibility in Ainsworth & Sussmann. II: ch.20, p. 661.
Esser K. 1971: Breeding systems in Fungi and their significance for genetic Recombination. Molec. Gen. Genetics, 110: 86-100.
Esser K. & Blaich R. 1973: Heterogenetic incompatibility in plants and animals. Adv. in Genetics, 17: 107-152.
Fries N. & Jonasson L. 1941: Über die Interfertilität verschiedener Stämme von Polyporus abietinus (Dicks.) Fr. Svensk. Bot. Tidskr., 35: 177-193.
Garza-Chapa R. & Anderson N.A. 1966: Behavior of single-basidiospore isolates and heterocaryons of Rhizoctonia solani from flax. Phytopathol., 56: 1260-1268.
Gaut I.P.C. 1969: Identity of the Fungal symbiont of Sirex noctilio. Aust. J. Biol. Sci., 22: 905-914.
Gilmour J.S.L. & Heslop-Harrison J. 1954: The deme terminology and the units of micro-evolutionary change. Genetica, 24: 147-161.
Ginns J. 1976: Merulius s.s. and s.l., taxonomic disposition and identification of species. Can. J. Bot. 54: 100-167.
Hallenberg N. & Hallingbäck T. 1974: Interfertility and Polarity in Amylostereum laevigatum (Fr.) Boid. and Peniophora lycii (Pers.) Höhn & Litsch. Göteborgs Svampklubb Arsskrift, 18-21.
Harmsen L. 1960: Taxonomic and cultural Studies on brown spored species of genus Merulius. Friesia, 6 (4), 234-277.
Jahn H. 1971: Stereoide Pilze in Europa (Stereaceae Pil. emend Parm. u.a., Hymenochaete) mit besonderer Berücksichtigung ihres Vorkommen in der Bundesrepublik Deutschland. Westfälische Pilzbriefe, 8: 69-176.
Jorgensen E. 1954: A method for the study of mycelial anastomoses. Friesia, 5: 75-79.
Kemp R.F.O. 1975: Breeding biology of Coprinus species in the section Lanatuli. Trans. Brit. Mycol. Soc. 65: 375-388.
Kniep H. 1920: Über morphologische und physiologische Geschlechtsdifferenzierung (Untersuchungen an Basidiomyceten). Verh. Physik-mediz. Ges. Würzburg, 46: 1-18.
Kniep H. 1922: Über Geschlechtsbestimmung und Reduktionsteilung (Untersuchungen an Basidiomyzeten). Verh. phys. med. Ges. Würzburg., 47: 1-28.

Kniep H. 1923: Über erbliche Änderungen von Geschlechtsfaktoren bei
    Pilzen. Zeitschr. ind. Abst. u. vererbgsl., 31: 170-
    183.
Kniep H. 1928: Die Sexualität der niederen Pflanzen. 544 p.,221 fig.
Laibach F. 1919: Untersuchungen über einige Septoria. Ber. deutsch.
    Bot. Ges., 37: 247.
Lamoure D. 1960: Recherches cytologiques et expérimentales sur l'Am-
    phithallie et la Parthénogenèse chez les Agaricales.
    Evolution nucléaire dans la baside des formes bispo-
    riques. Thèse Doc. Etat, Lyon, 115 p. 11 pl.
Lanquetin P. 1973a: Interfertilités et polarités chez les Scytino-
    stroma sans boucles. (Basidiomycètes Lachnocladia-
    ceae). Le naturaliste canadien, 100: 33-49.
Lanquetin P. 1973b: Utilisation des cultures dans la systématique des
    Vararia Karst. subg. Dichostereum (Pilat) Boid. (Ba-
    sidiomycètes Lachnocladiaceae). Bull. soc. Linn.
    Lyon, 42: 167-192.
Leger J.C. 1968a: Les anastomoses végétatives dans le genre Stereum
    Hill. ex S.F. Gray sensu stricto. Bull. Soc. Mycol.
    France., 84: 333-342.
Leger J.C. 1968b: Utilisation des anastomoses végétatives dans la re-
    connaissance des affinités interspécifiques: étude du
    genre Peniophora Cke sensu stricto (Basidiomycètes).
    Bull. Soc. Mycol. France., 84: 505-511.
Leger J.C. 1974: Recherches expérimentales des affinités interspéci-
    fiques par les méthodes électrophorétiques et immuno-
    logiques chez les Basidiomycètes: Etude spéciale du
    genre Peniophora cke. Thèse Doct. Etat. Lyon, 142 p.
Leger J.C. 1976: Analyse électrophorétique et taxinomie numérique
    dans le genre Peniophora Cooke (Basidiomycètes).
    Bull. Soc. Mycol. France, 92: 377-392.
Lemke P.A. 1969: A reevaluation of Homothallism Heterothallism and
    the species concept in Sistotrema Brinkmanii. Mycolo-
    gia, 61: 57-76.
Maas Geesteranus R.A. & Lanquetin P. 1975: Observations sur quelques
    champignons hydnoïdes de l'Afrique. Persoonia, 8: 1-
    22.
Macrae R. 1967: Pairing incompatibility and other distinctions among
    Hirschioporus (Polyporus) abietinus, H. fuscoviola-
    ceus, and H. laricinus. Can. J. Bot., 45: 1371-1398.

Mc Kay H.H. 1959: Cultural basis for Maintaining Polyporus cinnabarinus and Polyporus sanguineus as two distinct species. Mycologia, 51: 465-473.

Mc Kay H.H. 1962: Interfertility study of Polyporus palustris and other brown rot species occuring on conifers and hardwoods. Plant. Disease Rep. 46: 26-29.

Mc Keen C.G. 1952: Studies of Canadian Thelephoraceae IX. A cultural and taxonomic study of three species of Peniophora. Can. J. Bot., 30: 764-787.

Madhosing H.C. 1964: A serological comparison of isolates of Fomes roseus and Fomes subroseus. Can. J. Bot., 42: 1677-1683.

Madhosingh C. 1969: Serological relationships in the Pycnoporus complex Canad. J. Microbiol., 15: 1344-1347.

Madhosingh C. & Ginns J. 1974: Serological relationships between Fomes fraxinophilus and F. ellisianus. Canad. J. Microbiol., 20: 1399-1401.

Madhosingh C. & Ginns J. 1975: Serological relationships between Gloeophyllum trabeum and Gloeophyllum saepiarium. Canad. J. Microbiol., 21: 412-414.

Maxwell M.B. 1954: Studies of Canadian Thelephoraceae. XI Conidium production in the Thelephoraceae. Can. J. Bot., 32: 259-280.

Mounce I. & Macrae R. 1936: The Behaviour of Paired monosporous mycelia of Lenzites saepidria (Wulf.) Fr. L. trabea (Pers.) Fr. L. thermophile Falck and Trametes americana Overh. Canad. J. of Res. C 14: 215-221.

Mounce I. & Macrae R. 1937: The Behaviour of paired monosporous mycelia of Fomes roseus (Alb. & Schw.) Cooke and Fomes subroseus (Weir) overh. Canad. J. Res. C 15: 154-161.

Neuhauser K.S. & Gilbertson R.L. 1971: Some aspects of bipolar heterothallism in Fomes cajanderi. Mycologia, 63: 722-735.

Nobles M.K. 1943: A contribution toward a clarification of the Trametes serialis complex. Can. J. Bot., 21: 211-234.

Nobles M.K. 1956: Studies of Wood-inhabiting Hymenomycetes. III Stereum pini and species of Peniophora sect. Coloratae on conifers in Canada. Can. J. Bot., 34: 104-130.

Nobles M.K. 1958: Cultural characters as a guide to the taxonomy and phylogeny of the Polyporaceae. Can. J. Bot., 36:

883-926.

Nobles M.K. 1965: Identification of cultures of wood-inhabiting Hymenomycetes. Can. J. Bot., 43: 1097-1139.

Nobles M.K. & Frew B.P. 1962: Studies in wood-inhabiting Hymenomycetes. V. The genus Pycnoporus Karst. Can. J. Bot., 40: 987-1016.

Olive L.S. 1956: Genetics of Sordaria fimicola. I. Ascospore color mutants. Amer. journ. Bot., 43: 97-107.

Robak H. 1942: Cultural studies in some Norvegian wood-destroying fungi. Meddel. Vestlandets Forst. 7: 1-248.

Roed H. 1969: On the relationship between Typhula graminum Karst. and Typhula incarnata Lasch. ex Fr. Friesia, 9: 219-225.

Skolko A.J. 1944: A cultural and cytological investigation of a two-spored Basidiomycete, Aleurodiscus canadensis n. sp. Can. J. Res., C 22: 251-271.

Talbot P.H.B. 1958: Studies of some South African Resupinate Hymenomycetes. II. Bothalia, 7: 131-187.

Talbot P.H.B. 1964: Taxonomy of the Fungus Associated with Sirex Noctilio. Australian. J. Bot., 12: 46-52.

Ulrich R.C. 1973: Sexuality, incompatibility and intersterility in the biology of the Sistotrema brinkmanii. Mycologia, 65: 1234-1249.

Ulrich R.C. & Raper J.R. 1974: Number and distribution of bipolar incompatibility factors in Sistotrema brinkmanii. Amer. Naturalist, 108: 507-518.

Ulrich R.C. & Raper J.R. 1975: Primary Homothallism. Relation to Heterothallism in the regulation of sexual morphogenesis in Sistotrema. Genetics, 80: 311-321.

Vandendries R. 1923: Nouvelles recherches sur la sexualité des Basidiomycetes. Bull. Soc. Roy. Bot. Belg., 56: 1-25.

Vandendries R. 1933: De la valeur du barrage sexuel, comme critérium dans l'analyse d'une sporée tétrapolaire de basidiomycète: Pleurotus ostreatus. Genetica, 15: 202-212.

Vandendries R. 1933: Nouveau critère spécifique des Basidiomycètes. Bull. Soc. Linn. Lyon, 1-4.

Van Uden N. 1951: Einige einfache Methoden zum Studium der Pilzmorphologie im Allgemeinen und der vegetativen Anastomosen im Besonderen. Archiv. Dermat. Syphilis, 193: 468-484.

Verrall A.F. 1937: Variation in Fomes igniarius (L.) Gill. Minnesota
    Agr. Exp. Stat., Techn. Bull No 117: 3-41.
Weresub L.K. & Gibson S. 1960: "Stereum Pini" in North America. Can.
    J. Bot., 38: 833-867.
White L.T. 1951: Studies of Canadian Thelephoraceae. VIII. Corticium
    galactinum (Fr.) Burt, Can. J. Bot., 29: 279-296.
Whitney H.S. & Parmeter J.R. 1963: Synthesis of heterocaryons in Rhizoctonia solani Kühn. Can. J. Bot., 41: 879-886.

TABLE 1. NUCLEAR BEHAVIOR CLASSED ACCORDING TO THE GROWING IMPORTANCE OF THE PLURINUCLEATE PHASE

	Germination	Haplont	Diplont	Subiculum	Subhymenium	Basidia
Normal	---	---	==	==	==	==
Subnormal	xxx	---	==	==	==	==
Heterocytic	xxx	xxx	==	==	==	==
Astatocenocytic	xxx	xxx	x==	x==	==	==
Holocenocytic	xxx	xxx	xxx	x==	xxx	==

- - - : Unimucleate   == : Binucleate   xxx : Multinucleate

Cultures des Auriculariales   321

Légendes des Tableaux

Tableau I   - extrait de Boidin (1971).
Tableau II   - extrait de Mc Kay (1962). On notera que les cultures
                 isolées du pêcher par Peterson et indéterminées, c'est
                 à dire les 19 portant une deuxième référence numérique
                 entre parenthèses, sont très clairement les unes in-
                 terfertiles avec P. palustris, les autres avec le P.
                 sp. et sont ainsi déterminées expérimentalement.
Tableau III   - extrait de Macrae (1967). Confrontations entre des
                 Hirschioporus abietinus (=poroïd), fusco-violaceus
                 (=Irpicoïd) et laricinus (=lamellate). On notera 2
                 sous-unités parmi les poroïdes américains, tous deux
                 plus ou moins interfertiles avec les poroïdes euro-
                 péens.
Tableau IV   - extrait de Lanquetin (1973b). Résultats d'interferti-
                 lités chez les Vararia subg. Dichostereum permettant
                 la nette distinction de 5 espèces désignées en haut du
                 tableau. Les numéros des souches sont accompagnés des
                 déterminations initiales des récoltes croisées.
Tableau V   - extrait de Mc Keen (1952). Il montre d'une part
                 l'existence de 3 espèces affines totalement intersté-
                 riles, d'autre part celle de 2 sous-unités chez P.
                 (Hyphoderma) mutata.
Tableau VI   - extrait de Mounce et Macrae (1938). On notera l'exis-
                 tence de deux groupes américains A et B pratiquement
                 interstériles entre eux et le comportement des Fomes
                 pinicola étrangers.
Tableau VII   - extrait de Leger (1968a). Anastomoses végétatives dans
                 le genre Stereum; les flèches indiquent les anastomo-
                 ses. *) Stereum rameale (Bres.) Fr. selon Jahn (1971)
Tableau VIII   - extrait de la Thèse de J.C. Leger (1974). Profil pro-
                 téique de 22 espèces de Peniophora.

Table 2. Results of pairing monosporous mycelia.

[Complex table with rotated column headers. Row labels (left column): MS-48, Mad. 4787, FP 57010, FP 94152-R, FP 97400-S, FP 103991-R, FP 105322-S, FP 105323-S, FP 105323-B, P-13a-R, Roth 8, Roth 10, H455-R, 4828, 5096-48, 5096-52, FP 48293-S, FP 71096, FP 72012-Sp, LCF 585, FP 59162-R, FP 104996-R.

Column groups:
- On conifer: MS-48, Mad. 4787, FP 57010, FP 94152-R, FP 97400-S, FP 103991-R, FP 105322-S, FP 105323-S, FP 105323-B, P-13a-R
- On hardwood: Roth 8, Roth 10, H455-R, 4828, 5096-48, 5096-52, FP 48293-S, FP 71096, FP 72012-Sp, LCF 585, FP 59162-R, FP 104996-R, FP 103951-Sp
- Polyporus palustris on peach trees: FP 105428-R, FP 105424-R, FP 105391-S, FP 105390-S, FP 105393-S, FP 105394-Sp, FP 105395-S, FP 105395-Sp, FP 105396-S, FP 105397-S, FP 105398-S, FP 105399-S, FP 105400-S, FP 105401-S, FP 105402-S, FP 105403-S
- Polyporus sp. on peach: FP 105421-S, FP 105413-S, FP 105407-S
- Polyporus sp. on other hardwood: Brazil-15, FP 50336-R, FP 105261-R
- Polyporus spraguei on hardwood: FP 47476-S, FP 24658-S, FP 59068-S, FP 103257-S, FP 103258-S

Cell entries: "+" indicates compatible pairing, "o" indicates incompatible pairing, blank indicates no test. The "+" marks generally fill the conifer, hardwood, and peach tree columns for the top rows, while the right-hand columns (Polyporus sp. and Polyporus spraguei) show predominantly "o".]

Tableau II

* The numbers in parentheses are those of D. H. Peterson.

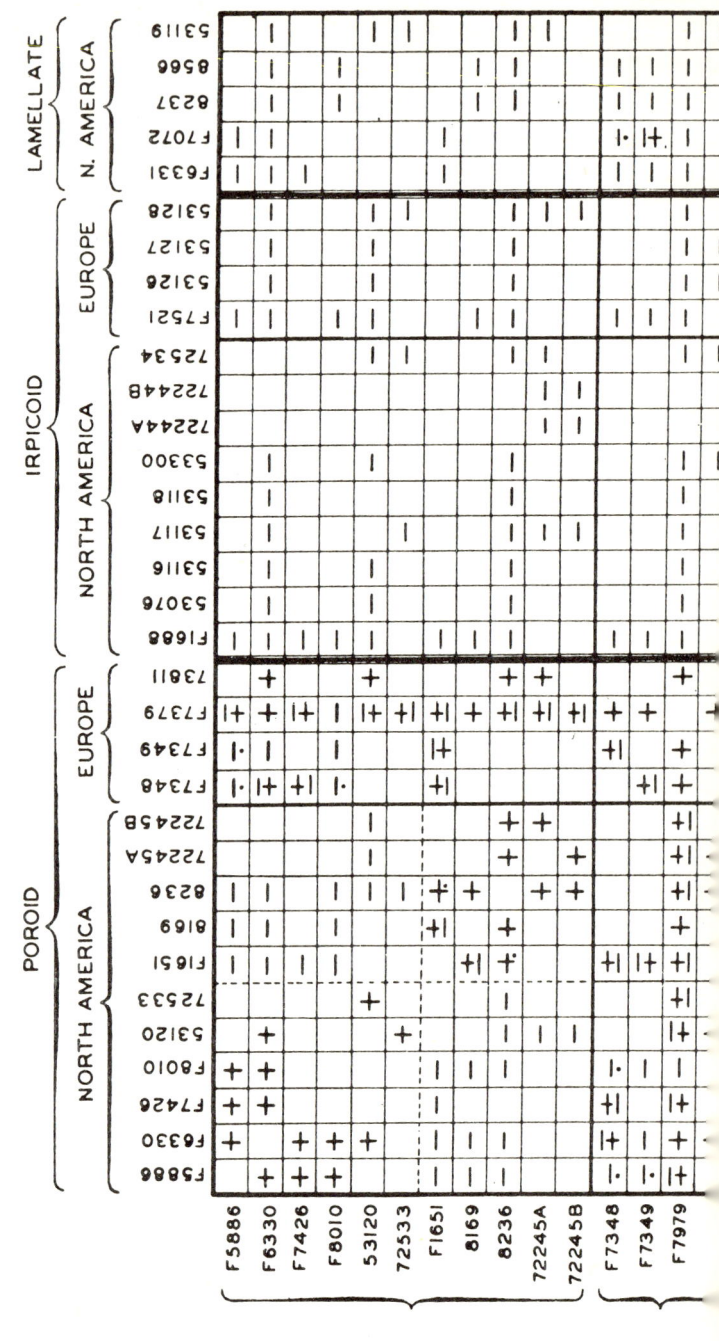

Tableau III

																	+	+	+	+	+				+	+	+	+	−	−	−	−	−
53076		−			−				−					−		+		+	+	+	+				+	+	±	+					−
53116		−			−				−					−		+	+		+	+	+				+	+	+	+					−
53117		−				−			−	−	−			−		+	+	+		+	+	+	+	+	±	+	+	+					−
53118		−							−					−		+	+	+	+		+				+	+	+	±					−
53300		−			−				−					−	−	+	+	+	+	+					±	+	+	+					−
72244A										−	−								+				+					+					−
72244B										−	−								+			+						+					
72534					−	−			−	−				−	−				+									+					−
F7521	−	−		−	−			−	−			−	−	−		+	+	+	±	+	±					+	+	+	−	−	−	−	−
53126		−			−				−					−	−	+	+	+	+	+	+				+		+	±					−
53127		−			−				−					−	−	+	±	+	+	+	+				+	+		+					−
53128		−			−	−			−	−	−			−		+	+	+	+	±	+	+	+	+	+	±	+						−
F6331	−	−	−				−					−	−	−		−									−					+	+	+	
F7072	−	−					−					−̣	∓	−		−									−				+		+	+	
8237		−		−				−	−			−	−	−		−									−				+	+		+˙	
8566		−		−				−	−			−	−	−		−									−				+	+	+˙		
53119		−			−	−			−	−				−	−	−	−	−	−	−	−	−		−	−	−	−	−					

FIG. 15. Table showing the results of intercollection pairings of monokaryons of the poroid, irpicoid, and lamellate forms of the *Hirschioporus abietinus* aggregate. The sign in each square indicates the occurrence of clamp connections in one series of pairings as follows: (+) in every pairing, (−) in none of the pairings, (±) in half or more than half the pairings, (∓) in less than half the pairings, (+˙) in every pairing but one, (−̣) in only one pairing.

	V.EFFUSCATA									V.sp.			V.PALLESCENS					
	C.B.S.	DAOM 22666/1	DAOM 22666/2	DAOM 31941/1	DAOM 31941/2	LY 6697/A$_1$B$_1$	LY 6697/A$_1$B$_2$	LY 6697/A$_2$B$_2$	LY 6697/A$_2$B$_1$	G 2458/1	G 2458/4	G 2458/6	FP 90120/A$_1$B$_1$	FP 90120/A$_1$B$_2$	FP 90120/A$_2$B$_2$	FP 90120/A$_2$B$_1$	DAOM 31244/1	DAOM 31244/3
C.B.S.		+	+	+	+	+	+		+				–	–	–	–		
DAOM 22666/1	+			+	+	+	+						–	–	–	–		
DAOM 22666/2				+	+	+	+											
DAOM 31941/1	+					+	+	+	+	+			–	–	–	–		
DAOM 31941/2						+	+	+	+		+							
LY 6697/A$_1$B$_1$	+	+	+	+	+			–	–	+	F	+	+	+	–	–	–	
LY 6697/A$_1$B$_2$	+	+	+	+	+			–	–	F	+	+	+	+	–	–	–	
LY 6697/A$_2$B$_2$	+	+	+	+	+	–	–			+	F	–	–	+	+	+	–	–
LY 6697/A$_2$B$_1$	+	+	+	+	+	–	–			F	+	–	–	+	+	+	–	–
G 2458/1		+				+	+	+	+				+	–	–	–	–	
G 2458/4	+					+	+	+	+				+	+	–	–	–	
G 2458/6				+		+	+	+	+				–	+	–	–	–	–
FP 90120/A$_1$B$_1$	–	–		–		–	–	–	–	–	–	–					–	+
FP 90120/A$_1$B$_2$	–	–		–		–	–	–	–	–	–	–					–	–
FP 90120/A$_2$B$_2$	–	–		–		–	–	–	–	–	–	–					+	–
FP 90120/A$_2$B$_1$	–	–		–		–	–	–	–	–	–	–					–	+
DAOM 31244/1								–			–	–	+	+	+	+		
DAOM 31244/3													+	+	+	+		
DAOM 31244/5								–			–	–	+	+	+	+		
FP 70855/1						–	–	–	–		–	–	–	–	–	–		
FP 70855/2						–	–	–	–		–	–	–	–	–	–		
LY 6627/A$_1$B$_1$						–	–	–			–	–	–	–	–	–		
LY 6627/A$_1$B$_2$						–	–	–			–	–	–	–	–	–		
LY 6627/A$_2$B$_2$						–	–	–			–	–	–	–	–	–		
LY 6627/A$_2$B$_1$						–	–	–			–	–	–	–	–	–		
G 2395/1								–	–		–	–	–	–	–	–		
G 2395/3								–	–		–	–	–	–				
G 2395/7								–	–		–							
LY 6242/1	–	–																
LY 6242/5				–		–	–	–			–	–	–	–	–	–		
LY 6242/8				–														
LY 6242/10	–			–		–	–	–			–	–	–	–	–	–		
LY 7097/1						–	–	–			–	–	–	–	–	–		
LY 7097/3	–					–	–	–			–	–	–	–	–			
LY 7097/5						–	–	–			–	–	–	–	–	–		
LY 7097/7						–	–	–			–	–		–		–		
T 433/A$_1$B$_1$						–	–	–			–	–	–	–	–	–		
T 433/A$_1$B$_2$						–	–	–			–	–	–	–	–	–		
T 433/A$_2$B$_2$						–	–	–			–	–	–	–	–	–		
T 433/A$_2$B$_1$						–	–	–			–	–	–	–	–	–		

Tableau IV

V.AFF.DURA		V.cf. PENIOPHOROIDES							V.RAMULOSA								V.RHODOSPORA			
FP 70855/1	FP 70855/2	LY 6627/A$_1$B$_1$	LY 6627/A$_1$B$_2$	LY 6627/A$_2$B$_2$	LY 6627/A$_2$B$_1$	G 2395/1	G 2395/3	G 23957	LY 6242/1	LY 6242/5	LY 6242/8	LY 6242/10	LY 7097/1	LY 7097/3	LY 7097/5	LY 7097/7	T 433/A$_1$B$_1$	T 433/A$_1$B$_2$	T 433/A$_2$B$_2$	T 433/A$_2$B$_1$
-	-	-	-						-	-										
-	-	-	-						-		-									
-	-	-	-	-	-	-			-	-	-	-	-	-	-	-	-	-	-	-
-	-	-	-	-	-	-	-	-	-	-	-	-	-	-	-	-	-	-	-	-
		-	-	-	-	-	-	-	-	-										
-	-	-	-	-	-	-	-	-	-	-										
-	-	-	-	-	-	-	-	-	-	-	-	-	-	-	-	-	-	-	-	-
-	-	-	-	-	-	-	-	-	-	-	-	-	-	-	-	-	-	-	-	-
-	-	-	-	-	-		-	-	-	-										
-	-	-	-	-	-			-	-	-										
-	-	-			-			-	-	-			-	-			-	-	-	-
-	-			-					-	-			-	-			-	-	-	-
-	-																			
+		+	+	+	+		+	+	+	-	-			-	-	-	-	-	-	-
+		+	+	+	+		+	+	+	-	-						-	-	-	-
+	+	-	-	+	F	+	+	+	-	-	-	-	-	-	-	-	-	-	-	-
+	+	-	-	F	+	+	+	+	-	-	-	-	-	-	-	-	-	-	-	-
+	+	+	F	-	-	+	+	+	-	-	-	-	-	-	-	-	-	-	-	-
+	+	F	+	-	-	+	+	+	-	-	-	-	-	-	-	-	-	-	-	-
+	+	+	+	+	+		+	-	-	-			-	-			-	-	-	-
+	+	+	+	+	+	+		+	-	-			-	-			-	-	-	-
+	+	+	+	+	+	-	+											-	-		
		-	-	-	-				-	+	+	+	+	+	+					
-	-	-	-	-	-	-	-		-		+	+	+	+	+	+	-	-	-	-
		-	-	-	-				+	+	-		+	+	+	+				
-	-	-	-	-	-	-	-		+	+	-		+	+	+	+	-	-	-	-
		-	-	-	-		-		+	+	+	+		+	-	+	-	-	-	-
		-	-	-	-		-		+	+	+	+		+	+	-	-	-	-	-
		-	-	-	-				+	+	+	+	-	+		+	-	-	-	-
		-	-	-	-				+	+	+	+	+	-	+					
-		-	-	-	-	-			-	-	-	-	-				-	-	+	F
-		-	-	-	-	-			-	-	-	-	-				-	-	F	+
-		-	-	-	-	-			-	-	-	-	-				+	F	-	-
-		-	-	-	-	-			-	-	-	-	-				F	+	-	-

	PENIOPHORA HETEROCYSTIDIA								PENIOPHORA POPULNEA		
	F7233	10899	17216	17551	17568	17587	21301	21308	10740	17217	17559
F7233	±	+									
10899	+								−		
17216		±		+	+	+	+	+		−	−
17551			+			+	+	+		−	−
17568			+	+			+	+		−	−
17587			+	+			+	+		−	−
21301			+	+	+	+		+		−	−
21308			+	+	+	+	+			−	−
10740	−	−									
17217			−	−	−	−	−	−			+
17559			−	−	−	−	−	−		±	
A											
17214			−	−	−	−	−	−		−	−
17558			−	−	−	−	−	−		−	−
17569			−	−	−	−	−	−			−
21205			−	−	−	−	−	−			−
21300			−	−	−	−	−	−			−
22316											
B											
F7991	−	−							−		
17215			−	−	−	−	−	−		−	−
17547			−	−	−	−	−	−		−	−
17550			−	−	−	−	−	−			−
21621			−	−	−	−	−	−			−
21622			−	−	−	−	−	−			−
21624			−	−	−	−	−	−			−

Tableau V

I+	I+	I+	+I	I+	+I		+	+	+	+	+
+	I+	I+	+I	+I	+I		+	+	+	+		+
+I	I+	I+	+I	I+	I+		+	+	+		+I	+
I+	I+	I+	I	I+	I+		+	+		+	+	+
+I	+I	I+	+I	+I	I+		+		+	+	+	+-
+I	I	I	+I	I+	I+		+I	+	+	+	+	+
+	+	+	+	+			I+	I+	I+	I+	+I	+I
+	+	+	+		+		I+	+I	I+	I+	+I	I+
+	+	+		+	+		+I	+I	I	+I	+I	+I
+	+		+	+	+		I	I+	I+	I+	I+	I+
+	+I	+	+	+	+		I	+I	I+	I+	I+	I+
+I	+	+	+	+	+		+I	+I	I+	+I	+	I+

I	I	I	I	I			I	I	I	I	I	I
I	I							I	I				
					I							

I	I	I	I	I			I	I	I	I	I	I
I	I	I	I	I			I	I	I	I	I	I
					I							

17214, 17558, 17569, 21205, 21300, 22316, F7991, 17215, 17547, 17550, 21621, 21622, 21624

A: 17214–22316
B: 17550–21624

PENIOPHORA MUTATA

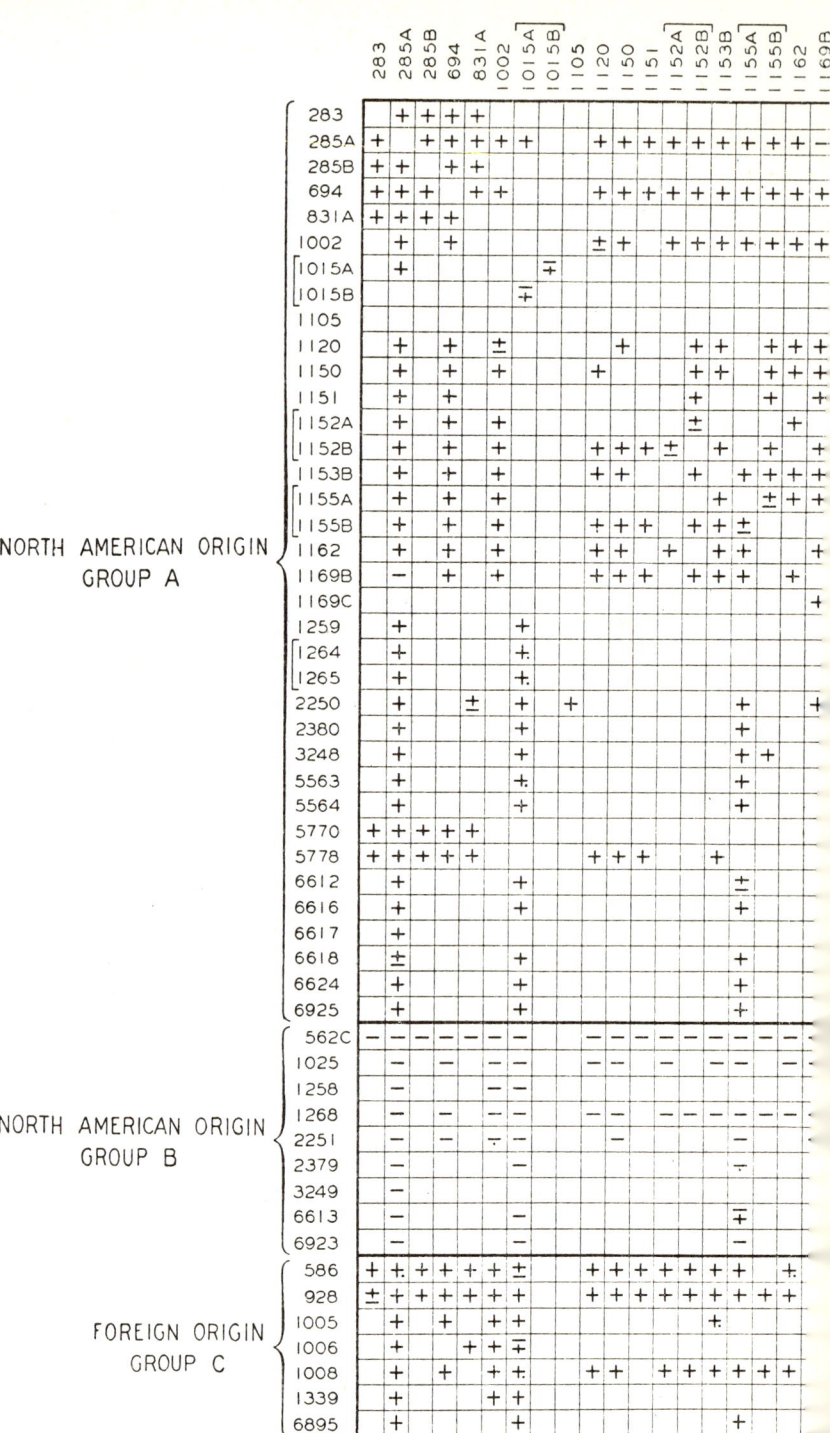

TABLE II. *The results of pairings of monosporous m*

Tableau VI

lia of Fomes pinicola from various hosts and localities.

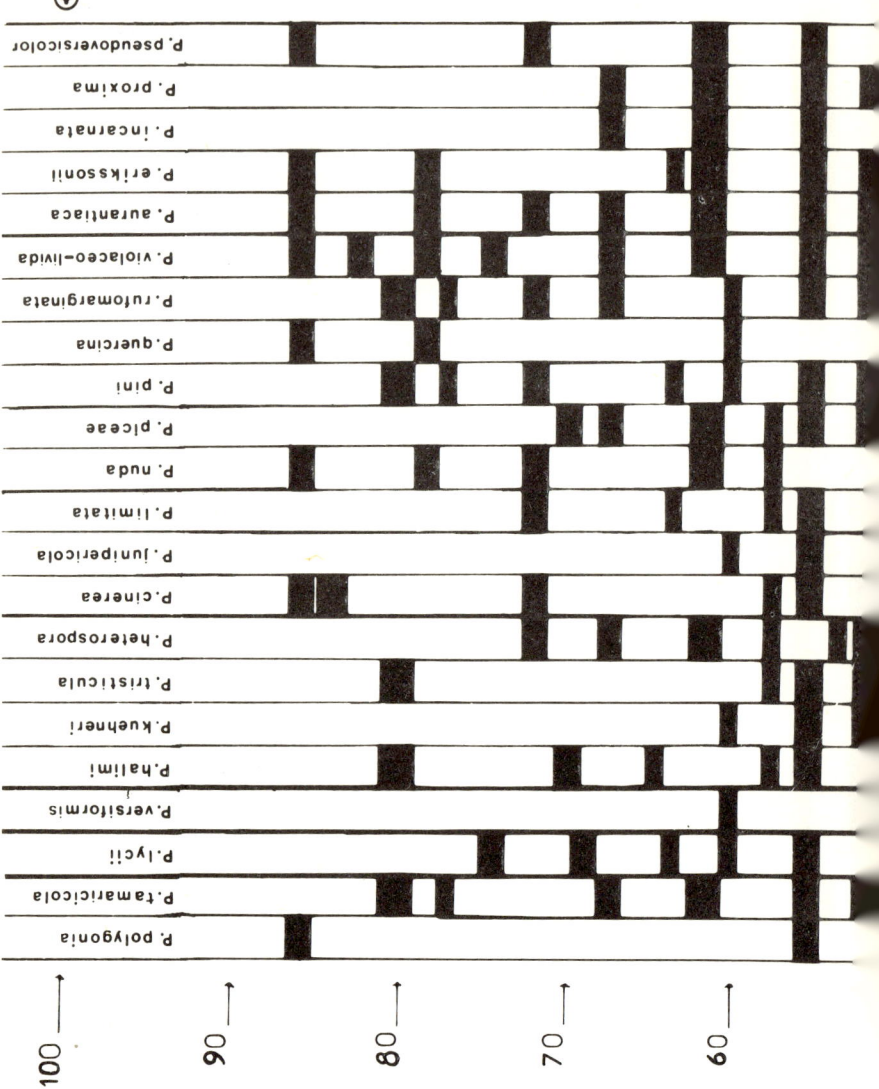

Tableau VIII

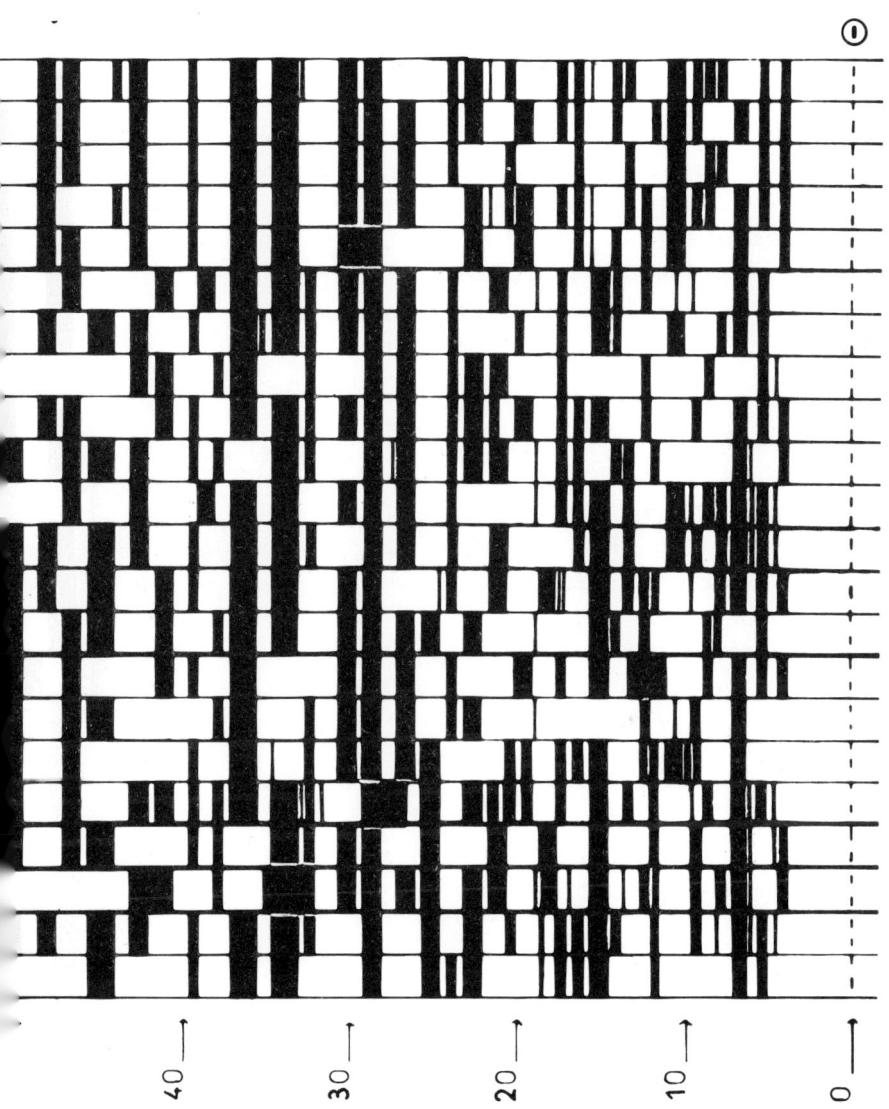

DISCUSSION FOLLOWING DR. BOIDIN'S PAPER

SINGER: Dans la partie A vous parlez de l'interfertilité. Je crois que maintenant tous les mycologues sont d'accord que ce sont des mêmes espèces s'il y a interfertilité totale. Maintenant j'ai des difficultés pour comprendre si l'interfertilité totale est limitée à la formation d'un mycélium secondaire et ma question est la suivante: Y-a-t-il des cas où on a déterminé si le mycélium a été sujet au gènes qui permettent de former les carpophores et s'il y avait caryogamie dans tous ces cas qui m'ont étonné un peu, par exemple, le cas des Auricularia. N'est il pas nécessaire de suivre le développement du mycélium?

BOIDIN: C'est une très bonne question. Vous savez sans doute que la majorité de ceux qui utilisent ces tests s'arrêtent à la production de boucles vraies. J'ai eu trop peu de temps pour vous dire dans quels cas la fructification et la descendance même avaient été obtenues. De mémoire, j'évoquerai les Lopharia mirabilis aux couleure et ornementations différentes; le croisement fructifie très bien et la descendance est parfaitement fertile avec les parents. Nous avons ici la certitude que caryogamie et méiose ont été normales. Mais, bien sûr, dans le cas très troublant des Auricularia polytricha de l'Assam et d'Australie, aux spores si différentes de taille, pour lesquel Duncan ne nous donne pas d'autres précisions que la présence de boucles, il serait indispensable de tout faire pour obtenir la fructification. En bref; il y a des cas où une fructification a été obtenue, mais pas toujours, loin de là!

BLAICH: I would like to add that this is the concept of Kemp which does not need the production of fruit bodies. Of course, we wish to obtain fruit bodies, but I think as a preliminary character it is sufficient if two nuclei may share the same metabolism.

SINGER: I do not know whether this is sufficient, because we cannot talk about interchange of genes in that case. It is possible that the carpophores are not formed because the genes do not act.

ESSER: Si on peut obtenir des fructification, c'est mieux. Le généticien pense que ce n'est pas indispensable.

KEMP: Could I answer Prof. Singer. I tried to indicate that the evolutionary decisions concerning speciation occurred at the level of the hyphal tip and were therefore largely concerned with plasmogamy. Caryogamy and Meiosis occur much later in the life cycle. If the decisions are taken at plasmogamy, at least in those species which have internal (or symplasmic) speciation, then I agree with Dr. Blaich that in most cases the formation of fruit bodies is not necessary.

ESSER: I would like to make a comment. I think there is fair agreement between the speakers of yesterday and today, that we could accept a genetic concept. And this comes close to your remarks, how can we proof, in absence of fertility, that this absence of fertility is caused by species differences. I think we have independently come to similar results. In your paper this morning you have shown many examples of crosses between species. That is the same thing that I have shown, and I am not quite sure whether in your cases we have seen barrage or other lines.

BOIDIN: Plusieurs auteurs signalent des phénomènes d'interaction, mais on ne peut savoir s'il s'agit du même phénomène que celui que vous appelez "barrage intraspécifique". D'ailleurs plusieurs auteurs font bien remarquer qu'il existe aussi pour eux de semblables réactions entre espèces voisines aussi bien qu'à l'intérieur de micro-unités inférieures à l'espèce. Il me semble qu'il serait très important pour l'avenir que l'attention soit portée avec précision sur ces phénomènes d'interaction. Il y a peut-être plusieurs sortes d'interactions à distinguer.

ESSER: I think we could agree on the following. We have many examples that in intra-species crossings there are isolating mechanisms instrumental caused by genic differences. These are more or less pronounced, they may cause total isolation, they may cause partial isolation, because fruit body production is late. And we must be able to differenciate, macroscopically or microscopically, between these intra-species interactions which we agree are the beginning of new species. We must distinguish between these interactions and the inter-species interaction which have been shown in various papers. I think all we could do, what the experimental mycologist could do, is to open the eyes of our collegues to write down records, to write down more precisely our observations. If they have done field studies, get the strains even if they do not have fruiting bodies, if they do not fruit in the laboratory, try, let the mycelia grow together and try to get this spectrum. I think that the hyphal interactions, or the hyphal-oidium interactions, are much more valuable than protein spectra.

BOIDIN: Si je vous comprends bien, ce qui est important ce sont les réactions mêmes des mycéliums au moment de la plasmogamie (réaction de refus, ou au contraire établissement du dicaryon) puis plus tard de la caryogamie...., c'est-à-dire la réponse objective de l'espèce elle-même. Les compléments d'information comme les spectres électrophorétiques sont d'intérêt moindre, et doivent être, comme les caractères morphologiques, interprétés (ils sont donc subjectifs!). Je suis de cet avis, et pense que ces compléments sont surtout nécessaires pour comparer des espèces homothalles.

ESSER: We must find out, and we must find as much as we can, criteria which allow us to explain the absence of interfertility. If we can explain the absence of interfertility then we have as much as we can, found possibilities to classify various morphologically closely related fungi and species.

BRESINSKY: I would like to repeat a question concerning these anastomoses which occur between different species of Stereum, would you think that there is any exchange of genetic material in form, perhaps, of plasmatic particles which could pass over?

ESSER: I do not know. We have to make experiments. I think there is a certain degree. We are very happy if we can observe fruit body production and fertile offspring. And if this does not work, then we have to look into all interactions, whether we have clamp connections, then the level below, whether we get hyphal anastomoses or aversion zones or even killing of hyphae. I think all three criteria, in a case like this, will help us to define species, at least to set a frame, a species frame on a genetic basies which we can more or less fit. You can fit it with your work, and some other people can fit it.

BLAICH: Je voudrais bien avoir une explication des basidiocarpes monocaryotiques. Vous avez donné un exemple de basidiocarpes en culture pure, monocaryotique. Quels sont les différences entre les basidiocarpes monocaryotiques en culture pure et les basidiocarpes en nature.

BOIDIN: Vous voulez parler, je crois, de la parthénogénèse, c'est-à-dire de la fructification d'un monocaryon dans la nature et au laboratoire. Le Dr Esser a projeté une page d'un livre récent où figure la liste des fructifications obtenues au laboratoire. Dans les cas que je connais, chaque fois que l'on obtient la fructification haploïde donc monocaryotique au laboratoire, toutes les spores formées par le basidiocarpe sont de même pôle et les mycéliums qui en sont issus sont parfaitement compatibles avec le pôle complémentaire. Par contre, si vous récoltez un carpophore parthénogénétique dans la nature, les cultures qu'il vous donne sont incompatibles avec celles obtenues à partir d'autres récoltes parthénogénétiques semblables à lui.

BLAICH: Et ce sont des vrais carpophores monocaryotiques dans la nature, et pas des cas d'amphithallie ou quelque chose similaire?

BOIDIN: L'amphithallie n'est pas toujours facile à reconnaître. Mais grâce à des colorations nucléaires, il est acquis que <u>Mycena galericulata</u> ainsi que beaucoup d'autres Mycènes et Hygrophores bisporiques sont parthénogénétiques.

BLAICH: Je croyais que les carpophores monocaryotiques soient connus seulement de culture pure.

KUEHNER: Non. Il est vrai que seuls des carpophores obtenus en culture pure à partir d'une seule spore d'une forme hétéothallique sont sûrement parthénogénétiques. Mais plusieurs espèces sont connues pour donner, dans la nature, deux sortes de carpophores: des carpophores à comportement nucléaire normal et des carpophores (toujours sans boucles) n'ayant qu'un seul noyau dans les articles du sous-hyménium et dans les plus jeunes basides, se comportant donc comme des carpophores parthénogénétiques obtenus en culture. Chez M. galericulata, ce sont les souches obtenues à partir de tels carpophores sauvages qui se sont révélées interstériles.

SINGER: I just wanted to give a reason for my first question. I asked Dr. Boidin, because there is Dr. Esser's definition: " Populations belong to different species when the failure to interbreed or to produce viable offsprings in nature is not caused by genetic parameters". I think that in order to apply this rule you have to know not only whether there is plasmogamy, but also whether there is caryogamy, otherwise no viable offspring in nature is possible. Is that right?

ESSER: That is right. I think in your cases and in my cases in the Ascomycetes we have an ABC-System. You cannot do this with two strains, you need a variety of strains, and then we have always one strain that crosses with an other, and this crosses with an other, and eventually you can go around. This, I admit, is a very hard job to do. I think what we want, from a genetic point of view, we want to have just limits to set, and we want to have examples, in German I would say "Vorbilder". We want to have "Vorbilder" where we can go and where we have to go.

BAS: Dr. Kemp a dit que dans 90% des cas de Coprinus, le concept d'espèce morphologique est confirmé par son travail. Et vous, et votre collaborateur, dans vos confrontations, avez-vous également une valeur comme cela? Ou est-ce que c'est une autre chose?

BOIDIN: Je pense même que l'on s'approche des 100%. Mis à part certains refus, mis sur ce compte de l'incompatibilité hétérogénique, les résultats correspondent presque toujours sinon au pronostic du morphologiste, du moins à l'une des hypothèses entre lesquelles il hésite à choisir. Il ne faut cependant jamais oublier que les espèces sont vivantes et en cours d'évolution, et qu'une espèce nouvelle récemment apparue qui est donc déjà décelable par l'incompatibilité peut ne différer morphologiquement que par des caractères peu marqués, ou du moins peu aisément remarquables.

BAS: Merci, je me sens un peu plus heureux maintenant.

WATLING: I agreee that the genetic species is an ideal. Surely we must achieve some compromise between the morphological and the genetic species; even in the flowering plants, we do not widely have this sort of information. We have to make some sort of compromise in order to deal with the material we are receiving on our laboratory desks every day.

ESSER: That is right, but on the other hand, in knowing what kind of interaction might take place, we have this in mind, that you look for these interactions, and the phenomena we know already will help us explain future observations. That is what I want to make clear. We know that something is present.

BLAICH: I think one thing should be mentionned. A species concept, however complicated it might be, is to be used by other biologists, as well as by taxonomists. Therefore a species concept should allow for people not familiar with this kind of thinking, to use it for their purposes. Therefore I think a species concept should be similar to the one proposed by Dr. Watling, with some macrospecies, sufficient for a chemist etc., and microspecies which can be reserved for people working in taxonomy.

KEMP: The approach I take (I hope I have some leanings towards taxonomy) is to confirm the findings of the taxonomist and not to disprove them.

SINGER: It seems to me that it is not right to have two taxonomies, one for taxonomists and one for children.

BLAICH: One is part of the other, and I think Dr. Watling did not intend his macrospecies to be for children.

SINGER: No, but he did mean something quite different from your angle. Excuse me, it was just a joke.

BLAICH: This taxonomy should be part of the other, of course.

SINGER: I agree with this. It seems to me, that we are precisely here for the purpose of understanding each other, and I think you expressed that very nicely; that both should try to understand the other camp. I would say, and you probably agree with me, that there are good mycologists and there are bad mycologists, and there are good geneticists, and there are bad geneticists, and when you used the word dogma, you probably had in mind a bad mycologist. Because I do not think we work with dogmas, I do not think we work for a religion, I think we have a scientific approach, and here it is for the geneticists to try to understand what the procedure of scientific taxonomy is.

ESSER: What I really mean: for most geneticists it is very hard to understand details of the many forms the taxonomists work with. Even within the field of genetics there exists a danger that people do not understand each other. It is evident that in biology exists the tendency to split up in small clubs. Few people belonging to a "so and so" club do understand each other, they speak a common language; but they do not any more understand the language spoken in other clubs. Therefore I was fascinated to come here to Lausanne to a meeting which brought together taxonomists, mycologists and geneticists who sit together and are able to bring their opinions "on the table".

SINGER: But what would you suggest that we could do to bring our working committements and the structure of our thinking in line with your necessities?

BRESINSKY: I have the feeling that some people object against diversity in nature. It is sure that taxonomists are not responsible for the diversity in nature. And the other point is: we have seen in this symposium that the taxonomic units, which were delimitated by the taxonomists show reasonable coincidence with genetic barriers, so the work of the taxonomists is not to bad. Why should we forget that we have diversity in nature? I do not understand you.

ESSER: No, no, let us find another track. There is diversity in nature, O.K., and taxonomists are not responsible for it, but the taxonomists should not be responsible for making diversity more complicated that it already is.

BRESINSKY: We do not. You have seen that there is a coincidence with genetic barriers and the taxonomic units are fairly reasonable.

ESSER: Certainly I have it, and that is what I am happy for. I am fighting for a coming together.

PETERSEN: What I hear you talking about is "something" _versus_ taxonomy, as though the taxonomist were one kind of person working on one kind of characters, in aposition to those other persons who are not doing taxonomy. I think the people we are talking about as omitted

Cultures des Auriculariales

from taxonomy are actually doing taxonomy. They are describing additional characters for everyone to use and they are sorting the taxa. I think Dr. Kemp gave us some characters this morning, and Mr. Blaich gave us some characters the other day. These are characters that somehow are not being used in the present discussion of "small t" taxonomy. Furthermore, I am afraid I do not understand where we arrived in the evolution of the discussion to talk about that "something" (I have no name for it) versus what we are all calling taxonomy. I am not sure what that "something" is, but I think I know what "small t" taxonomy is. But are you people not also doing "large t" taxonomy? I think I get the overtones that some of the experiments you are doing give us "small t" taxonomists some additional characters with which to sort our taxa. Am I not right?

ESSER: Yes, you are absolutely right. What we want is to keep in mind that there are other characters than maybe morphological characters.

KEMP: I would like to support Dr. Bresinsky. In fact, a great many of the groups we have looked at do have good characters which are used by the taxonomist. The limits of the taxa are confirmed by the living species, and we respect the views of the taxonomist. I think it is the species in culture that will bring us together and to respect what other workers are doing.

BAS: I believe that we should not see the morphologic species concept as being in contradiction with the genetic species concept. I think we should consider them as stages of knowledge. We are gradually expanding our knowledge. In some groups, perhaps, we cannot do anything else but trying to get to this morphologic species concept. In other groups we have the possibility to go beyond that stage. And so one concept is passing into the other.

Herbette Symposium on Species Concept in Hymenomycetes 1976

SPECIES- AND GENERIC-CONCEPTS IN THE CORTICIACEAE

F. Oberwinkler

During the past twenty years the number of genera within the non-gilled Hymenomycetes has rapidly increased. Even a corticiologist will find it difficult to survey the generic jungle in some groups. I am of the opinion that a shifting of valuing the characters of taxa has taken place by splitting old and establishing new genera. By this procedure the species concept may have been changed too.

Characters, restricted to comparative morphology, for delimiting species and genera in some selected taxa of the Corticiaceae should be treated in the following.

1. The Botryobasidium-Sistotrema-group with remarks on species of related genera.

Within the genus Botryobasidum (fig. 1) a group of species can be arranged which is characterized by hyphal septa without clamp connections and spores which are apically rounded (at least not biapiculate).

Species to be filed in this group may be keyed as follows by using spore data:
1 Spores more or less ellipsoid
  2 Spores 4-5 x 7.5-12 microns   B. obtusisporum J. Erikss.
  2'Spores 2.5-3.5 x 5-8 microns   B. laeve (J. Erikss.) Parm.
1'Spores more or less cylindric
  3 Spores 2.5-5 microns broad, not allantoid
    4 Spores not longer than 10.5 microns
      Spores 3-3.5 x 6-9 microns   B. aureum Parm.
      Spores 2.5-3.5 x 7-9 microns   B. conspersum J. Erikss.
      Spores 3-3.5 x 6-9 microns   B. robustior Pouz.& Holub.-Jech.
      Spores 3-4 x 8.5-10.5 microns   B. simile Pou.& Holub.-Jech.
    4'Spores 3-5 x 11-14-(17)microns B. danicum J. Erikss.& Hjortst.
  3'Spores 2-3 x 8-9 microns,   B. vagum (Berk.& Curt.) Rogers
      almost allantoid

There are three points to be discussed under "species concept" in
this case:

a) The species listed in the key are described (with one exception)
within the last two decades. Several mycologists who have studied -
like John Eriksson - fungi intensively and with great accuracy with
the light microscope, could support evidence that characters which
are easily to be overlooked may be of taxonomic value. Often these
investigations are carried out in so-called natural groups, that are
commonly identic with "new genera" like Botryobasidium (Donk 1931),
which may really be a forerunner in the generic boom. The genus Botryobasidium which is circumscribed by its peculiar basidial type and
the branching of the broad hyphae by right angles is an example for
shifting species concepts to the generic level. Originally B. subcoronatum, B. coronatum and B. solani where included in the genus. Probably because the type species (B. subcoronatum) is wellmarked, the
generic membership is easily to be recognized. This seems to me a
supposition for a detailed study which may reveal further, still not
known characteristics for new species concepts.

b) A main pool for these characteristics is bound with basidiospore
morphology. In the selected group it seems likely to separate four
taxa by spore measurements. It must be admitted that there are sometimes really small differences, but they seem to be highly relevant.

c) In addition to this the species concept in Botryobasidium is
strongly influenced by characters derived from asexual conidial
states (fig. 2). The species affiliated to key point 3 in realty not
only cover the same spore range but are very close in other details
of the sexual state. In contradiction to this the conidia of the
Oidium-type provide essential specific characteristics.

Botryobasidium agrees with Sistotrema in basidia which bear more than
four basidiospores. Moreover the urniform meiosporangium is the most
typical organ of species generally referred to Sistotrema (fig. 3,
4). At the moment all species which develop urniform basidia with six
to eight sterigmata are submitted to Sistotrema with the probable exception of Multiclavula Petersen. Species with basidia of this type
occur within a broad range of fruitbody types. The type species of
Sistotrema, S. confluens (fig. 3) develops basidiocarps which can be
more or less pileate; the hymenium is rather irregular but may be

sometimes porioid. An extremely frequent "species" is **S. brinkmannii** (fig. 4) which already is fertile by growing with some few basal hyphae. Under good environmental conditions the fungus may build up odontioid hymenia. At least for practical purposes the splitting of this taxon has as yet not been acceptable. Even small differences in micromorphology of basidia and basidiospores could apparently not been used for a definitive clearing of the species concept.

I have chosen these examples for demonstrating quite different situations in close relationships (Botryobasidium and Sistotrema). There is no doubt that this genus is in urgent need to be monographed. It is to be expected that more natural groups can be established by reevaluating types of hyphae, hyphal context, structure of the hymenium, spore morphology and probably also a more detailed basidium morphology.

A second point in this connection seems worthy to be discussed: Species that have the hyphal type of Botryobasidium but the meiosporangium type of Sistotrema. Eriksson & Ryvarden recently (1973) were opposed to this problem and made a decision in favouring the "value" of basidial morphology by indicating to transfer Botryobasidium heteronemum to Sistotrema. Though the pattern of characteristics is the same, by the "influence" of generic membership the species concept may be altered in so far as within Botryobasidium the bacidium will be a specific marker whilst grouped in Sistotrema this character is merged in the generic concept. Vice versa can be argumented for the hyphal type. Properly this is the point where a discussion on the generic concept had to start. But this is not the intention of this symposion.

Nevertheless I take the opportunity to cast a glance on some related genera which comprise only few species or which are even monotypic. In 1958 Eriksson erected the genus Sistotremastrum, based on Corticium suecicum (fig. 5), a species that lacks probasidial swellings at the base of the young basidium. In my opinion the genus is at present monotypic.

In comparison with Sistotrema I would challenge the uniformity of the basidial type for this genus too. Theoretically a number of hypothetic species could be expected by different but specifically fixed

spore types. On purpose I do not follow now the line towards Paullicorticium, as I did already in 1965. For that, some remarks on what
may be a species in further groups with fungi that develop urniform
basidia is added. The urniform basidium however is now of the four-
spored type. This can be accepted inter alia as a generic feature for
taxa proposed as genera like Waitea, Galzinia, Urnobasidium or Galziniella.

Isolated from soil in Australia, Warcup & Talbot (1962) cultivated a
fungus with microscopical details as shown in a section of the type
material in fig. 6. I will restrict the interpretation to one character, viz. the septate spores. As the genus is monotypic, logically
this enters the species- and the generic concept. The same is due to
Cejpomyces (Corticium terrigenum) whereas in Exobasidium, Septobasidium and genera of the Dacrymycetales this criterion can be used also
for generic and even suprageneric circumscription. In Platygloea uni-
spora (Auriculariales) it is a marker on specific level if the current classification is accepted. I would like to draw attention here-
with to the fact that on account of our present knowledge (or actually) equal morphological structures are valued differently in various
taxa. In this connection we can touch shortly the valorization of an
additional spore character: the germination by repetition. In Corticium cornigerum spore repetition refers to the specific description;
as a member of the genus Ceratobasidium this behaviour enters the generic concept; in Ceratobasidiaceae the family, in Tulasnellales the
order are additionally characterized and finally in Heterobasidiomycetes the "class" is circumscribed solely by this character. Concerning Waitea circinata I think a decision is outstandig whether the
spore type is purely a specific character.

Bourdot's genus Galzinia was based on a new species, Galzinia pedicellata (fig. 7). According to Eriksson & Ryvarden (1975) the basidial type is much more variable as from Bourdot's (or even Erikson's
1958) figure can be concluded. Nevertheless I am convinced that a
specific (and generic) marker is - at least a high percentage of -
the urniform basidium. I agree with Eriksson & Ryvarden (l.c.) that
"the repetition of basidia ... is thus of little value for the generic diagnosis" (in this genus and elsewhere, e.g. Hyphoderma, Subulicystidium or Hymenochaete).

On the other hand this basidial formation should be part for the specific description. In Repetobasidium (Eriksson 1958, Oberwinkler 1965) this peculiarity is a general generic phenomenon.

In the "urnobasidium-group" remain two genera which have been proposed by Parmasto (1968) as Urnobasidium (typified by Corticium sernanderi, fig. 8) and Galziniella (type species: G. pereximia, fig. 9).

In Urnobasidium the author stresses the combination of the urniform basidium (two- to four-spored) with the presence of gloeocystidia. Although young developmental stages of the Galziniella basidium are described as "subgalzinoidea" a phenomenon during spore formation is apparently emphasized to be of generic importance, i.e. the occurrence of one, two or three sterigmate basidia. After studying the type specimen I am convinced that these basidia are abnormally developed and moreover that the normal four-spored basidium is quite typical for this fungus. I am confirmed in this interpretation by the note of Parmasto (l.c.) himself that sterigmata lengthen up to 10 microns. This can be observed generally when the "normal" basidium- and sterigma-formation is suppressed and abnormally two or only one sterigma is formed. The differences in cystidial types seem to me more important. In Urnobasidium sernanderi a yellow tinted pigment characterizes the cystidia as gloeocystida, whereas morphologically similar structures in Galziniella are absolutely colourless (at least in herbarium material of the type collection). The specific differences of both fungi may be underlined by spore characters too. Altogether the mentioned details seem to me as worthy for a species concept but not at all for generic delimitation.

A second group of corticioid fungi which will be discussed in the following by stressing the dependence of species- and generic concepts, are the

2. athelioid "Corticia".

The term "athelioid" signifies a special type of fruitbody: the trama hyphae are very loosely interwoven (almost cobweb-like), so that the pellicular hymenium can easily be detached. The type species of Persoon's genus is Athelia epiphylla (fig. 10), a wide-spread species to which the generic features just mentioned can be fully assigned. By monographing the genus, Jülich (1972) elaborated a species-concept, which in some groups seems to be so narrow, that Eriksson & Ryvarden (1973) for example prefer to have a "wider interpretation" of the type species aggregate, that means that they recognize only A. epiphylla from the species in question. Using Jülich's key data, six taxa of this group are characterized as follows:

	Spores (microns)	Basidia (microns)
A. alnicola (B.&G.) Jülich	ellipsoid 6.5-8-8.5 x 3.6-4-4.4	15-20-25 x 5-8
A. epiphylla Pers.	cylindric, basally round (5.5)-6-7.5.(8) x 2.8-3.2	13-18 x 5-8
A. macrospora (B.&G.) Christ.	M.P.9-11-13.5 x 5-6-6.5	20-30-36 x 8-10
A. nivea Jülich	very broad ellipsoid to nearly cylindric 6.5-8-(9) x 4.3-5.2	15-18 x 5.5-6.5
A. ovata Jülich	8-9 x 3.8-4.2	16-18 x 5-7
A. tenuispora Jülich	long and small cylindric 8-10-(12) x 3.5-4	16-18 x 6.5-7.5

Although this is a somewhat simplified presentation (for full information see Jülich, l.c.) the author's restricted species concept is evident.

There is still another problem involved in athelioid fungi: the delimiting of genera, which in my opinion can only be discussed thoroughly by an accurate understanding of the included species.

As segregates from Athelia, Jülich has proposed several new genera. Leptosporomyces (type species Corticium galzinii Bourd., fig. 11) is originally defined by its "... sehr kurzen, zylindrischen Basidien und die kleinen, schmalen Sporen". Further the author gives some remarks on the delimitation. For example "Fibulomyces mutabilis - (fig. 12) - kann auch mit relativ kurzen Basidien angetroffen werden, jedoch sind auch stets längere Basidien vorhanden und ausserdem ist der Fruchtkörper dieser Art nicht dünn-häutchenförmig, sondern dicklich membranös". As for the basidia I am doubtful whether this is important. The difference in hymenial structure is obvious (compare fig. 11 and 12). The main difference results from a thin or a thickening hymenium. The torulose hyphae of the subhymenium in Fibulomyces mutabilis show the former steps of basidial formation. A fungus with a thickened hymenium therefore is relatively old. Whether a non thickened hymenium is a young developmental stage or a specific character has to be proven in any case very carefully. Practically it is already troublesome to check this character for purposes of specific interpretation. On the generic level I would hesitate to make a decision.

There is a third genus described by Jülich (l.c.) and derived as a satellite from Athelia s.ampl.: Confertobasidium, which is typified by Corticium olivaceoalbum Bourd. & Galz. (fig. 13). This is a species which agrees in basidial and spore morphology with Leptosporomyces and Fibulomyces. The hymenial configuration is of the Fibulomyces-type too. As a different character can be stated - as has already be done by Jülich and Eriksson & Ryvarden - "the brown pigmentation of the basal hyphae". The presence of rhizomorphs seems not to be of generic value because they are built up also by certain taxa of the closely related genus Fibulomyces. At the moment there is one species, the type of the genus. What is the specific and what the generic concept in this case? I have mentioned already divergent and corresponding characters between the three genera involved. As to Confertobasidium it is difficult for me to accept the genus concept. An analogue situation could be constructed in splitting Botryobasidium laeve and especially B. pruinatum from the existing genus. I think that both examples could be used for demonstrating that membrane pigmentation of basal hyphae has to enter the species concept.

By studying pleurobasidial species I became aware (1965) of Xenasma aurantiacum M.P. Christ. (fig. 14), a species which I made the type of the new genus Athelidium. At present I would prefer to strengthen other generic features than those mentioned in 1965. The hymenium is adnate to the substratum, therefore the fruitbody is not typically athelioid, i.e. a pellicular hymenium is lacking. Probably Eriksson & Ryvarden (1973) are correct in interpreting the basidia as Hyphoderma-like. This modified generic concept forces the exclusion of Xenasma pyriforme, as has already been practiced by Jülich, who transferred the species to Athelia. I am not sure whether this is correct. As to the species concept within Athelidium I would claim for taxa which may differ in spore dimensions for example. The above given generic description should not be challenged; if it is to be broadened, e.g. by including fruitbody pigmentation, has further to be proofed.

In connection with athelioid fungi some comments should be given on
two genera also recently proposed by Jülich (1972, 1974). Ceraceomyces is based on Corticium tessulatum Cooke (fig. 15), a species with
strongly thickening hymenium whose basidia are (narrowly) clavate.
Sometimes inconspicuous cystidioles occur in the hymenium; they have
an hymenial origin and may protrude the basidia only slightly. The
type of fruitbody, hyphal construction and context, growth of the hymenium, basidium- and cystidiole-morphology should be summarized for
circumscribing the genus. Additional characters, e.g. as colour and
surface structure of the hymenium, measurements of cystidial organs
and spores may be useful for specific arrangements.

A wide-spread species that is even known by non-corticiologists is
currently named as Corticium evolvens (fig. 16). I wondered if Jülich
includes this species in Ceraceomyces; but he raised it to generic
rank in erecting the genus Cylindrobasidium. The name already indicates the generic character, viz. the long cylindric basidia. In fact
these organs may have a length of 40-80 microns. In comparison with
Ceraceomyces tessulatus (basidia 25-35 microns long) this seems to me
not really important, because this only is a characteristic of dimen
sions not correlated to different structural features. Besides this
the morphology of the basidium itself is quite similar to that of the
Ceraceomyces meiosporangium. The identity of morphological criteria
as listed above (under Ceraceomyces) of the two species under discussion indicates, that the length of the basidium is in this case better used as a specific marker.

3. Pleurobasidial fungi
are a third group in which specific and generic characters may be
located. As to the pleurobasidium I would like to make some distinctions: There are fungi with gelatinous fruitbodies and a cathymenium,
where a sterile hyphal context is primarily developed. Secondly basidia are formed in (or partly in) this trama in the way which is called pleurobasidial. "Pleurobasidia" can also be observed in many
other basidiomycetes, especially at the growth line of the hymenium,
where generative hyphae are straightened and then give raise to laterally fixed basidia. This is clearly a developmental stage and should
not be merged in the cathymenial pleurobasidium type. When I proposed
the family Xenasmataceae I had in mind fungi which could be brought

Corticiaceae   339

together on the basis of this basidial character. Whether the given
scope was correct in details cannot be discussed in this connection.

The generic concept of Xenasma (based on Corticium rimicolum Karst.,
fig. 17) is, as to my present knowledge: with cystidia; spores warted, warts invisible in KOH. In adopting this circumscription, Peniophora subcalcea has to be excluded. The remaining group may be split-
ted into species on the basis of cystidial, basidial and spore morphology. For pleurobasidial species without cystidia and relatively
short basidia I have (l.c.) introduced the genus Xenasmatella as a
segregate from Xenasma. As type species Corticium subflavido-griseum
Litsch. has been designed. For this genus the generic name Phlebiella
Karst. (1890) has to be used. The type species, Phlebia vaga Fr.
(fig. 18) which is synonym to Cristella sulphurea, can be distinguished from Corticium subflavido-griseum only at the specific level,
i.e. the spores of the latter species are not warted in a ventrically
zone. This genus should be restricted to species with warted spores.
What species lie inside the scope of Phlebiella or Trechispora has to
be elaborated thoroughly, especially in recognizing cat- and euhyme-
nia. The species concept of Phlebiella will be strongly dependent of
spore morphology. For example two collections of a fungus (fig. 19)
extremely similar to Phlebiella subflavido-grisea show spores with
few warts in a slightly irregular position. This different spore type
can indicate what has to be understood as species concept in this genus.

I have restricted now Phlebiella to species with warted spores. In
Xenasmatella were included also taxa with smooth spores and those
species whose spores are smooth and amyloid. Without any comment on
related fungi Hauerslev (1974) recently has described the genus Melzericium, which is typified by Corticium udicolum Bourd. I do not
know the type, but judging from figures given by Hauerslev, I am convinced that the fungus develops a cathymenium and in addition is typically pleurobasidial. Probably Corticium rallum Jacks. could be a
synonym. When this is correct, the genus has to comprehend what I had
placed in Xenasmatella subgen. Amyloxenasma. Because we had and still
have no information of what causes the amyloid reaction in the spore
wall of fungi of this group I omitted to raise the taxon to generic
level. The species concept is here again clearly dependent on the delimitation of the genus. As members of Melzericium the species can be

distinguished by spore morphology mainly. A probably common species -
at least in Central Europe - which I have named Xenasmatella allantospora (fig. 20) is easily distinguished by the spore shape.

There remain at least two problems: a) where to place the smooth-
walled, non-amyloid species of the residual Xenasmatella group and
b) what is the status of Corticium subcretaceum Litsch.?

Without a drastic emendation of the genus, species like Corticium fi-
licinum and C. subnitens cannot be included in Melzericium. The situation is now quite the contrary as when I had to fix the scope of
Xenasmatella. Therefore I would prefer for the time being to unite
those non-amyloid species in a genus of its own. The species concept
is similar to that of Melzericium; the species may be separated therefore especially by spore characters. At present I know two species
(FO 5600, 12994, fig. 21, 22), which are not yet described. They differ from the already known species by spores which are relatively
small (2-2.5 microns broad). Finally the spore shape is the distinguishing character between the species themselves.

Corticium subcretaceum (fig. 23) is apparently cathymenial. However
the species has no pleurobasidia and the spores are inamyloid. A
clear solution would be to establish a new genus for this species.
But this would again be one of many genera which in realty are
"simple" species.

A fourth chapter is added to demonstrate that within a fairly well
delimited genus, the separation of species by methods of comparative
morphology is far from easy.

    4. Species of the genus Subulicystidium.

    When Peniophora - in a natural sense - was restricted to Bourdot &
Galzin's sect. Coloratae, a vast field for establishing new genera
with cystidiate species had to be covered. I remember a discussion
with Dr. Donk in 1967 when he predicted a genus to be based on Peniophora longispora. Cystidial morphology in this species has always
been recognized as the leading character and there has never been any
doubt in accurate identification. Meanwhile the scope of the well
known species is challenged. Cunningham (1955, 1963), being fully

Corticiaceae

aware of the close relationship to Peniophora longispora, proposed his species P. nikau in favouring spore characters. Talbot & Green (1958) were of the opinion that a short spored fungus of South Africa might represent a variety of P. longispora only. Finally, Jülich (1969, 1975) revised the genus in recognizing the taxa mentioned as separate species. He has shown that scanning electron microscopy can be used for a better understanding of micro-details. But in separating the species, he uses conventional criteria, i.e. characters to be detected by light microscopy. At the moment I can refer only to this method too, which yields the following data: fruitbody corticioid with loosely interwoven and distinct hyphae which are constantly clamped; the branching of the hyphae reminds to what occurs in species of Hyphodontia or Amphinema; cystidia seem to be uniform (Jülich 1975); the basidia may be called suburniform and at best be compared with the meiosporangia of Hyphodontia-species; as variable characters remain the size and form of basidiospores whose range runs from very long and small to elliptical. For the discussion of the species concept in this genus I have selected nine samples which I would like to demonstrate in detail.

The type species of the genus Subulicystidium Parm., S. longisporum (fig. 24), has somewhat curved spores which measure 2-2.5 x 13-15 microns in the type collection. The actual range of spore dimensions may not be covered by this data. - Subulicyctidium brachysporum (fig. 25) spores of the authentical material show 2.5-3 x 6.5-8 microns. - Subulicystidium nikau spores of the type specimen (fig. 26) vary within 3.5-4.5 x 6.5-8.5 microns.

A fungus collected in the botanical garden in Munich (FO 17359, fig. 27) has extremely small and long spores (1.5 x 15-20 microns) as a basis for separating it form S. longisporum. A juxtaposition with the type species would confirm the idea: the trama is scanty; the cystidia are very small and show a tiny incrustation in observing them with the light microscope; finally the basidia which have 1/2 or 2/3 of the length of the S. longisporum basidium may be occasionally developped in a "pleurobasidial" manner (compare the "pleurobasidial group", point 3.!). Summarizing these characters, a "new species" could be described, which is collected twice (second sample FO 17371). In this case however I think that we need more information especially on the development of S. longisporum. The few trama hyphae with a scarce hyphal-branching and consequently poor hymenium may indicate the very beginning of the fungus fructification. In this view the cystidia may not be fully developed. For the time being I cannot interprete the spore type. So far I think that we are not yet prepared to answer the question "what is a species?" in this complex.

Two fungi, collected in Venezuela (FO 13010, 13728) may be merged in S. longisporum agg. too (fig. 28). They differ however from typical S. longisporum by stout basidia and straight spores. Here again it seems to me that the descriptive status is not at all sufficient for the decision of what may be a different species.

Two collections, again from Venezuela (FO 15970, 16014), show fungi (fig. 29) whose spore length is precisely amidst (2.5-3 x 9-10 microns) of those of S. longisporum and S. brachysporum. Again I would prefer not to fix the status of this fungus because information on the variability of spore length is not available at present time.

Against my own intention I have brought together a mixture of organisms (or morphological characters) in the immediate neighbourhood of S. longisporum (or its clear-cut data available from the type collection). On the whole there is little evidence that the situation can be cleared by the methods presented. Rather I think that additional information, e.g. available from electron microscopy or cultural experiments is urgently needed to solve these problems.

Judging from Jülich (1975) S. brachysporum is known only from Africa (three collections, South Africa, Nigeria, Sierra Leone). We have collected five fungi in Venezuela (FO 13761, 13795, 14104, 14781, 16091) which correspond to fig. 30 and which show similar spore length and width. The shape of the spores however seems to be constantly different, i.e. sligthly curved. In general a difference like this one in spore form is used - at least in corticiology - to separate species. I think that this can be accepted without hesitation. Whether there exist additional "specific" characters cannot be decided as yet, because, for example the basal incrustation of the basidia is not known to be a constant feature.

I have studied a further South American specimen (fig. 31) which agrees in all respects with S. nikau, except the incrustation of the cystidia, which seems to be (at this magnification one should be careful in interpreting micro-details!) fairly irregular, at least in comparison with those structures of other Subulicystidium species. I am not yet prepared to give any decisive comment on the systematic rank of this fungus. Finally there remains a Subulicystidium (fig. 32), also collected in Venezuela (FO 12778) with spore characters that may indicate an additional clear-cut species: the spores are mostly navicular and vary within the range of 5 x 10-12 microns. I am convinced that this is sufficient for delimiting a species.

N e w   c o m b i n a t i o n s   a n d   n e w   s p e c i e s

Ceraceomyces evolvens (Fr.ex Fr.) Oberw. comb.nov. (basionym: Thelephora evolvens Fr.ex. Fr., Syst.mycol. 1, 1821)
Phlebiella insperata (Jacks.) Oberw. comb.nov. (basionym: Corticium insperatum Jacks. Canad.J.Res. C 28, 1950)
Phlebiella subflavido-grisea (Litsch.) Oberw. comb.nov. (basionym: Corticium subflavido-griseum Litsch. Ann.Myc. 39, 1941)
Phlebiella tulasnelloidea (v.Höhn.&Litsch.) Oberw. comb.nov. (basionym: Corticium tulasnelloideum v.Höhn.& Litsch. Sitzb. Akad.Wiss. Wien, Math.-nat.Kl. 117, 1908)
Subulicystidium meridense Oberw. spec.nov.
Fructificatio effusa, resupinata; hyphae distinctae hyalinae, nodosiseptatae; cystidia subulata, incrassate tunicata, incrustata; basidia cylindracea, suburniformia; basidiosporae ellipsoideae adusque suballantoideae, 2.5-3 x 6-7 microns, hyalinae, tenuitunicata, tunicis levibus, non amyloideis.
Hab. Ad ligna putrida. Venezuela.
Typus: FO 13761, Herb. F. Oberwinkler
Fig. 30

Subulicystidium naviculatum Oberw. spec. nov.
Fructificatio effusa, resupinata; hyphae distinctae, hyalinae, nodosi-septatae; cystidia subulata, tunicis undulatis, incrustata; basidia cylindracea, suburniformia; basidiosporae naviculatae, 4.5-5 x 10-12 microns, hyalinae, tenuituicatae, tunicis levibus, non amyloideis. Hab. Ad ligna putrida. Venezuela.
Typus: FO 12778, Herb. F. Oberwinkler
Fig. 32

Literatur.

Cunningham, G.H.: The Thelephoraceae of Australia and New Zealand.
N. 2. Dep.Sc. indust.Res.Bull. 145: 359 pp.
Donk, M.A. 1931: Revision der niederländischen Heterobasidiomycetae und Homobasidiomycetae-Aphyllophoraceae. Meded.Neder.mycol.Ver. 18-20
Eriksson, J. 1958: Studies in the Heterobasidiomycetes and Homobasidiomycetes-Aphyllophorales of Muddus National Park in North Sweden, Symb.Bot.Ups. 16: 1-172
Eriksson, J. & L. Ryvarden, 1973: The Corticiaceae of North Europe. Vol. 2
Eriksson, J. & L. Ryvarden, 1975: The Corticiaceae of North Europe. Vol. 3
Haversler, K. 1974: New or rare resupinate fungi. Friesia 10: 315-322
Jülich, W. 1969: Über die Gattungen Piloderma gen.nov. und Subulicystidium Parm. Ber.deutsch.Bot.Ges. 81: 414-421
Jülich, W. 1972: Monographie der Athelieae (Corticiaceae, Basidiomycetes), Willdenowia, Beih. 7
Jülich, W. 1974: The genera of Hyphodermoideae Corticiaceae). Persoonia 8: 59-97
Jülich, W. 1975: Studien an Cystiden-I. Subulicystidium Parm. Persoonia 8: 187-190
Karsten, P. 1890: Fragmenta mycologica XXXI. Hedwigia 29: 270-273
Oberwinkler, F. 1965: Primitive Basidiomyceten. Sydowia 19: 1-72
Parmasto, E. 1968: Conspectus Systematis Corticiacearum. Tartu
Talbot, P.H.B. & Green, 1958: Bothalia 7: 148-149
Warcup, J.H. & P.H.B. Talbot, 1962: Ecology and identity of mycelia isolated from soil. Trans.Brit.Myc.Soc. 45: 495-518

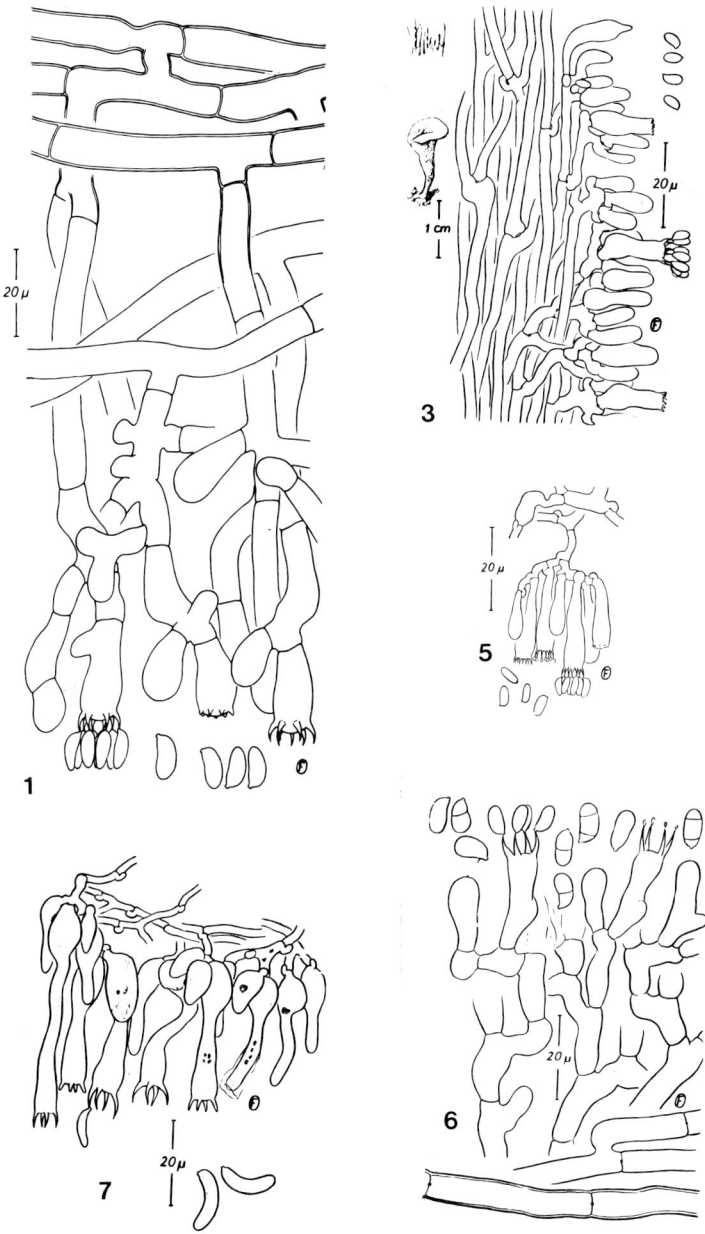

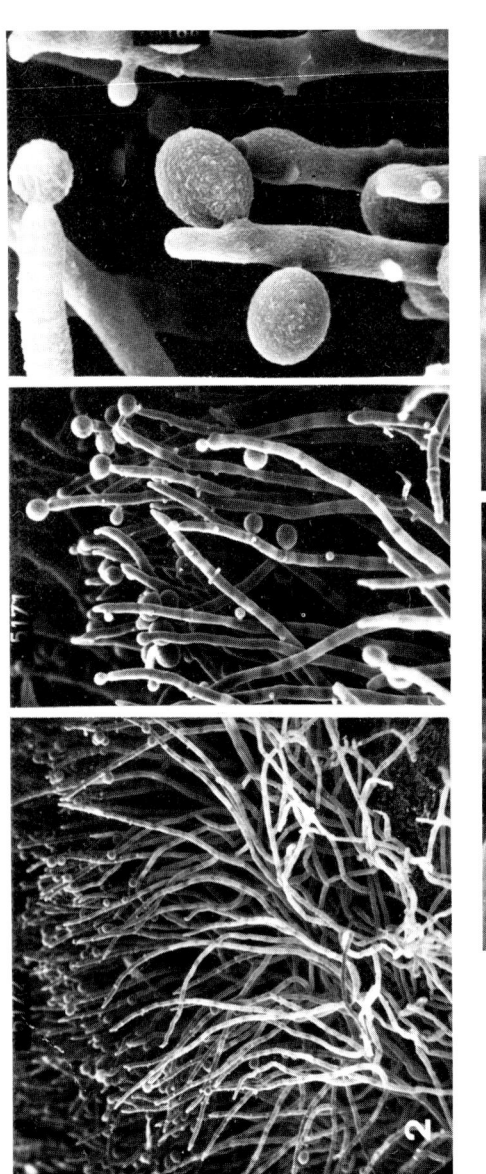
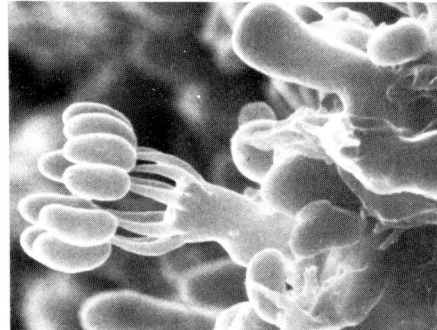
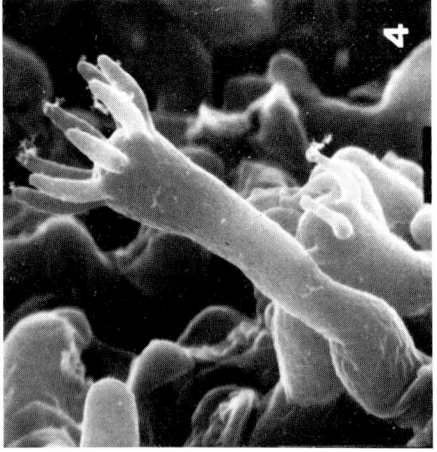

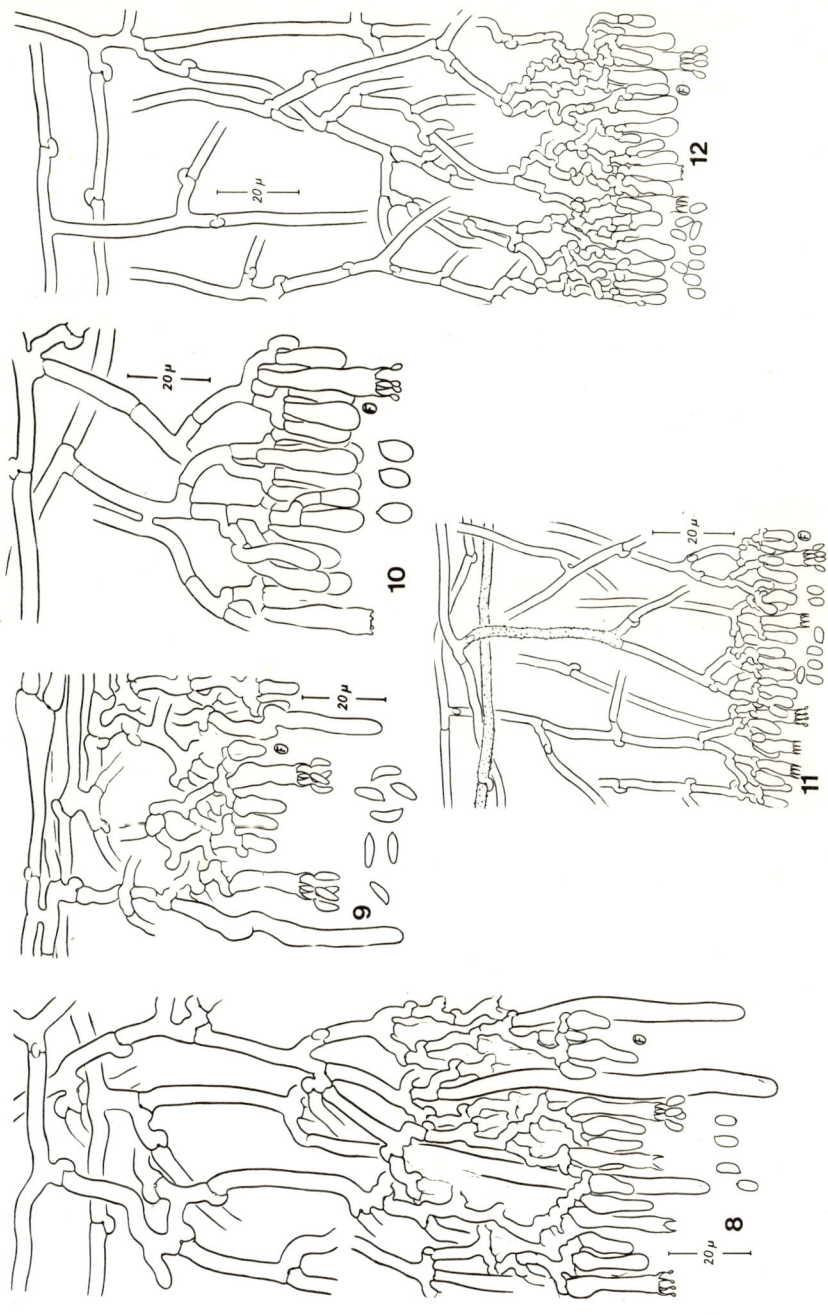

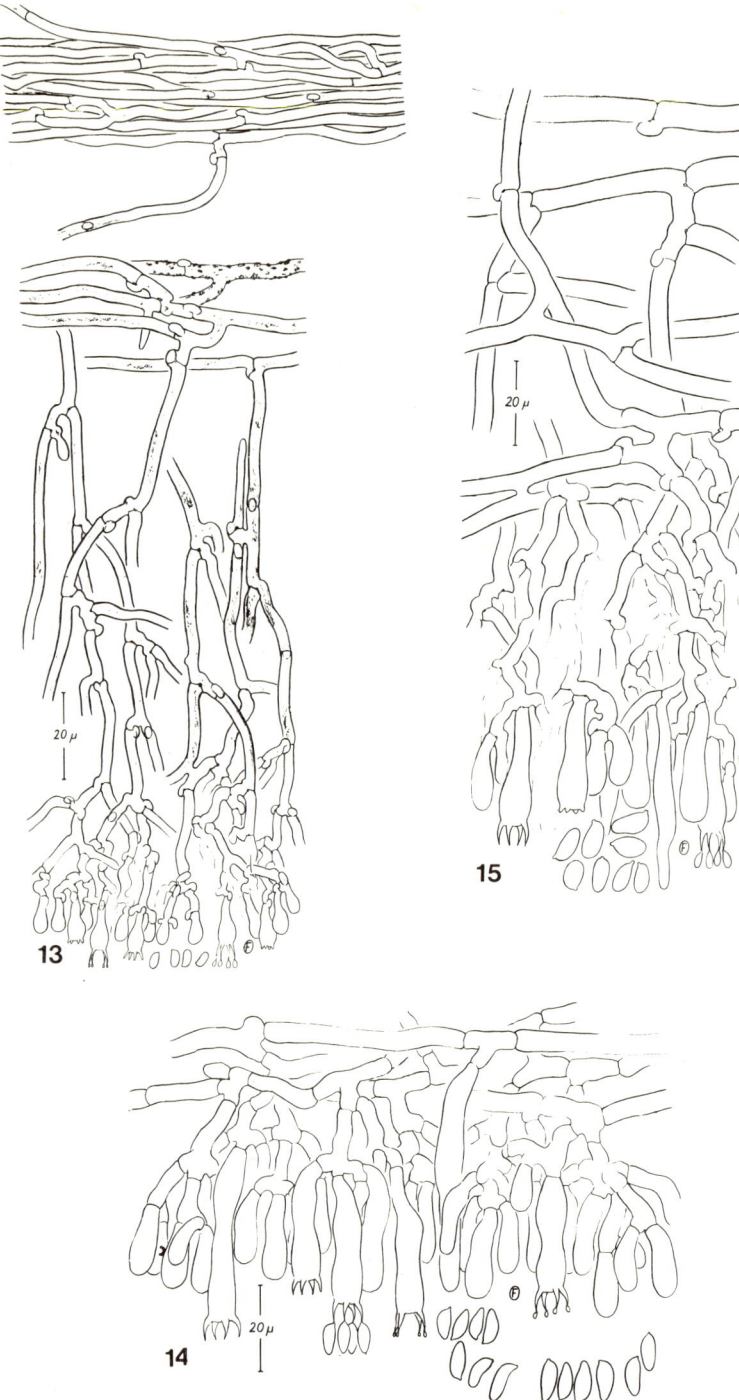

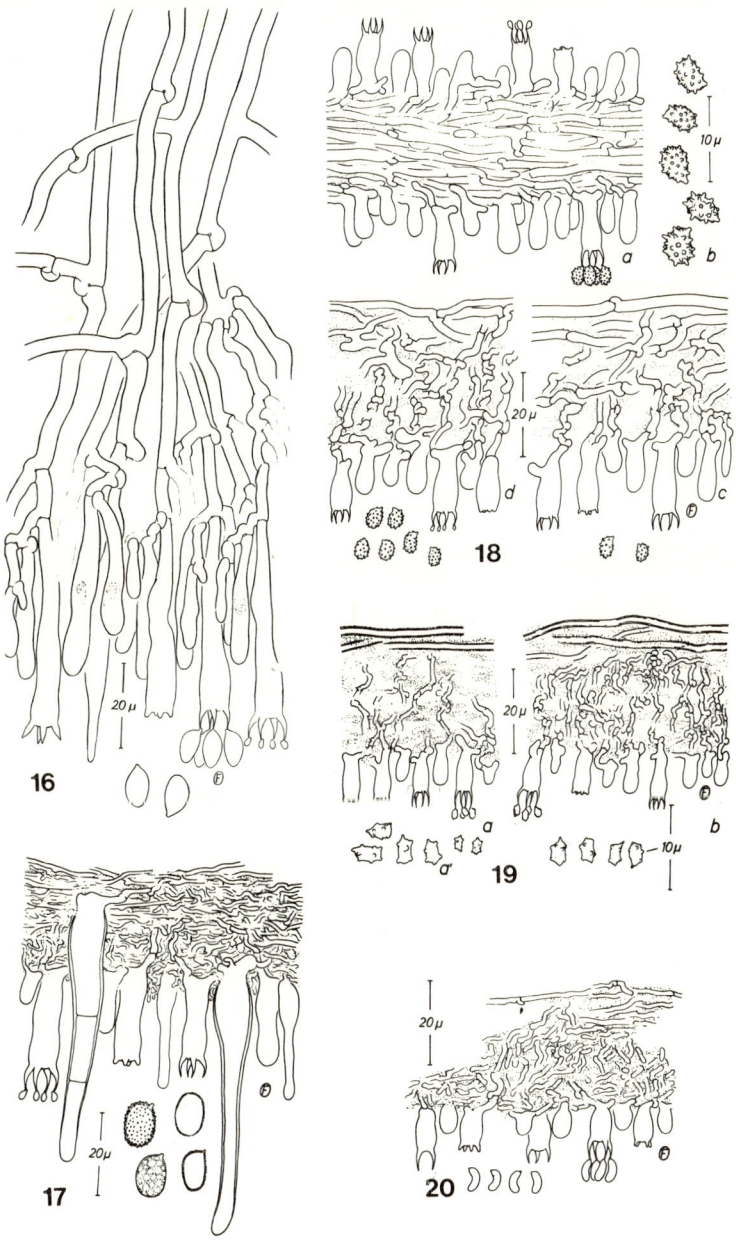

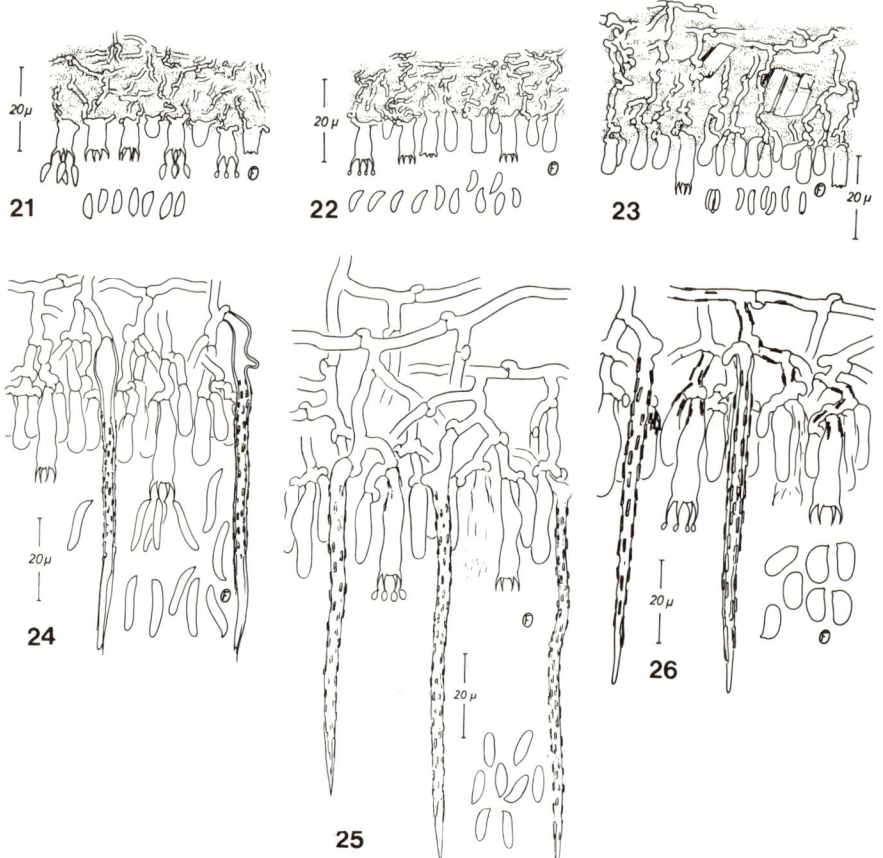

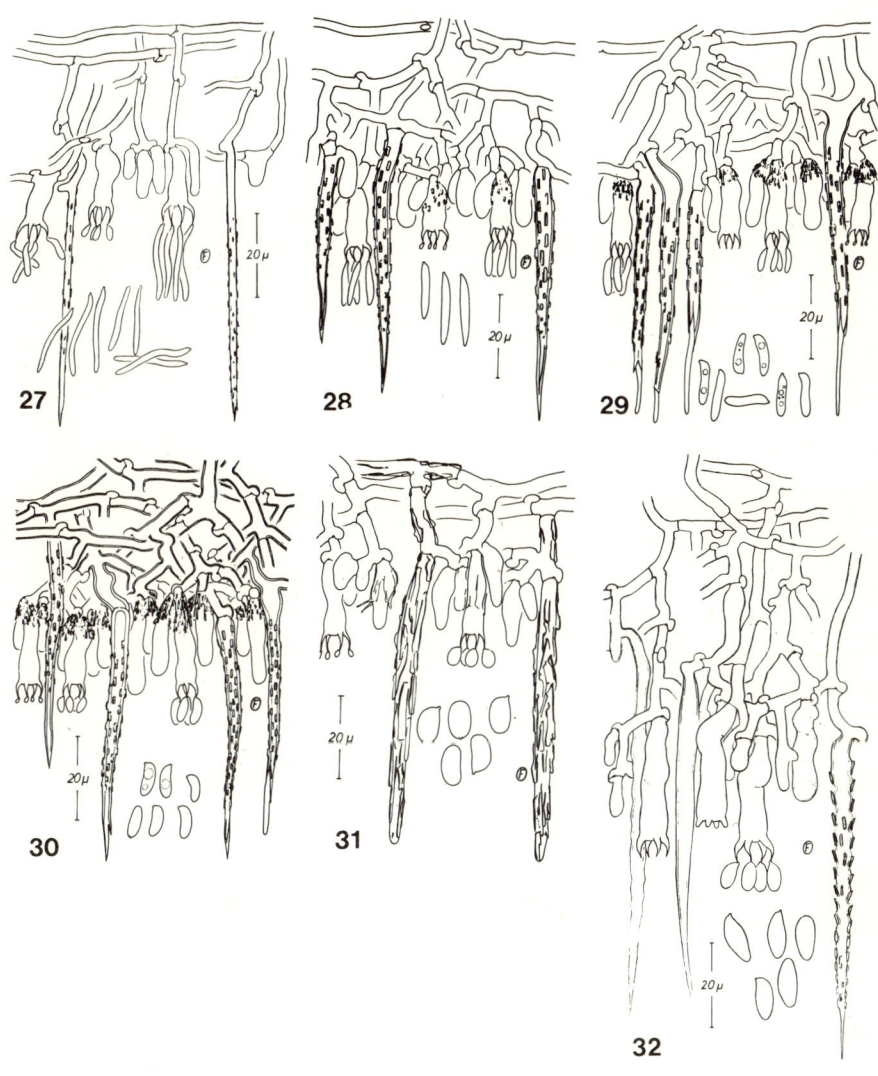

fig. 1: Botryobasidium obtusisporum J. Erikss.
fig. 2: Oidium conspersum (Link) Linder
fig. 3: Sistotrema confluens Pers. ex Fr.
fig. 4: Sistotrema brinkmannii (Bres.) J. Erikss.
fig. 5: Sistotremastrum suecicum Litsch. ex J. Erikss.
fig. 6: Waitea circinata Warcup & Talbot
fig. 7: Galzinia pedicellata Bourd.
fig. 8: Urnobasidium sernanderi (Litsch.) Parm.
fig. 9: Galziniella pereximia Parm.
fig. 10: Athelia epiphylla Pers.
fig. 11: Leptosporomyces galzinii (Bourd.) Jülich
fig. 12: Fibulomyces mutabilis (Bres.) Jülich
fig. 13: Confertobasidium olivaceoalbum (Bourd. & Galz.) Jülich
fig. 14: Athelidium aurantiacum (M.P.Christ.) Oberw.
fig. 15: Ceraceomyces tessulatus (Cooke) Jülich
fig. 16: Cylindrobasidium evolvens (Fr. ex Fr.) Jülich
fig. 17: Xenasma rimicolum (Karst.) Donk
fig. 18: Phlebiella vaga (Fr.) Karst.
fig. 19: Phlebiella sp.
fig. 20: Xenasmatella allantospora Oberw.
fig. 21: FO 5600
fig. 22: FO 12994
fig. 23: Corticium subcretaceum Litsch.
fig. 24: Subulicystidium longisporum (Pat.) Parm.
fig. 25: Subulicystidium brachysporum (Green & Talbot) Jülich
fig. 26: Subulicystidium nikau (G.H.Cunn.) Jülich
fig. 27: Subulicystidium sp. FO 17359
fig. 28: Subulicystidium sp. FO 13010
fig. 29: Subulicystidium sp. FO 15970
fig. 30: Subulicystidium meridense Oberw. FO 13761
fig. 31: Subulicystidium sp. FO 14338 a
fig. 32: Subulicystidium naviculatum Oberw. FO 12778

DISCUSSION FOLLOWING DR. OBERWINKLER'S PAPER

CLEMENCON: My experience is that Agaricologists and Aphyllophorologists coexist in mutual taxonomic ignorance. You confirmed this impression by showing that morphological taxonomists in Aphyllophorales are building up their own genus and species concepts which are not in agreement with what we have seen up to now. I understand your contribution as a kind of a warning to mycologists not to let it go this way. We should make an effort to work out a species concept that works at least for all homobasidiomycetes. We should not allow a gap to be developped between us. Am I right?

OBERWINKLER: It is really difficult, because the Corticiologists are going in an other direction.

BLAICH: There is one difficulty for all morphological taxonomy of which I was not aware yet. That is the value of the characters. I think it would be sufficient, e.g. for these conidial forms, if you could show in one or two cases by culturing whether this differences are correlated with a genetic separation of the species. And then I would be quite satisfied and admit that this might be generalized for the other types. Even if it is not true in all cases it would be sufficient.

OBERWINKLER: I absolutely agree with you.

BLAICH: You must not mix in the generic problem. It is an art to create higher taxa. But before you accumulate morphological data without knowing onto which level you are working you should decide by some experiments what morphological differences are characteristic for the species level. It is not necessary to do this for all species. Two examples would be sufficient. In Sistotrema there are some examples which you could use.

OBERWINKLER: Well, I am not sure.

BLAICH: I do not know either, but just one test, and the geneticist would be satisfied.

OBERWINKLER: I think this is a kind of generalization to which I cannot agree.

BLAICH: Because taxonomy has the concept to represent evolution I think, not just to put some species into drawers. I think that the unit of evolution is the species, and then we should know exactly this level on which evolution goes on.

OBERWINKLER: Of course, but I do not see any contradiction.

BLAICH: It might be that you are working on the genus level what you consider to be species, and the species problem lies still deeper, because there are minor differences which you are not aware yet.

OBERWINKLER: Might be, but I think that on the basis of comparable morphology that is not possible. I agree fully with you that we should make more experiments, but my impression is quite the contrary to that one.

HORAK: I am an outsider both in _Corticium_-like fungi and genetics. Could it be possible, e.g. to stimulate a further step in cell division in basidia to get a _Sistotrema_?

OBERWINKLER: This is a problem of cytology and caryology and I do not know whether there are some possibilities.

BRESINSKY: Just a little supplementum. I think if we have in taxonomy a worker who is to a high degree skillful to make this morphological analyses and to make these wonderful drawings, I think it would be a waste of time to put this man on any other type of work, experimental work or culture work. I think these analyses and these drawings will certainly survive all of us, but to some extent I would doubt whether a protein profile would survive.

SMITH: I think we left an important detail at of the whole discussion. The taxonomists work with organisms that are surviving and evolving in nature, and the species as he studies it is a product of this evolution. Now when you go into the laboratory, you change the conditions, and you have an artificial system. You are not really qualified to make taxonomic decisions with what you find until you have thoroughly investigated the whole genetics and chemistry of your organism and can come to the taxonomist with a clean cut analysis of this current section of a living continuum that a hundred years from now might be different from the one that you analyse now. Our species are naturally occurring populations in which change is continuous through each generation. Evolution in most instances is the process of steady change.

ESSER: I bet you can't prove it!

SMITH: You just look around you!

OBERWINKLER: I think we again reached a point where mutual understanding becomes very difficult. I would like to remind you that you can study material that is now 200 years old, and you go out and collect material in good condition and can see that the characters are, at least in a morphological-analytical sense, very much the same.

SMITH: In very short periods of time there are likely to be very minor mutations that would not show up. The taxonomic aspect of it, your morphology and anatomy, will be the last to change.

SINGER: It is always useful, at this level, to go out and find other characters. I think this is very important. You want more characters. Because this is the only way for you as a taxonomist to decide some of your problems. Now, if the geneticists can give you an additional class of criteria, I think you should be content and accept that. So why should mutual understanding become difficult?

BOIDIN: J'ai trois points à soulever. Premièrement, j'ai, comme vous, remarqué une très grande diversité de la spore chez les <u>Subulicystidium</u> tropicaux; malheureusement leur culture ne semble pas possible à l'heure actuelle; je vous enverrai volontiers tout le matériel dont je dispose avec pour chaque récolte une sporée faite le jour même. Le deuxième point touche à la notion de genre; je suis bien convaincu, et vous l'êtes certainement, que le genre <u>Corticium</u> était extrêmement hétérogène, mais en exagérant à peine, je crains que dans quelques jours il y ait un genre par espèce; alors nous aurons supprimé la notion de genre, et tout se résumera à notre souci actuel, l'espèce! Je crois que le seul intérêt de la multiplication des genres est (à coté de la satisfaction de voir son nom immortalisé, mais qui osera l'avouer?) d'obliger les autres mycologues à des observations plus fines des détails microscopiques et des structures; je ne sais si c'est le but recherché. Troisièmement pourquoi, dans votre tableau, deuxième colonne figure Hyphodontia avec sa baside un peu étranglée, et pas Hyphoderma qui a aussi la baside étranglée? Je rappelle que, à ce jour, tous les Hyphodontia se sont montrés tétrapolaires, et tous les Hyphoderma bipolaires (avec quelques souches homothalles). C'est très net, bien que ce caractère ne soit évidemment pas morphologique. Un genre est bipolaire, l'autre est tétrapolaire, sans exception. Qu'est-ce qui définit, au fond, la baside urniforme selon les morphologistes? La baside urniforme la plus typique, celle des Sistotrema, est stichobasidiée. Il en est de même chez Sistotremastrum niveo-cremeum. Est-ce que l'on connaît la position des fuseaux chez Galzinia? Je pense qu'il serait important de le savoir, par ce que la disposition stichobasidiée étant rarissime, elle a sans doute une grande valeur.

Herbette Symposium on Species Concept in Hymenomycetes 1976

INCIDENCE DES CARACTERES NON MORPHOLOGIQUES SUR LA NOTION
D'ESPECE ET AUTRES TAXA CHEZ LES MACROMYCETES

H. Romagnesi

Il est vrai, comme on le dit souvent, que le concept de l'espèce varie beaucoup d'un mycologue à l'autre, mais il dépend sans doute moins de l'individu que de l'époque considérée. En effet, contrairement à ce qui s'est passé en Phanérogamie par exemple, où les caractères accessibles à l'oeil nu ou tout au plus renforcé par la loupe, se sont, en gros, montrés suffisants pour définir les espèces et établir une classification naturelle, il n'en a pas été de même pour la Mycologie, dont l'histoire, beaucoup plus complexe, est celle de l'utilisation de moyens de plus en plus perfectionnés et sophistiqués.

Dans la période ancienne, où le mycologue ne se servait guère, outre ses sens, que de la loupe, il était inévitable que l'espèce fût conçue très largement: sous un même nom, on groupait des formes de physionomie en apparence identique, ou ne différant entre elles que des caractères légitimement considérés comme négligeables, ou tout au plus variétaux.

Quand l'usage de plus en plus généralisé du microscope optique (d'abord pour la spore, ensuite pour les éléments hyméniens, enfin pour les revêtements et les tissus) eut permis d'accéder à la structure plus profonde des espèces fongiques, on s'est aperçu que, sous cette apparence d'identité, ou derrière des variations infimes, se cachaient des différences assez importantes pour justifier sans conteste une distinction spécifique.

Malgré les immenses progrès que ces nouvelles méthodes d'investigation microscopique (et nous y comprenons la microscopie électronique) ont fait réaliser à notre connaissance de l'univers fongique, il semble bien que nous nous trouvons aujourd'hui dans une situation analogue à celle qu'ont connue nos prédécesseurs d'il y a cent cinquante ans. En effet, on commence depuis peu à soumettre les espèces définies uniquement à l'aide de leurs caractères morphologiques (macro et microscopiques) à des tests d'ordre chimique, écologique ou biologique, qui mettent souvent en lumière, à l'intérieur d'un ensemble de formes morphologiquement identiques ou peu différentes, des variations comparables à celles qu'autrefois le microscope avait révélées. D'ailleurs, avant même l'intervention de ces méthodes nouvelles, beaucoup de Systématiciens s'étaient rendu compte que les procédés traditionnels de leur science aboutissaient parfois à des impasses, à des échecs dans la définition des espèces; c'est encore aujourd'hui le cas dans certains genres très compliqués, où se pose perpétuellement le problème de savoir, si, oui ou non, de légères différences morphologiques entre deux lots de carpophores suffisent pour justifier leur isolement spécifique; on s'est trouvé ainsi confronté à des situations inextricables, où le concept d'espèce prenait un aspect tout à fait subjectif, fondé sur des intuitions et non sur des faits objectifs, donc très flou et indéfiniment contestable.

Sur le plan spécifique, certains résultats ont été obtenus d'abord par des caractères ressortissant à la chimie: l'odorat et le goût sont les seuls de nos sens à nous fournir quelques indications sur eux, et on sait combien ils sont imparfaits, rudimentaires même. Mais ç'a été tout d'abord d'une manière très grossière et empirique, au moyen de simples réactions colorées provoquées par le contact des revêtements et de la chair avec certains corps tels que les bases fortes, l'ammoniaque, les acides, les sels de fer, le phénol, ou les réactifs des phénoloxydases; mais, sauf dans ce dernier cas, la signification de ces réactions restait tout à fait inconnue. Grâce à elles cependant, nombre d'espèces de Russules et de Cortinaires par exemples ont été clairement définies, et, dans ces genres en particulier, le concept de l'espèce s'en est trouvé à la fois rétréci et renforcé. Depuis, de grands progrès ont été réalisés grâce aux divers procédés de l'analyse chimique, recherche directe de certains corps, spectroscopie, et surtout chromatographie: une science nouvelle est en train de prendre une grande importance, la chimiotaxinomie, dont les premiers résultats sont très prometteurs, surtout dans la recherche des pigments, par exemple les différents carotènes chez les Discomycètes, divers corps colorés chez les Cortinaires du groupe Dermocybe; chez les Russules, il semble bien que les espèces puissent parfois n'être caractérisées que par leur combinaison pigmentaire, et chez les Inocybes, la présence de muscarine et la nature des acides aminés libres contribueront peut-être à débrouiller ce genre inextricable.

L'écologie elle-même vient de commencer à apporter sa contribution - quoique encore très modeste, et ayant un caractère d'appoint, d'adjuvant, par rapport aux autres critères. On sait que chez les Micromycètes, il existe des parasites que rien ne distingue dans la morphologie, mais qui n'attaquent chacun qu'un hôte déterminé, incapables qu'ils sont de vivre sur un autre. Chez les Macromycètes, où la notion d'écologie est loin d'avoir la même rigueur, vu leur ubiquité beaucoup plus grande, on peut cependant citer quelques cas, comme celui de <u>Phellinus Hartigi</u>, propre aux conifères, mais difficile à distinguer macro et microscopiquement des espèces des feuillus. Pourtant, les particularités des mycéliums en culture ont démontré qu'il s'agissait bien d'une espèce distincte; il s'agit certes là d'un caractère physiologique, mais qui se traduit dans la morphologie du mycélium, et peut donc être regardé comme morphologique.

En matière de Biologie, ce sont les caractères cytologiques qui, dès le début de ce siècle, ont été les premiers utilisés, mais surtout sur le plan de la taxinomie et non de la spécification. C'est ainsi que la définition de la famille des Cantharellacées s'est trouvée précisée et enrichie par l'orientation du fuseau de la division des noyaux dans les basides jeunes. Aujourd'hui, leur nombre dans les spores, et surtout le comportement nucléaire des mycéliums en culture pure prennent une place très importante dans la définition de taxa de rang divers, y compris celle de certaines espèces, par exemple chez les Polyporellus.

Ainsi, l'apport de la chimie, de l'écologie, de la cytologie sont en train d'enrichir l'arsenal des critères spécificateurs que le systématicien moderne a à sa disposition. Cependant, la solution du problème de l'espèce ne s'en est trouvé rapprochée que dans une certaine mesure, car comment savoir si les nouvelles différences découvertes constituent des caractères vraiment spécifiques, et non variétaux ou même individuels? Prenons l'exemple de l'Homme: les quatre groupes sanguins du début du siècle étaient devenus en 1950 29'252, et aujourd'hui, on connait 2'717'245'240 phénotypes; l'intervention d'autres critères chimiques doublerait ce nombre, si bien qu'on en arrive à penser que chaque individu possède, au point de vue chimique, un sang qui lui est propre, comme les empreintes digitales. On comprend l'imprudence qu'il y aurait à attribuer trop d'importance à la découverte d'une originalité de cet ordre chez tel ou tel groupe de carpophores fongiques. De même, pour l'écologie, on a constaté par exemple chez des espèces qui ne se rencontrent que sur la terre brûlée, que leurs spores germaient très bien sur des milieux de culture artificielle qui ne comportaient pas la moindre trace de carbone à l'état libre, et, bien mieux, que le <u>Pleurotus Eryngii</u> (ainsi que ses variétés) qui ne croît dans la nature que sur les souches d'Ombellifères, produisait des carpophores normaux en laboratoire sur un compost d'une tout autre composition.

C'est pourquoi, c'est dans un autre domaine de la Biologie, celui de la sexualité, qu'on peut espérer trouver un moyen plus objectif de définir l'espèce. Malheureusement, son champ d'application est encore très limité: il s'agit en effet du critère d'interfertilité entre haplontes obtenus en culture pure (et, accessoirement, de la faculté entre mycéliums diploïdes d'une même espèce de s'anastomoser quand ils proviennent d'individus différents; mais en pratique, ce critère est d'application difficile). Or, on ne sait pas encore, malheureusement, faire germer les spores de la majorité des Macromycètes, surtout en cultures monospermes.

Ce critère semble pour le moment avoir une valeur plus grande chez les champignons dits supérieurs que dans les autres groupes végétaux: on n'a jamais réussi à hybrider des mycéliums d'espèce différentes, même les plus voisines. Certes, on peut faire de grandes réserves: il est a priori paradoxal que les gros champignons soient les seuls êtres vivants où il n'y ait pas d'hybrides interspécifiques; de plus en comparaisons de ce qui se passe à tout moment dans l'immensité de la nature, le nombre de nos expériences peut légitimement paraître dérisoire, donc peu concluant. Cependant, c'est un fait d'expérience, que les haplontes d'espèces extrêmement affines, ou même obtenue à partir de carpophores où aucune différence morphologique (macro ou microscopique) ne peut être relevée, refusent parfois obstinément de copuler, alors qu'ils le font facilement dans d'autre cas. Nous ne parlons ici que d'individus hétérozygotes, car, pour des individus issus à l'origine d'un même mycélium secondaire, il arrive que les haplontes perdent la faculté copulatoire lorqu'ils sont séparés par un certain intervalle entre les générations successives, et cela tout le long de la lignée, sans qu'on puisse y marquer une limite ponctuelle précise.

Il n'empêche que l'on ne saurait négliger complètement la possibilité
d'hybridation interspécifique chez les Macromycètes. Mais il est certain qu'une interfertilité constitue une très grande probabilité
d'indépendance spécifique. Quant à l'interstérilité, il faut tenir
compte de l'existence possible de quelque facteur empêchant la copulation entre individus de la même espèce; dans ce cas, il peut très
bien se passer quelque chose au niveau des noyaux, sans qu'il y ait
forcément répercussion sur les caractères actuellement accessibles du
phénotype, de quelque ordre qu'ils soient.

C'est pourquoi, à l'heure actuelle et dans l'état de nos connaissances, on ne saurait se fier à ces seuls critères d'interfertilité et
d'interstérilité pour décider si deux groupes de carpophores sont ou
non "conspécifiques". Mais on ne peut nier qu'il s'agit d'un caractère extrêmement important, qui marque sans aucun doute une limite
nette dans la hiérarchie des taxa, une limite très fine et située
assez bas. D'ailleurs, en pratique, il a déjà donné de remarquables
résultats.

Nous ne citons que les travaux bien connus de Quintanilha, de Morten
Lange, de Yen, et surtout de l'école lyonnaise avec Kühner, J.Boidin,
Mlle Lamoure, Mme David, Mme Galland, Berthier, etc... Ils ont permis
de délimiter correctement les espèces de certains groupes, Coprins
Micacei, Drosophila Q.(Psathyrella ss. lato), Polyporellus Clitocybe
et Omphalia, diverses Théléphoracées, notamment les Vararia
(Asterostromella B.-G.)etc...

Ayant nous même assuré la collection et l'étude botanique du matériel
sur lequel ont travaillé d'abord le professeur Quintanilha, puis récemment, Mme Galland (Contribution à l'étude du genre Psathyrella
Fr.) Q. (Agaricales), Thèse no 170 de Sc. Nat., 1973), nous avons
toutefois constaté que de sérieuses difficultés existaient dans le
maniement de ces critères. Certes, dans certains cas, les interfertilités ont remarquablement confirmé les distinctions spécifiques que
nous avions faites à partir des caractères morphologiques. Mais dans
d'autres, il n'y avait pas coïncidence avec nos propres conclusions:
tantôt, nous avions soupçonné certains lots interfertiles, en apparence assez différents botaniquement parlant, d'être de bonnes espèces, tantôt nous en avions assimilé d'autres, faute de différences
palpables entre eux, et dont les haplontes refusaient obstinément de

copuler; pourtant toutes les précautions avaient été prises, notamment les quatres groupes sexuels de ces champignons tétrapolaires ayant été isolés, et le nombre d'expérience répété.

Mais une remarque s'impose: dans le cas d'interfertilités imprévues, jamais il ne s'agissait d'espèces unanimement acceptées par les auteurs antérieurs (à une exception près), mais distinguées provisoirement par nous-même, et les différences morphologiques relevées étaient relativement faibles: l'exception citée concerne _marcescibilis_ et _lactea_, distinguées par Lange, mais que nos propres observations nous avaient conduit à assimiler spécifiquement, ce qu'a confirmé leur interstérilité. En revanche, même après avoir été informé de l'interfertilité de certains lots, nous n'avons pu découvrir entre eux de différences morphologiques importantes.

Ces difficultés ne sauraient nous dispenser de proposer une solution, toute provisoire qu'elle puisse être, au problème de la spécification. A l'heure actuelle, et tant que la possibilité d'hybridation entre espèces unanimement reconnues comme telles n'aura pas été démontrée, on peut admettre l'interfertilité comme un critère suffisant de "conspécificité". Si des différences, en premier lieu d'ordre morphologique, mais aussi chimique, écologique ou biologique, sont reconnues comme nettes et constantes entre individus interfertiles, on pourra provisoirement les attribuer à des taxa de rang inférieur à l'espèce. On rencontre ici les notions de forme, variété et sous-espèce; elles ne reposaient à ce jour que sur des appréciations toutes subjectives, mais il est sans doute temps qu'elles commencent à prendre un aspect plus objectif chaque fois que ce sera matériellement possible.

Comme nous l'avons proposé dans notre communication au VI° Congrès européen de Mycologie (Avignon, octobre 1974), publiée dans le Bulletin de la Société mycologique de France 91, 4: 108, 1975, nous recommandons de considérer comme des _formes_ un ensemble d'individus se séparant du type par un ou plusieurs caractères de tous ordres dont le déterminisme réside dans l'action de facteurs externes, donc de l'environnement; si le mycélium qui les a produits cesse d'être soumis à ces facteurs, les différences disparaissent dans les carpophores.

Dans les _variétés_ au contraire, les différences en questions devront avoir un support héréditaire, le plus souvent de nature mendélienne très probablement; la notion de variété rejoint ici celle de race en Zoologie par exemple.

Reste la <u>sous-espèce</u>: en Mycologie, elle pose des problèmes, car, historiquement, les mycologues ont usé de ce terme pour désigner des espèces considérées comme extrêmement voisines, de la valeur desquelles ils doutaient quelque peu: Konrad et Maublanc, Bourdot et Galzin, ont ainsi distingué nombre de sous-espèces, qui ont été ultérieurement soit restituées au rang d'espèces distinctes, soit au contraire réduites au rang de forme ou de variété, quand elles n'ont pas été purement et simplement synonymisées. C'est pourquoi il semble raisonnable de conserver au concept de sous-espèce le sens qu'il a dans les autres branches de l'Histoire Naturelle, c'est-à-dire de le réserver aux cas où une ou plusieurs différences constantes coïncident avec une aire géographique déterminée. D'ailleurs, on ne connaît pas d'exemples certains de sous-espèce en mycologie: ceux auxquels on songe, par exemple ceux des <u>Amanita caesarea</u> et <u>muscaria</u> européennes et américaines n'ont pu être soumis au critère d'interfertilité, et il s'agit peut-être d'espèces distinctes. De plus, bien que théoriquement, la sous-espèce dût être l'échelon supérieur des taxa infraspécifiques, il se peut qu'elle se rapproche bien davantage de la variété ou même de la forme, car rien ne prouve qu'un mycélium transplanté d'une aire à une autre (chose impossible à réaliser pour le moment), les carpophores produits ne perdraient pas leurs particularités respectives.

Il est vrai, malheureusement, que ces considérations ne sont valables que dans les cas, encore rares, où il est possible d'obtenir à partir des spores, des cultures monospermes de mycélium. Pour les autres, l'immense majorité, nos définitions demeurent toutes théoriques. Mais comme nous l'écrivions (Bulletin de la Société mycologique de France 91, 3: 140, 1975): "On peut réussir, grâce à la pratique de ce test (l'interfertilité) à perfectionner son jugement en matière de spécification dans un sens conforme à la réalité biologique. On peut aussi espérer en tirer profit même dans les groupes où l'on n'arrive pas à obtenir de cultures monospermes en laboratoire. L'expérience acquise dans l'un peut être précieuse pur l'étude d'un autre."

Mais les problèmes les plus épineux sont posés par le test d'interstérilité: devons-nous en déduire que tout haplonte qui - toutes précautions prises - refuse de copuler avec un autre appartient nécessairement à une espèce différente? Il serait téméraire de le croire.

Dès avant la seconde guerre mondiale, des biologistes américains avaient constaté des cas d'interstérilité entre espèces que les systématiciens avaient cru identiques, aucune différence n'ayant été relevée dans la macro ou la microscopie; d'autres fois, les systématiciens avaient établi des distinctions spécifiques ou même génériques chez certains Corticiés à l'aide de tels ou tels caractères, et de leur côté, les biologistes avaient été conduits à reconnaître l'hétérogénéité du groupe; malheureusement, les distinctions des uns et des autres ne coïncidaient pas du tout. Pourtant, le fait que les premiers avaient pressenti qu'ils avaient affaire à un ensemble multispécifique se trouvait d'une certaine manière confirmée par la biologie, et demeurait des plus troublants. N'était-ce pas que les caractères vraiment spécifiques étaient tellement légers et si peu frappants qu'ils avaient échappé aux systématiciens, ou qu'ils les avaient sous-estimés?

Certains faits ont corroboré quelque peu cette hypothèse. Parmi les travaux récents, ceux de J. Boidin et de D. Lamoure sont particulièrement intéressants. Par exemple, il s'est avéré chez deux espèces de Peniophora, les individus interstériles se distinguaient uniquement par la constance ou la dispersion des boucles aux cloisons des hyphes, et par l'habitat, dans un cas sur Abies, dans l'autre sur Picea. Dans les Omphalia du groupe pyxidata, la seule différence morphologique constatée était une silhouette légèrement plus angulée du sommet des spores (vues en masse, si non le caractère était imperceptible), et, là encore, les deux espèces ne se rencontraient pas dans le même biotope. Quel systématicien aurait eu l'audace de fonder sur de telles bases des distinctions spécifiques?

Dans un autre ordre d'idée, il a été vérifié dans plusieurs genres que les formes bisporiques et tétrasporiques, hétérothalles ou amphithalles, étaient régulièrement interstériles, alors que rien ne les distinguait botaniquement. On ne peut dire en effet que la bisporicité soit un caractère morphologique comme les autres: c'est un caractère biologique particulier seulement parce qu'il est décelable par la simple observation au microscope optique et n'exige même pas de coloration spéciale. C'est pourquoi nous avons proposé (loc. cit.) d'appliquer à ces formes le nouveau nom de pseudovariété: c'est en somme l'inverse de la variété telle que nous la concevons, puisque, identique morphologiquement, elle est interstérile avec le type, alors que c'est le contraire pour la variété.

Chez les Polyporellus du groupe brumalis, Mme David a montré que l'interstérilité était liée à des différences dans le comportement nucléaire des mycéliums, ainsi d'ailleurs qu'à de très nettes différences morphologiques, dimensions des pores et des spores, qui varient parallèlement.

Tous ces faits démontrent que, lorsqu'on réussit à établir une relation entre l'interstérilité et un autre caractère, celui-ci n'est pas forcément d'ordre morphologique, mais qu'il peut être d'ordre biologique ou écologique; de plus, quand il y a différence morphologique, elle peut être en apparence tout à fait insignifiante.

On n'a pas encore à l'heure actuelle d'exemple où l'interstérilité serait uniquement liée à un caractère chimique. Mais il est plus que vraisemblable que ce sera réalisé dans un avenir plus ou moins proche.

Il reste cependant le cas résiduel où aucune différence, de quelque ordre que ce soit, n'a pu encore être décelée entre des lots de carpophores interstériles. J. Berthier et d'autres avant lui l'ont montré (Typhula, etc...) et nous-même l'avons constaté chez les Drosophiles du groupe Candolleana. Etant donné la remarquable constance du phénomène pour des groupes donnés, comme l'ont prouvé les travaux de Mme Galland, on peut certainement penser que l'on n'a pas encore découvert le ou les caractères liés à ce refus de copulation. Comme nous l'avons fait observer à plusieurs reprises, saurions-nous distinguer beaucoup d'espèces de Cortinaires Scauri si notre oeil, comme celui de certains animaux peu évolués, était incapable de percevoir les couleurs? Peut-être nous manque-t-il un sixième sens pour découvrir le caractère déterminant pour la spécification. Cette hypothèse nous paraît gratuite: nous admettons qu'au sein d'une même espèce, pour une cause ou une autre, certains individus produisent des haplontes qui n'ont pas la possibilité de copuler avec certains autres. Bien qu'il s'agisse probablement de phénomènes d'un ordre différent, le fait a été constaté chez les Phanérogames (Chrysanthemum leucanthemum).

Nous en conclurons que, pour le moment, le critère d'interstérilité a une valeur moindre que celui d'interfertilité. On ne le regardera que comme une marque de différence spécifique que si, parallèlement, il s'accompagne d'une ou plusieurs différences principalement morphologiques (à notre avis, la morphologie doit avoir ici un rôle dominant), et accessoirement d'un ordre chimique, biologique ou écologique. Dans le cas de complète similitude morphologique mais de différences d'ordre biologique comme la bi- ou tétrasporicité ou la thallie, on aura affaire à des pseudovariétés. Pour le moment, aucune expérience ne nous permet malheureusement de savoir si des différences d'ordre chimique et écologique justifieraient __à elles seules__ un isolement spécifique. Dans le premier cas, on peut prévoir avec une forte dose de vraisemblance une réponse positive. Dans le second, l'écologie, il est possible que nous ayons à modifier la conception que nous avons exposée ci-dessus de la sous-espèce, que, comme nous l'avons dit, nous postulons interfertile avec le type, comme c'est souvent le cas pour les Phanérogames, mais sans aucune preuve expérimentale. Si notre opinion se trouvait un jour controuvée, il faudrait renoncer à cette conception, et la sous-espèce se définirait alors par une interstérilité jointe à une différence d'ordre chorologique. Mais présentement, cela nous paraît prématuré.

Soulignons bien pour terminer le rôle prépondérant que nous laissons encore à la morphologie; pour le moment, les autres disciplines ne peuvent avoir qu'un rôle d'adjuvant; mais il n'est pas exclu que l'avenir leur accorde une importance grandissante au point d'égaler peut-être un jour celle de la morphologie.

Toutefois, il faut se garder de toute illusion: non seulement les critères que nous proposons sont à ce jour impraticables à la grande majorité des espèces fongiques, car il s'écoulera encore de longues années avant qu'on réussisse à faire germer artificiellement les spores de la plupart des champignons supérieurs, mais encore, même dans cette situation idéale, jamais l'on ne réussira à dégager la notion d'espèce de toute trace de subjectivité. La nature se moque de nos classifications et de nos concepts qu'elle ignore avec mépris: celui du genre, sans aucun doute, le plus arbitraire de tous: certes, il en est d'excellents, et qui ont une existence réelle, comme les __Melano-__

leuca, les Pluteus, les genres monospécifiques aberrants comme les Rhodotus; mais la prolifération absolument délirante de "genera nova" inutiles qui pullulent par exemple chez les Polypores et autres Aphyllophorales dans les publications modernes, et chez les Pyrénomycètes chez les anciennes, le démontre très fâcheusement. Or, la notion d'espèce ne peut, elle non plus, échapper totalement à l'arbitraire et à la convention. On ne peut espérer mieux que de lui apporter un peu plus de précision en usant des méthodes de plus en plus variées, en faisant une soigneuse critique de leurs résultats, et en essayant de s'accommoder au moins mal possible des contradictions auxquelles elles conduisent trop souvent à notre gré de cartésiens. Le problème de l'espèce, pourtant l'un des plus fondamentaux de la Biologie, ne sera sans doute jamais complètement résolu.

DISCUSSION FOLLOWING DR. ROMAGNESI'S PAPER

SINGER: Pour la pseudovariété je voudrais savoir si Mr. Romagnesi la considère comme terme théorique, où s'il propose un nouveau taxon dans la hiérarchie de la nomenclature?

ROMAGNESI: Oui, certainement. Mr. Boidin nous a parlé d'une subspecies amphithallica pour un Vararia. Mais il nous n'admettons pas en ce sens le terme sous-espèce, mais nous préférons un terme spécial.

SINGER: Il sera nécessaire de le proposer.

ROMAGNESI: Oui, parfaitement.

CLEMENCON: Est-ce que vous voyez cette nouvelle catégorie au même rang que la variété, sorte de ramification, ou bien voyez-vous un échalonnage?

ROMAGNESI: Je verrais plutôt un échalonnage. La pseudovariété n'est même pas une variété. C'est quelque chose d'inférieur.

CLEMENCON: Et pourquoi ce n'est pas une forme?

ROMAGNESI: Parce qu'elle n'est pas soumise à des facteurs externes.

CLEMENCON: Je vois, vous insérez la pseudovariété entre variété et forme.

WATLING: I think it comes out of this discussion that there is a necessity for another unit at this sort of level. I think that a pseudovariety may be what I tried to express as a microspecies.

ROMAGNESI: Je ne crois pas. Je crois que la micro-espèce est plus large que la pseudovariété. La pseudovariété n'est même pas une microespèce, elle n'a pas d'autre fondement que biologique.

SMITH: Is it what we call a form?

CLEMENCON: No, a form is influenced by environmental factors and it is not genetically fixed, whereas a pseudovariety is the level just above it, already genetically fixed.

SMITH: Why not take a broader view of the term variety? I find the term pseudovariety a little bit of a contradiction, it is opposed to variety.

KUEHNER: Je pense qu'il est difficile de comparer les "pseudovariétés" de Mr. Romagnesi aux "microspecies" de Mr. Watling; la méthode expérimentale permet seule de reconnaître si l'on ou non affaire à une pseudovariété; il n'est pas forcément nécessaire de l'utiliser pour délimiter les "microspecies".

BOIDIN: Une petite remarque, sur le problème sous-espèce, pseudovariété. Si j'ai bien compris, votre sous-espèce est supposée interfertile, et elle est uniquement créée par une séparation géographique, liée à quelques petites différences morphologiques; la pseudovariété, telle que vous la concevez, est interstérile et il y a donc une séparation génétique qui s'exprime plus ou moins par des différences morphologiques souvent légères. Aussi si l'on doit placer votre "pseudovariété" dans la hiérarchie des taxons infraspécifiques, elle ne peut être en dessous de la sous-espèce, mais au dessus car la séparation génétique est totale et sans doute définitive, alors qu'entre vos sous-espèces il n'y a qu'une séparation géographique qui peut ne pas être définitive.

ROMAGNESI: Je suis tout-à-fait d'accord avec vous. J'ai dit moi-même, vous vous en souvenez, que les sous-espèces sont supérieures, mais au fond, vous avez raison sur ce point. Je ne me suis pas posé la question s'il fallait la mettre en-dessus ou en-dessous, du point de vue morphologique il est certain que c'est en-dessous de la forme.

SMITH: I think it is interesting to see that in most of the discussion in this symposium we do agree that we need infra-species categories.

Herbette Symposium on Species Concept in Hymenomycetes 1976

SPECIES CONCEPT IN HIGHER BASIDIOMYCETES:
TAXONOMY, BIOLOGY AND NOMENCLATURE

Ronald H. Petersen

The word species, as applied to higher fungi, can, it seems to me, include at least three separate connotations. They are as follows: a) species as a taxonomic unit; b) species as a biological unit and c) species as a nomenclatural rank.

Species as a taxonomic unit.
For the purposes of this paper, I would explain taxonomy as the disposition of describable specimens in a logical order. Such words must be read without biases - that is, disposition is meant to encompass both "lumping" and "splitting", logical need not imply a "natural" system, and order should not be construed as limited to a single row or line. I would further have to qualify what passes as practical taxonomy from the above definition. Taxonomists of fleshy basidiomycetes, I suspect, do not consider a "species concept" at all, but utilize almost exclusively a single character field ("morphology"), exhibited by a single portion of the organism (the fruit body), and the results often seem only coincidentally related to a "natural" system reflecting phylogenetic relationships.

If the concept of species arose in a period between Hieronymus Bock and Kaspar Bauhin, then the concept was based solely on characters observable with the raw senses, perhaps aided somewhat by a lens. This idea of species, reinforced by Tournefort's organization into genera, was absolutely entrenched long before conscious manipulation of plants was applied to their classification. The theories of Mendel were separated by two centuries from the solidification of the species concept, as were von Leeuwenhoek and De Bary, who began the manipulation of microbes. In short, with fleshy fungi treated as extensions of the flora, the species concept governing flowering plants also was applied to these fungus forms, and by the time fungi were grown purposefully by people, taxonomy was already widely separated as a discipline from physiology and its new partner, genetics. The schism has remained, and when most taxonomists of fleshy fungi talk about "taxonomic literature", it is almost always in apposition to biochemical, physiological or genetic literature, and almost always deals with "morphological" characters. In short, morphological, fruit body taxonomy historically has been etched in marble.

Over the years, to be sure, a layer of technology was interpolated between the observer's senses and the specimen. Karsten, Patouillard, Maire and others introduced the serious use of the microscope, after which Melzer and Bataille were among those who applied chemicals and observed their reactions with fungi. Corner introduced and developed the description of hyphal systems. In the first step toward taxonomic manipulation of the fungi, Mildred Nobles exploited cultural characters of the dikaryotic mycelium, but the effort was not born of an attempt to use another character field for taxonomy of polypores, but to cope with hundreds of unnamed cultures submitted by foresters from all over Canada.

At the same time, it must be admitted that taxonomy of fleshy basidiomycetes, of course, lags far behind taxonomy of some other fungus groups in the extent of manipulation in classification. In several groups of Phycomycetes, especially in the Fungi Imperfecti, and in selected families of Ascomycetes, growth of the fungi in pure culture is almost a prerequisite before meaningful taxonomy can be begun. In the yeasts, not only is axenic culture necessary, but taxonomy has been extended to fermentation and assimilation tests. Conversely, it may be argued that microfungi exhibit microcharacters, observable most clearly under cultural conditions, and that macrofungi exhibit enough fruit body characters to get the taxonomic job done. This is, of course, true to some extent, and it is this principle - just enough to get the job done - that has encouraged taxonomists of fleshy Basidiomycetes to evade investigation of the "species concept" and concentrate on separating "taxa" one from another, using only the limited characters of the fruit body.

Now, if this discussion appears to be peripheral to the concept of a species, then a discussion of psychology, explored however gingerly, will seem even less germane. But quite to the contrary: if "species" has been the invention of the taxonomist's mind, then to know the kind of mind which does taxonomy is to be better acquainted with that mind's invention.

I would submit that the taxonomist, in Freudian terms, must surely
have a healthy (or unhealthy) dose of anal fixation. What mixture of
genes and environment produce such a personality is beyond our subject, but anyone versed in personality development can quickly identify the adult traits reflecting the fixation and the childhood phase
in which crayons, toys, and all other objects were taxonomized into
logical groups - by color, by size, by subject - so that all cars are
in one pile and all trucks in another; all books sorted according to
shape; etc, etc... Persons intent with such arranging also tend to
project this into the world at large, attempting to bring order to
their milieu - to know their place in an ordered world. Moreover,
certain predictions can be made about such fixated personalities.
First: such a person will be a hoarder, a collector, a gatherer, a
preserver. Second: such a person will be more concerned with things,
possessions, objects, than with people and social amenities.
Third: the task of sorting, arranging, indexing, will be more important than close study of the objects themselves. Fourth: in many such
people, criticism and/or scepticism is accepted as threatening, and
competitors not as welcome additions but as potentially controlling
forces, to be kept at bay.

Borne of such a personality, how will a "species" be conceived?
First: it will be easily sorted, both away from dissimilar objects
and together with similar. Second: it will likely not be investigated
too exhaustively, for the task of sorting or arranging is more important than deep knowledge of the object itself. Third: it will be
stable, passed on from one taxonomic personality to another. Fourth:
it will not be open to much criticism, not because of its innate qualities, but because of the personality of its inventor. Fifth: it is
liable to be based on some inanimate portion of the organism, requiring little manipulation for identification.

So higher fungus taxonomy in the historical perspective has evolved
in the hands of a certain personality type, and has, therefore, not
been concerned with the manipulation necessary for deep investigation
of a "species concept". Instead, practicality, convenience, and personality have retained fruit body morphology as the alpha and omega
of species as taxonomic units.

Actually, such a species concept is quite adequate for many uses. As a shorthand identification between taxonomists, and as a useful communication to foresters, plant pathologists, mycophagists and others interested in applied uses, the taxonomic unit is an operational designation filling an obvious need.

S p e c i e s   a s   a   b i o l o g i c a l   u n i t .
But to conceive of "species" as r e s t r i c t e d to the above concept is only to ask very limited questions about it, namely: What is its name; and into which taxonomic pigeonholes does it fit. Other workers, with other backgrounds and personalities, would ask very different questions about "species", such as: What is its function in the ecosystem; With what other population representatives does it interbreed; What, in short, is its "biology"?

A few examples might be offered.

Papers by Duncan (1972) and Duncan & MacDonald (1967) on Auricularia species are particularly interesting. Within A. auricula (Hook.) Underw., three strains were identified by biometrical and interfertility data. Two occurred within North America, could be separated by substrate preference, and were 26% interfertile. The European strain was totally intersterile with the North American strain on deciduous trees, and only 2% interfertile with the North American strain on coniferous wood.

Moreover, in A. polytricha (Mont.) Sacc., a southern United States strain appears genetically isolated from all others, a Hawaiian strain virtually totally interfertile with all others (except the above), and African, Indian, and Australasian strains partially interfertile amongst themselves. Does the North American strain constitute a new taxon, and if so, at what rank?

Perhaps just as startling, however, was the strain of A. polytricha from Assam, which, while totally interfertile with strains from Bengal, Australia and Hawaii, and partially interfertile with strains from Kenya and New Zealand, was morphologically so dissimilar as to suggest it as a separate taxon. Such data are applicable also to work by Orson Miller (private communication) and coworkers on Omphalotus. Strains of O. olearius (Fr.) Singer from Italy were mated against

isolates of Clitocybe "illudens" (Schw.) Sacc. from Michigan and
Virginia, and found completely interfertile. All strains were mated
against isolates of Monadelphus subilludens Murr. from the southern
United States, and again were completely interfertile. While usually
considered morphologically distinct, all three were genetically completely compatible.

Thus, on the one hand, the idea that a species is "the sum total of
interbreeding individuals, and, hence, the most inclusive unit of
normal biparental reproduction" (Grand, 1971:19) must be questioned.
But in higher plant systematics, such events as described above come
under more precise terminology. That strain of Auricularia polytricha
genetically isolated but morphologically indistinguishable, would be
called a sibling species to the other strains. Those complexes of
Auricularia auricula and Omphalotus species, where morphological or
taxonomic taxa interbred as though one, would be referred to as syn-
gameons. Basidiomycete taxonomists have not yet utilized such terminology, but may well do so in the future.

A second example. Fiasson, Le Breton & Arpin (1968) furnished a con-
cise chart showing biosynthetic pathways of carotenoid pigments.
Using such a chart and knowledge of the taxa of Cantharellus, several
important problems can be raised concerning species concepts in that
genus.

First: the various steps in carotene synthesis are, of course, enzy-
matically mediated. Moreover, as shown in other fungi (Dacrymyces
palmatus, for instance) carotenes can be photo-induced, and the resultant molecules photo-oxidized. It is not uncommon to find significantly paler pigmentation on portions of the pileus of C. cibarius
Fr. protected by leaf mold and, therefore, shaded from the sun and
air. In other fungi pigmentless mutants have been produced, and this
process may well take place in nature (viz. C. subalbidus Smith,
etc.). So a genetic or physiological block of a single enzyme can
have profound effects on coloration, and so on taxonomy. This may be
seen in subgenus Leptocantharellus (e.g. C. infundibuliformis Scop.
per Fr.; C. tubaeformis Schaeff. per Fr.; C. melanoxeros Desm., C.
ignicolor Pet.; etc.) in which neurosporene is deposited, but not
,-carotene, as compared with subgenus Cantharellus (C. cibarius; C.
minor Peck; C. friesii Quel.; C. cinnabarinus Schw.; etc.) in which

γ-carotene is deposited, but hardly neurosporene. In fact, this information was most important in the final correct placement of C. melanoxeros (Fiasson, 1973).

Second: Apparently, as in many other fungi, pathways are influenced by physiology in localized tissues within the fruit body. This is surely no surprise, but indicates that enzymatic reactions may be suppressed in local tissue even though the biosynthetic pathways are present genotypically. For example, C. subalbidus exhibits white pileus, flesh and stipe, but pinkish salmon hymenium. Clearly the enzymes required for conversion of colorless precursors into carotene molecules are present, but are repressed in all tissues but hymenium. This process not only requires the biosynthetic enzymes, but the complex physiological process of local repression and/or derepression.

Parenthetically, a totally albino form could result from a single block of the pathway near its inception, or represent a system in which the pathway was never developed. Taxonomically, such forms, morphologically similar, would be dealt with similarly, but biologically would represent very different systems, probably phylogenetically widely separated.

Third: Fiasson & Arpin (1967) reported on the pigments of C. tubaeformis, including -carotene and a dihydroxy- -carotene (yellow), neurosporene (orange) and lycopene (red). A blend of such pigments would result in colors from golden yellow to orange red, dictated by the ratio of one pigment to the others. Absence of one or another pigment would even more drastically change the phenotypic coloration. Examination of many specimens of the C. tubaeformis complex convinces me that several pigments are produced not only generally in the fruit body, but within specific tissues (pigmented fruit body but white spore print) and at various times during ontogeny (pigment developing very early in pileus formation or delayed until adolescence). Moreover, each pigment may be significantly repressed independently, either genetically or physiologically, to alter the ratio of pigments and, therefore, change the overall color. Such suspicions (which can hardly be confirmed until the group is cultured and fruited under standard conditions) make any determination of species concept in this group very uncertain. Again parenthetically, these intermediate color forms seem most numerous in the Pacific area of North America, while in Europe they appear much fewer, and the taxa much more distinguishable.

Fourth: if it is true that related taxa should exhibit characters in common, and if this principle starts (not ends) in biochemical characters, then Cantharellus should be related to other carotene-producing groups. Arpin & Fiasson (1971) have advanced Gerronema as a likely candidate for this reason. Conversely (Cibula, private communication) Hygrophorus must be rejected on the same basis. Within Hygrophorus, however, very similar problems are likely to be identified as the biology of that genus is uncovered.

Yet another example. Specimens of Amanita citrina (Schaeff.) Gilb. were routinely cultured in our lab in 1974 and 1975. Of several cultures, one exhibited significantly different culture mat characters (Campbell & Petersen, 1975), enough to suggest a re-examination of the original fruit bodies. David Jenkins, then in our lab working on Amanita, corrected the identification to A. citrina var. lavendula Vesely, distinguished from the type variety only by its dull lavender universal veil fragments and annulus margin. More recently, however, Anita Mahoney, also of our lab, has performed many thin layer chromotograms on amanita toxins from the genus in our area, and has found that while A. citrina produces small amounts of one or two amanita toxins, var. lavendula produces all of the toxins also produced by A. phalloides (fruit bodies from Europe used as standard). So var. lovendula differs in three characters: fruit body morphology (volva and annulus color), culture micromorphology, and biochemistry. Should this combination of characters be used to elevate var. lavendula to species rank?

One final example. Within Clavulinopsis, two infrageneric complexes were identified (among others) by Corner (1950). One produced spores with very large apiculi, the other produced spores with small, papillate apiculi. Petersen (1968) subsequently found that some species produced a characteristic green reaction with ferric salts on the hymenium, while others did not. Still later, the pigments in several taxa with small-apiculate spores were found to be carotenes (Fiasson, et al., 1969). The large apiculate-spored taxa were known not to produce carotenes, but the best estimate (Jameson, 1973) seems to be substituted xanthophylls. In short, until a very short time ago, it appeared that the two complexes were becoming clearly separate on micromorphology and biochemistry. Most recently, Marsha Hubbard, working in our lab, examined basidial nuclear behavior in taxa with small-apiculate spores (chiastic behavior had already been reported for the large apiculate-spored group). While she found behavior to be chiastic, in every case the eventual nuclear number was eight, with four residual nuclei remaining in the basidium after spore discharge. In the large apiculate-spored group, only four nuclei are formed, all passing into the young basidiospores. I am sorely tempted to resurrect <u>Donkella</u> Doty for the large apiculate-spored group on at least three characters: spore morphology, biochemical pigmentation, and basidial nuclear number. Such a mixture of characters at the genus level would surely not conform to the idea of convenience as a taxonomic unit.

What can be said about biological species versus taxonomic species? Taxonomic species reflect phenotypically expressed gene pools, while biological species emphasize non-phenotypically expressed genes, perhaps, although the interface between genotype and phenotype is anything but distinct.

S p e c i e s   a s   a   n o m e n c l a t u r a l   u n i t .
The International Code of Botanical Nomenclature historically has forcefully declined to consider the "species concept" as part of its mission. In fact, the ICBN assumes that it will be consulted only after two operations are completed, namely, all taxonomic decisions, and formulation of all "taxon concepts" (including all pertinent ranks). In short, it assumes that the taxonomist knows what species are, and has arrived at conclusions concerning their arrangement. So we should not expect to find any discussion of genotype, phenotype, interfertility, and the like, in the ICBN.

What we do find, however, is a hierarchy of nomenclatural ranks, ranging from physiological races to kingdom, and because the various ranks are available for nomenclatural status, some thought must be given to the differences between, say, a variety, a species, and a genus.

Surely the simplest solution to this hierarchal ranking is to assign rank by numbers of differing genes - two gene differences for a form, three for a variety, four for a species, more for a genus, etc., etc. Such arbitrary assignment neatly solves the problem, but does not grasp the most serious issue raised in the discussion above namely, are morphological, "phenotypic" genes somehow more "valuable" in defining the essence of "species", or are "genotypic", less easily interpreted genes of more or equal value? While we may perceive three phenotypic character differences (caused by only three genes?) between two specimens, how many unperceived gene differences could be discerned if biochemical, genetic and cytological phenomena were investigated? The ramifications of such questions are endless, and render meaningless the number of gene differences as a basis for assignment of nomenclatural rank.

What then?

Perhaps an answer must be forthcoming from a historical, political and nomenclatural perspective. First: the concept of a species chiefly defined as morphological is deeply rooted in history. To radically change that concept would require rethinking by, and retraining of, many taxonomists. If, before proposing a new species, it were necessary to understand and describe its physiology, genetics and ecological role, the task of collecting, preserving and cataloging the taxa of higher fungi would slow to an agonizingly ponderous pace, albeit based on a much more solid foundation. Already several ecosystems are in danger of extinction, and the fungi of such regions must be cataloged as quickly as possible. Second: although some of the scientific constituency recognizes the discrepancies between the "taxonomic unit" and the "biological unit" concepts, most are equally familiar with "species" based on morphology, and are able to tolerate the concept. In fact, our colleagues in related fields, especially more applied areas, would be frustrated and angry if, before a positive identification for a fungus could be furnished, exhaustive tests for "unexpressed" gene differences had to be performed. I would contend that even those workers who seriously question a species concept based on morphology would be hard pressed to adopt any other within practicality. Finally: within nomenclature, the "type method" has become firmly established as the means by which we ascertain the "true characters" of a taxon. But living type specimens have always been outside the ICBN - expressly forbidden (although this may come under serious question in the near future) - and tests for most "unexpressed" gene differences become impossible given the usual small quantity of the type specimen and the restrictions on its manipulation. When the type specimen was killed for preservation, the species concept was essentially frozen in time.

It is impossible, therefore, to escape one species concept based on morphology, on phenotypy - essential for communication to the floricist, the forester, the plant pathologist, the mycophagist and others - and readily recognized with a minimum of manipulation.

The ICBN makes another assumption: decisions involving the taxonomy and nomenclature of a group of plants (including fungi) will not be made by newspaper reporters, clothing salesmen or swimming instructors, but by competent, well-trained botanists with taxonomic background. By concommitantly purposefully avoiding definitions of nomenclatural ranks, these definitions are left to the minds of those taxonomic botanists. Here they must remain, for the quality and quantity of morphological difference dictating a species of one fungus group may well not be congruent to those defining a species in another, and such inconsistencies are not bad, if clarity and convenience in taxonomy are to be valued.

At the same time, accepting the inevitable species concept just outlined, there is every reason to pursue the investigation of each taxon until it is well understood - morphologically, cytologically, genetically, biochemically, ecologically. We must welcome the best methodology and terminology from other disciplines and biological kingdoms so that we can better express the phenomena uncovered, and with an eye toward better and better understanding of the phylogenetic and ecological relationships of the fungi involved. When non-morphological data can appropriately be put to good taxonomic use, new taxa should be proposed and old taxa synonymized or lowered in rank. Although a practical species concept might well start with morphological characters or phenotypic genes, less readily perceived gene differences must be given significant weight in any refinement of the superficial morphological concept.

Summary

Historically, taxa were conceived of necessity as morphological. Taxonomy as a discipline attracted a certain personality type which reinforced this species concept, and it has been perpetuated until the present. But practical taxonomy hardly deals with "species" as a biological unit, and such investigations into the "biology" of a taxon raise immediate, serious questions concerning the validity of the "species" in its "taxonomic unit" sense. Furthermore, for several, often non-scientific reasons, there appears little hope of drastically changing the prevailing species concept, but by adopting the best methodology and terminology from other disciplines and organismic groups, the "taxonomic unit" concept may be used as a foundation on which to build a superstructure of "biological" data and conclusions.

Literature

Arpin N. & J.L. Fiasson 1971: The pigments of basidiomycetes: their chemotaxonomic interest. in R.H. Petersen, Ed. Evolution in the higher Basidiomycetes, Univ. Tennessee Press, Knoxville. pp. 63-98.

Campbell M.P. & R.H. Petersen 1975: Cultural characters of certain Amanita taxa. Mycotaxon 1: 239-258.

Corner E.J.H. 1950: A monograph of Clavaria and allied genera. Ann. Bot. Mem. 1: 740 pp.

Duncan E.G. 1972: Microevolution in Auricularia polytricha. Mycologia 64: 394-404.

Duncan E.G. & J.A. MacDonald 1967: Microevolution in Auricularia auricula. Mycologia 59: 803-818.

Fiasson J.L. 1973: Les caroténoïdes de Cantharellus ianthinoxanthus (R. Maire) Kühner et sa position taxinomique. C.R. Acad. Sci. Paris 276: 3219-3220.

Fiasson J.L. & N. Arpin 1967: Recherches chimiotaxinomiques sur les champignons. V. - Sur les caroténoïdes mineurs de "Cantharellus tubaeformis" Fr. Bull. Soc. Chimie Biolog. 49: 537-542.

Fiasson J.L., P. Le Breton & N. Arpin 1968: Les caroténoïdes des champignons. Bull. Soc. Natural. Archeol. l'Ain. 82: 47-67.

Fiasson J.L., R.H. Petersen, M.P. Bouchez & N. Arpin 1969: Contribution biochimique a la connaissance taxinomique de certains champignons cantharelloïdes et clavarioïdes. Rev. Mycol. 34: 357-364.

Grant, Verne 1971: Plant Speciation. Columbia Univ. Press. N.Y. x + 435 p.

Jameson A. 1973: Characterization of the pigments of Clavulinopsis fusiformis. M.S. Thesis, ined., Univ. Tennessee. vi + 51 pp.

Petersen R.H. 1968: The genus Clavulinopsis in North America. Mycologia Mem. 2: 39 pp.

DISCUSSION FOLLOWING DR. PETERSEN'S PAPER

SMITH: I have several comments to make. I would first like to take up the historical. I think this applies rather appropriately, here, because the impression seems to be that the bilogical aspects of species are something, in a way, recent developments, and this needs, blanket contradiction. Kauffmann, Farlow and Couch, in North America, and at least Kauffmann and Farlow were greatly influenceed by the professors in European Laboratories, so I have a feeling that so me of their views probably came from this side of the Atlantic, they tought their students the biological approach. We grew our specimens, we studied their life cycle, and we learned what characters were used to identify them. In my early work on Mycenas, which was never published because it did not seem that cultural work of that type at that time was not generally considered was worth publication. I grew my Mycenas, I studied them bilogically, chemistry was a lot simpler then than it is now, but Kauffmann's students all cultured their material and studied every aspect of it. So I want to emphasize very plainly, this bidlogical approach is absolutely nothing new in mycology. I admit that after world war II it was neglected and developed under different names. Now, second, I would like to go to the matter of pigment-less mutants and Cantharellus subalbidus. Ron, did you mean that that C. subalbidus is just a mutant of C. cibarius?

PETERSEN: No.

SMITH: It is a valid species in its own right. The point I wanted to make is that C. cibarius and C. subalbidus have been found to have different amounts of sterol. I am not even sure that C. subalbidus descended directly from C. cibarius. Its fruiting body has a much different ratio of sterile to fertile tissue. As to the study of albinos, one must recognize certain restrictions in defining an albino. It must be white over-all including the spore deposit. A white Psathyrella with colored spores is not an albino. The best known albino in the Strophariaceae is the one Dr. McKnight studied. It had white spores, but was interfertile with the parent stock having normally pigmented spores.
These were all men that were prominent in the development of taxonomy, Couch and Kauffmann in particular.

PETERSEN: Yes. On the other hand, I would respond in this way. The last fifty years have formed a really significant part of the history of the species concept. I think this perhaps is inaccurate in that a morphological species concept has been around for several hundred years now. In terms of the biology of the organisms, I am quite aware that several taxonomists of former years (e.g., John Couch and others) surely were interested in, and concerned with, experimental mycology. On the other hand, one finds relatively little influence of experimentation on taxonomy. Kaufmann's agaric species, Couch's species of Septobasidium, Coker's species of clavarias and hydnums, show very little reflection of biological concepts that they knew about. They cultured fungi, but I do not see that as influencing the morphological or taxonomic unit of species, especially at that time.

SMITH: Well, it is a case of following the format of the times rather than the philosophy of the concept. That would be my answer.

PETERSEN: The second point you made dealt with Cantharellus subalbidus. In both places it was mentioned in the text, it was accepted as a species.

SINGER: Dr. Petersen's presentation was so clear and so good, that I do not have any question. I just want to make a nasty remark. Ron, don't you think that your taxonomy of the psychology of the taxonomists may fall into the same traps into which taxonomy of the fungi falls?

PETERSEN: It certainly is, as Kees Bas said, skating on thin ice. It is certainly a difficult task to taxonomize taxonomists.

CLEMENCON: You certainly know that you are not the first one to do this. Linné classified all taxonomists known to him according to a military system. And he was the general.

ROMAGNESI: Il faudrait s'assurer si le caractère de la couleur et de la pigmentation est un caractère génétiquement fixé. Gilbert a autrefois fait l'hypothèse que c'était le climat méditerranéen qui provoquait une différence dans la coloration du chapeau chez Clitocybe olearius et illudens. Mais c'est une hypothèse certainement gratuite, aucune preuve expérimentale n'a été donnée.

PETERSEN: I might comment on that particular point. Some years ago I was very interested in the Hydnum repandum, H. rufescens, H. flavescens group. Visiting at Leiden, Maas Geesteranus told me that in Scandinavia it was very easy to distinguish between H. repandum and H. rufescens, but in southern France, it was very difficult. And being young and sceptic, I did not believe him readily. Now I have travelled a bit more into southern Germany, and to the middle of France, and in parts of the United States. He is absolutely correct. Fruitbody pigmentation in different geographic areas leads to confusion in this genus. I think this is a very difficult problem.

ROMAGNESI: D'autre part, je voudrais vous dire quelque chose au sujet de l'Amanita citrina variété asteropus; pour moi, d'après les définitions que j'ai données tout-à l'heure, je suis d'accord avec vous pour dire que ce serait une espèce différente. Tandis que s'il se révèle que Clitocybe illudens et olearius sont vraiment une espèce climatique cela tombe exactement dans la définition d'une nouvelle variété.

ESSER: I do not know whether I understood you right. Do you attribute to this pigment differences in the albinos and other pigment differences, any taxonomic rank?

PETERSEN: I would have to respond that I do not know the pedigree of each particular albino. Why is it white? Did it never have any of the enzymes which take these colorless precursers and give them chromophores? Or is only a single enzyme missing, perhaps due to a very "recent" mutation?

ESSER: Do you this phenotype give a rank in taxonomy?

PETERSEN: I think it would be safe to say yes. Certain albino taxa, of known pedigree, could be given species rank.

ESSER: Taxonomy should be something which is not restricted to fungi. There are albinos in the human race, and there are many of genetically fixed metabolic diseases in man, like diabetes and others, so you would, according to your scheme, consider an albino human being a different species, and a negro as a different species.

EN: I understand your comment. Of course, I cannot respect it very much. I think you are talking about biological units, while I am duscussing taxonomic units. I do not know the pedigree of the fruit-body that I collect in nature. I do not know whether the mycelium mutated once, yesterday at 4:00 PM, or whether it never had the enzyme system at all. But the taxonomic unit concept says I must deal with that fruitbody, and I deal with it taxonomically.

KUEHNER: Mes collaborateurs MM. Arpin et Fiasson, qui travaillent sur la chimie des champignons, ont une formation naturaliste de base, mais il n'ont fait que des études de Sciences naturelles de type classique; si les questions de Systématique sont l'objectif de leurs préoccupations de chimistes, les problèmes de nomenclature leur sont passablement étrangers. Je ne me souviens plus s'ils ont écrit que les Gerronema peuvent être rapprochés des Chanterelles à cause de la présence de caroténoïdes; s'ils l'ont fait c'est en pensant aux Ag. chrysophyllus et venustissimus et non à l'ensemble des Gerronema. Gerronema n'est que le résidu de la coupure friesienne Omphalia après élimination de quelques espèces, en particulier des espèces à pigmentation incrustante, seules rangées par Mr. Singer dans le genre Omphalina. Si la coupure Omphalia était hétérogène, la coupure Gerronema l'est encore beaucoup trop à mon avis; en particulier il me parait impossible de rapprocher dans un même genre Gerronema, comme le fait Mr. Singer, les Ag. chrysophyllus et fibulus; si la première espèce est riche en caroténoïdes, il est peu probable que ce soit le cas pour fibulus puisque l'essentiel de sa pigmentation est vacuolaire. Pensez vous, Mr. Singer que votre genre Gerronema soit naturel?

SINGER: Il est très possible qu'il y ait une certaine hétérogénéité dans ce genre. C'est l'avis aussi de mon ami Moser et il est très possible qu'il a raison. Mais il y a un grand nombre d'espèces intermédiaires rencontrées dans les tropiques.

SMITH: I wanted to take up the matter that Dr. Esser brought up. Answering the question on the taxonomic position of the albinos. In my own case I simply refer to them as albinos. And the requirement to me for an albino is that the collection is fertile, produces spores, and the spores lack pigments as well as all parts of the fruiting body. But we have a complicated problem. We have various degrees of sterility in naturally occurring fruiting bodies. They may be almost entirely sterile or nearly completely fertile, and the pigmentation of the basidiocarp may be different in each. Sterility in the Strophariaceae is often correlated with over-development of yellow pigments.

BLAICH: I just want to modify some genetic concepts. Obviously both morphological taxonomists and geneticists accept the species as the smallest evolutionary unit. Is that correct?

MANY: Oh no!

SINGER: Well, some do.

SMITH: That will be right in the sense that the subspecies, variety and form are included in your word species.

BLAICH: A lot of taxonomists think that this concept might be well approached by morphological means, and as we have seen, in most cases they are perfectly right. However also the genetic concept of non-crossability should not contradict the morphological concept. It is quite clear that the evolution of genetic barriers and morphological mutations do not necessarily coincide. The fixation of morphological mutations needs some kind of isolation mechanism. This isolation might be brought about e.g. by geographical separation, or by ecologic separation. Geneticists should then not insist on conspecificity in species isolated by mechanisms other than genetic barriers, which evolved divergently but are still cross-fertile. E.g. if you have two species, one from Australia, and one from America, and they are quite different, so that everybody would recognize them as different species. If these species are cross-fertile, that will not necessarily mean that there are the same species.

SINGER: Do you have such an example. Does that exist?

BLAICH: I think there were some examples cited. That different species which normally do not get into contact, may cross under artificial conditions, and are described as conspecific. A genetic isolation mechanism will certainly lead eventually to speciation. But speciation due to other isolation mechanisms must not necessarily be accompanied by genetic isolation. So I think that even Esser's concept here allows for this fact. He talks of "failure to breed", but not "caused by genetic parameters". It does not exclude other isolation mechanisms. But if I am not right in considering the species as the smallest evolutionary unit I should like someone who has another concept to explain why species are not the smallest unit.

Herbette Symposium on Species Concept in Hymenomycetes 1976

THE SPECIES CONCEPT IN AGARICALES AND ITS ADAPTATION TO TAXONOMY

Rolf Singer

The species concept in Agaricales has been discussed from the theoretical side in several papers which are infrequently cited and hardly ever taken up as a rigorous basis in practical taxonomic work excepting, sometimes, by the authors of the respective theoretical papers. This has several reasons. The most serious one is the lack of pertinent data to reconcile theory with taxonomy. Lacking experimental data on the heredity of certain characters, degree of interfertility in two related putative species or infraspecific taxa, chromosome numbers, area studies and ecology of the taxa involved, often also sufficient observations on variability of different populations, the taxonomist relies exclusively on descriptive data, whereby the microscopical data are often those of a few dried specimens. Generalizations are more often than not dependent on intuition which, in value, varies according to the talent and experience of the observer.

Can we say that therefore the species concept of the majority of authors is based on data insufficient to lend theoretically satisfactory standing to the taxa published or recognized? And does, if this were so, the taxonomical literature, up to the present time, present an art (as has been suggested) rather than a strictly scientific endeavor with well defined units?

The first question no doubt calls for an affirmative answer. The second cannot be answered with an unqualified yes or no. While it is impossible, at the present time, to solve all questions related to the adaptation of current taxonomy to theory, it is unreasonable to devaluate the conclusions of the taxonomist as unscientific since the methods used are - if limited - certainly scientific in the sense that they are based on a scientifically valid approach (unless, of course, the author is negligent or illogical). Consequently, like all scientific conclusions, they are not final but subject to changes brought about by the evolution of science itself (cf. my short discussion of this matter, 1975, p iv). As I interpret Vasilkov's (1958) point of view, it coincides with mine although that author appears to be exaggerating our relative ignorance. Such a pessimistic view should be modified in the light of recent taxonomic work, particularly in the groups specifically mentioned by Vasilkov as examples for insufficient knowledge. At the same time one cannot but applaud his statement that the most important task of the agaricologist is the study in all taxa of those characters which thus far have been neglected or omitted - or not yet discovered - and that a more meaningful discussion of the species concept will be a consequence of the fulfillment of that task by mycologists. An example for such studies may be found in Demoulin (1968) for Lycoperdineae, and a similar attempt has been made by Singer (1950).

The species concept of monographers is neither fixed and unchanging, nor consistently independent of theory, nor is it uniform at any given time. It can easily be seen how the species concept of the leading taxonomists has changed over the years, say, from that of Fries to that of recent monographs. As has been correctly stated by Kreisel (1974), nearly all newer monographs are based on a narrow species concept which is also accepted by users of a wider genus and family concept like A.H. Smith (1968). This development is a consequence of more and more detailed observations by more and more sophisticated techniques and a wider array of characters available, for example the use of chemical analysis and EM observations, more experimental
    cytological and ecological data. Even the quantitative approach by numerical taxonomists begins to contribute to the same effect.

Yet, in spite of all that, the differences of opinion as to what is a "good" species, what a race or variety or subspecies, what an ecologically determined ecotype, and what a forma, is often hotly disputed, even among first-class taxonomists. This proves that not only is there a semantic problem but that available data are not sufficient or sufficiently utilized.

In fundamental agreement with my own (1943) definition, in Agaricales, a species is the lowest and first genetically fixed step in evolution, producing a sum of individuals having a specific number of heritable characters in common which determine their ecological niche and, eventually, a definable area. In Agaricales at least we have the further criterium of intersterility between species - a rule we find consistently valid in our group (if not in some others, e. gr. the Ustilaginales). Since full interspecific fertility of primary mycelia is according to most authors (for example Nobles 1962), an indication of conspecificity, we may be led to assume that the opposite - intersterility between strains - must indicate a sexual hiatus of sufficient significance to permit specific separation. The concept of "geographic races" as introduced by Vandendries (1933), obviously a misleading term at any rate, where different populations of different geographic origin may have a sexuality barrage, has, in the Agaricales, originated with Vandendries' work on Agaricales, e.gr. <u>Pleurotus ostreatus</u> (Jacqu.ex Fr.) Kummer. Since this species is now strongly reduced and, in its wider sense, clearly recognized as a collective species (cf.Romagnesi 1969), the existence of "geographic races" in Vandendries's sense may well be doubted. And where it seemingly finds an expression, taxonomists should be forwarned to look for minor correlated morphological, cytological or chemical differences, or differing host adaptations, confirming the existence of a minor hiatus between the sexually incompatible populations rather than to assume a single homogeneous macrospecies.

In the Agaricales, as in other groups of fungi, we observe that different climatic or physiological (host and mycorrhizal partner) conditions produce races, geographical races or what I have called "mycoecotypes". Since these are less sharply divided from each other by a minor hiatus and are supposed to be reproductively not fully separated, hybrids being possible as a result of reduced interfertility, transitions between populations of one and the other such race are observable. I have in the past considered these races as subspecies of the type form, but believe that this status should be invoked sparingly and only in cases where the available data permit the assumption of races of this kind. In the splitter's hand, these races may be considered species. The arguments for the use of subspecific taxa in this sense are (a) the observable fact that there is no sharp division between these and the formae speciales (as recognized in the Uredinales) although typical formae speciales are, to my knowledge, unknown in Agaricales, (b) the restricted interfertility between races where populations "touch" which would set them apart from the species as defined by complete intersterility; (c) the smaller hiatus between races as compared with other species of the same supraspecific taxon, a phenomenon which should be expressed in giving them a different hierarchic level than the species. The splitter's arguments are more on the practical side. To recognize these races as such is often - not to say mostly - too difficult a problem with the data the taxonomist has at hand so that the decision is largely judgmental or inferred, and to recognize these subspecies as species avoids many problems (i.a. the question whether partial intersterility is due to senescence of cultures).

In order to indicate a good example of justified use of the term subspecies for races, we shall cite the case of a pair of taxa in the Aphyllophorales: Peniophora pini ssp. pini and P. pini ssp. duplex (Burt) Weresub & Gibson (1960). The transferring authors showed that between the two taxa there is reduced interfertility with evidence of erratic and sparse matings of monospore culture hyphae; there are morphological differences on a definitely lower level than those shown to exist between either taxon and related species viz. P. pseudopini Weresub & Gibson. In this case, we have a European and an American geographic race. As for the Agaricales, a paper by Singer & Kuthan (in print) suggests that we have a similar situation between Chroogomphus helveticus ssp. helveticus and C. helveticus ssp. tatrensis (Pilat) Sing. & Kuthan, but here the race differentiation is on the basis of mycorrhizal partner rather than climate, and the difficulties of obtaining viable monospore cultures have made it impossible for us to provide statistics of mating frequency. Sundermann (1975) wants the use of subspecies restricted to geographical races but this merely reflects the reluctance of most phanerogamists (Sundermann uses Ophrys as his leading example) to take into account the problems of and the literature on lower organisms. Vasilkov (1954) proposes to replace the term mycoecotype (Singer 1943) by the narrower term "forma mycorrhizica" which, however, for reasons which merely restate the lack of precise data in cases studied by him, was abondoned later (Vasilkov 1958). Some authors follow a school of phanerogamic systematists who call variety what we here, in the spirit of Wettstein's definition, called subspecies. The present author has consistently used the status variety for those taxa which differ from the type form in but minor characteristics but cannot be disposed of because there is not enough factual evidence to determine their exact status - as forma, subspecies in the sense indicated above, or some sort of microspecies. It may be suspected that, if partial reproductive intersterility or intraspecific barrages exist, the resulting "microspecies", are a first step of speciation and may coincide with Sundermann's (1975) prae-species (p.624), a term of predictable usefulness in genera such as those of the obviously habitat-induced varieties in the stirps _Laccaria_ _laccata_ (where we observe seriate alignments or clusters of "microspecies") and perhaps also in the stirps Ostreatus in _Pleurotus_. But for the time being, the application of the term variety as a provisorium is in practice very useful and almost inavoidable. It may be virtually synonymous with such terms as "putative variety" (Salisbury in Huxley 1940), but not with Sundermann's definition of variety.

Until and unless the status of a much larger number of published taxa can be defined by additional data and a consensus regarding the use of the appropriate taxa for definable categories can be achieved, at least among mycologists, we cannot and should not try to coordinate the use of the terms forma, variety, subspecies, species, or for that matter "ecotype" or "praespecies", and I am far from convinced that the attempt to unify the use of these terms and categories is at present promising or advantageous. All that can and must be achieved is to impress mycotaxonomists that the words species, subspecies, variety forma etc. should be used consistently and with a declared meaning. The mycological literature is full with examples of varieties being raised to species, a new status being proposed for a subspecies to become a variety and vice versa without an attempt to explain what the particular author accepts as his definition of these words. One should never forget to ask oneself whether such a change has any scientific meaning or serves for a better understanding of the taxonomic group involved.

But even when all pertinent facts have become available, it will be obvious that not all the taxa which correspond fully to the definitions given for the term and taxon we call species are equivalent with each other. Their sexuality and evolutive history may be so totally different that the standing and value of a species in one genus or family may be difficult to compare with the standing and value of a species in another group; even within a single group, like _Conocybe_ or _Omphalina_ we may have totally different phenomena in the speciation leading to homothallic respectively apogamic, parthenogenetic forms than we have in amphithallic or heterothallic species. For the phanerogamic systematist, there can be no doubt about the difference of species concept when dealing with such species as the majority of the European Alchemillas or Taraxacums.

It is tempting to introduce into this picture of multiple species

concepts some additional, modern cytogenetic data originating from
generations of fungus organisms (rarely Agaricales) maintained in pure culture solely in the laboratory. It may even be anticipated that
criticism of the principles enumerated above may come from that side.
Far from wishing to minimize experimental results of this sort, the
taxonomist must insist on the condition modifying his view - that the
definitions and conclusions relate exclusively to organisms exposed
to the competitive interplay of gene pools in nature. He is not or
only indirectly concerned with the genetic possibilities or the patterns of sectoring of an organism under artificial conditions and in
isolation from the ecosystem of which it is a part. Many gene combinations, although capable of producing living organisms in vitro, are
unsuited for survival in nature, have no ecological niche except in
the laboratory and have no natural geographic area. What valid terms
to apply to such "cultivars" is a problem not within the range of the
present discussion.

Frequently, especially in the literature of the first part of this
century, attempts have been made to decide which and how many characters, and of what category, should be exhibited by a species, or by
every one of the infraspecific taxa of Agaricales, whereby it would
seem that some characters are particularly and invariably important,
others generally unimportant. Anybody who has been studying many
groups of Higher Basidiomycetes knows, however, that specific characters cannot be weighted in a generally valid scheme. Evolution has
taken various courses, some emphasizing, developing or reducing, certain characters or character complexes while maintaining others. Some
characters may have been more often, others more rarely, indicators
of a certain direction of speciation; none has an overall value.
Practically, this means that it is idle to prepare a monograph by
first deciding which are the characters variable within a genus or
family, and then separating all species according to the pattern and
distribution of character states in the whole group because some of
these character combinations may well be diagnostic in one stirps,
series, section or subgenus and not at all in another, even a closely
related one where these patterns do not separate species but individuals.

Amyloidity of cell walls is a good example. While it is indeed provable that in naturally delimited genera like <u>Amanita</u>, <u>Cystoderma</u>,

Dermoloma, Fayodia, Mycena, and others, species or sections or subgenera with both amyloid and inamyloid spore walls are included, it is equally true that entire genera of undisputed naturalness, even families, have either all inamyloid or all amyloid spore walls. Needless to add that aside from a certain variability in Pseudohiatula, the species of Agaricales have either amyloid or inamyloid spores (some inamyloid spores may sometimes be mixed in where the majority of spores is amyloid) or else pseudoamyloid spores (here again some inamyloid ones may occasionally be mixed in) in the sense that the presence of amyloid or pseudoamyloid spore walls is quite constant for any given species. Amyloidity is, of course, as aptly pointed out by A. H. Smith (1968), not always the same, chemically, but different types of amyloidity do not erratically occur in a single species. Consequently, we may say that amyloidity is sometimes a highly diagnostic family character, sometimes merely a specific character; but no intrinsic overall value can be attributed to it. Likewise the characteristic of free lamellae is extremely valuable on the family level in the Pluteaceae, but in Mycena and Psathyrella it varies from specific to sectional character and at best becomes an individual character in many other genera (e. gr. Russula).

Our discussion may not have any relevance for those who, like Möbius (1886) feel that there are as many species as there are different species proposed by their authors, which means simply that species are man-made units invented for practical classificatory purposes. This point of view frees the taxonomist from any consideration of affinity, the estimate of hiatus, isolation, and propagation patterns. My impression is that such an attitude has been abandoned by most modern workers who believe that the species is a product of the laws of evolution and has a definite, objective existence in nature, even though it has developed, step by step.

Literature cited.

Demoulin, V. 1968: Art- und Rassenproblem bei Pilzen. Int. Symp.
Biol. Ges. DDR, pp. 111-116.
Huxley, J.S. 1940: The New Systematics. Oxford.
Kreisel, H. 1974: Die Gattungs- und Artkonzeption bei Grosspilzen in
A. Vent (ed.), Widerspiegelung der Binnenstruktur
und Dynamik der Art in der Botanik, Berlin,
pp. 117-127.
Möbius, K. 1886: Die Bildung, Geltung und Bezeichnung der Artbegriffe
und ihr Verhältnis zur Abstammungslehre.
Zool. Jahrb. 1. Jena
Nobles, M.K. & B.P. Frew, 1962: Studies in wood-inhabiting Hymenomy-
cetes V. Can. Journ. Bot. 40: 897-1016.
Romagnesi, H. 1969: Sur les Pleurotes du groupe ostreatus (ostreo-
myces Pilat). Bull. Soc. Myc. Fr. 85: 305-314.
Singer, R. 1943: Das System der Agaricales III. Ann. Myc. 41: 1-189.
Singer, R. 1950: Naucoria Fr. i blizkie rody v SSSR. Trud. Bot. Inst.
II, 6: 402-498.
Singer, R. 1975: The Agaricales in modern taxonomy 3rd ed. Cramer,
Vaduz.
Smith, A.H. 1968: Specietion in Higher Fungi in relation to modern
generic concepts. Mycologia 60: 742-755.
Sundermann, H. 1975: Zum Problem der Definition taxonomischer Kate-
gorien ... Taxon 24: 615-627.
Vandendries, R. 1933: De la valeur du barrage sexuel ... Pleurotus
ostreatus. Genetic 15: 202-232.
Vasilkov, B.P. 1954: Opyt izucheniya vida u shliapovnykh gribov ...
Botan. Journ. 39: 5.
Vasilkov, B.P. 1958: in Problema vida v Botanike I: 85-101.
Moskva-Leningrad.
Weresub, L.K. & S. Gibson 1960: "Stereum pini" in North America.
Can. Journ. Bot. 38: 833-867.

DISCUSSION FOLLOWING DR. SINGER'S PAPER

SINGER: I have added to the original manuscript a few words which perhaps more clearly link our past discussions to my views. I would like to read that to you.
It seems to me that the experimental approach, including the genetic approach, to the species problem has for systematics a twofold usefulness, and a possible impact. In the first place it can help explain mechanisms of evolution in the stage of initiation of specietion. It provides us with data such as the earliest recognizable units of micro-evolution, in the words of Duncan, or races characterized by heterogenous incompatibility, according to Boidin; secondly the experimental approach in general will provide the taxonomist with additional criteria, an advantage which should always be warmly welcomed. Let us not forget, nevertheless, that aside from these two admittidly extremely important contributions to systematics and the species concept there is still the wide field of an incoordinated vast and intricate interplay of such things as climatic changes, migration, qualitative changes of ecosystems, competition among microorganisms and other interactions of organisms in general. All these factors have been and are to a certain extent shaping the species and its hiatus in nature. In spite of such useful inventions as the computer and systems analysis this formation of species and their hiatus can not be studied experimentally for a long time to come, and possibly for ever. Simply because a model that should be constructed would be too vast, too impredictable or too complex. Thus the taxonomist has to go on, for the foreseable future, observing, collecting, describing and comparing species, probably soon with the help of some partially quantitative and more objective methods, but still basically in his old pedestrian way. Where this is necessary we should always aim at more outside data, and what I have explained just now, I hope, will make it understandable when I say: Let us have not only more experimental data, but also more ecology. This last sentence is meant to draw attention to the ecological part of the definition of species which I have attempted to give.

BAS: I want to ask you a question that is somewhat beside the main subject of your lecture. I have the impression that in connection with infraspecific taxa we need a rank that can serve as a kind of waste basket, in which we can put the things we do not know very well but still wish to distinguish. Why do you use for this the rank of variety and not that of form? A "forma" is rather generally accepted as something that we cannot define very well. I think we need the rank of variety for things that are genetically fixed.

SINGER: Well, the reason for that is a purely practical one. It seems to me, and I feel from all I heard that everybody seems to agree, that forma comprises a sum of individuals with non-heritable characteristics, that are within a complex within which we have interfertility. I realize there are probably some who use the term forma for some other purposes, but it seems somehow extravagant to waste one of these terms unless there is a reasonable coincidence of use. So what is left is "variety" and "subspecies". Now, I am a Wettstain student and I was brought up in the idea of geographical and climatic "races" being called subspecies, and I feel that many botanists agree with that, whereas for variety, and I am not talking about taxonomic hierarchy, for variety everybody seems to have some different definition. Since this is so, it might just as well be used for what you call the wastebasket. I am very strongly of the opinion that we need such a thing. However, I do not know whether I made myself fully clear. I do not think the time has come where we attribute to everyone of these ranks a definite meaning. We have to have a definite meaning for what we have in our minds when we describe subspecies, varieties, forms, and every author must think of it and state "what do I understand variety to mean"? When the authors define their use of the ranks, then we will understand each other, whether we use the same terms or others.

THIERS: In this connection, Rolf, in your comment about <u>Chroogomphus</u> <u>helveticus</u>, which is separated into two subspecies based upon mycorrhizal associations, is that representative of what your concept of subspecies might be?

SINGER: Well, it might be, yes. I would take critisism from anybody who makes the necessary next steps in confronting monosporic cultures. I am personally convinced that somebody will eventually be able to culture them just like the other fungi, but we have to find out. I do not think that a specialist's time should suddenly be used for a problem which has no immediate bearing on his problems at the present time.

SMITH: Rolf and I have discussed the use of subspecies versus variety ever since we became acquainted. The reason I use the name variety is that the term subspecies has (as in Zoology) been used for a population geographically isolated. Let us face it, we do not know enough about the distribution of higher Hymeno-mycetes to really know, when we describe a taxon, we do not know its distribution, rather only a few places where it has been collected - often only in areas where some investigator is actively working. It is a geographic location for our collections where persons happened to have been. Now if we can say that a continent has been thoroughly collected, the flora all documented, then we would know whether we can use subspecies in a classical sense, for a geographical hiatus, but we do not know that, and so I recognize these variants as "varieties" which I think are genetically based; and the emphasis is thus on the genome rather than on the distribution.

SINGER: I think we are much closer to valid conclusions than you think. We have published a few things together, and we have published varieties and certain other taxa, and made it quite clear in what sense we used it.

SMITH: But what I mean, is the lack of knowledge of distribution.

SINGER: Yes, but you know that whenever the lack of knowledge became alarming enough I used the term variety and you subscribed to it. I think we should use the term subspecies in my sense very sparingly and only if we have enough evidence and the right to use it. My main point is that every author is free to use his set of taxa as long as he defines them. I am not in any way trying to convince anybody to do it my way.

KEMP: How do you use the term microspecies? The problem, as I see it, is that we have at the extremes two forms of speciation. One could be entirely cytoplasmic and the other could be based on geographical separation. Do you feel that microspecies (populations which are completely intersterile but lacking morphological differences), are the beginning of allopatric speciation?

SINGER: I am sorry, when I was reading. I could not always express the quotation marks which I have put on "microspecies". I think microspecies is a very useful term for anyone who does not want to be very specific. It has too wide a scope of applications. It may, in the end, perhaps be divided into pseudovariety and praespecies. I liked the word praespecies because it contains a notion of the evolution of the species, this being a step toward a species, as we understand it. Now, if we talk about the ones that are intersterile but that have no morphological characters to really separate them, in that case I do not think they are worthy of any taxonomic term. They are extremely interesting, and I think we should not try to press them into that system of variety, subspecies and species.

KEMP: Can I follow up on this point. I agree entirely that we should not press them into varieties or similar taxa, because these species have already made the complete and final step in becoming separate one from another, and to me microspecies are as fully qualified to be species as any others, because they are fully separate.

SINGER: Agreed. But this will depend on the definition of species we finally agree upon.

WATLING: It is suggested on the basis of this analysis, experimental work in the laboratory, it will be possible to circumscribe biological species based on the action of conidia and/or detailed microscopic analysis in correlation with field data, which may not be possible from field data alone. These taxa, I would suggest, should be termed microspecies; these can be clustered into a fairly homogeneous group which would be designated a macrospecies. A stirps would be a cluster of macrospecies.

SINGER: I have often used the word "microspecies" in that sense. Then I agree perfectly, I think this is right as long as we talk about biological species, which you see is not within my definition, here, because, I think I made that clear, I talk about the species as it occurs in nature and not in the laboratory.

Herbette Symposium on Species Concept in Hymenomycetes 1976

INDIVIDUAL PROPOSITIONS OF A SPECIES CONCEPT PRESENTED BY THE PARTI-
CIPANTS OF THE SYMPOSIUM, AND FINAL DISCUSSION IN ORDER TO ARRIVE AT
A COMMON CONCEPT.

INDIVIDUAL DEFINITIONS:

1. BRESINSKY:
   Different species concepts for different purposes are to be admitted: 1. To survey and establish diversity of organisms for all
   kind of biological research: A species comprises all individuals
   which exhibit identical or continuously varying or distributed
   characters. 2. To understand the biological species: A species
   comprises all individuals not kept apart by any type of isolation
   mechanism.

2. ESSER:
   Populations (races) belong to different species when the failure
   to interbreed and to produce viable offspring is not caused by genetic parameters operating in the completion of the sexual cycle.

3. BOIDIN:
   1) Des populations (même isolées géographiquement ou écologique-
   ment) appartiennent à la même espèce si elles sont capables de "se
   croiser" et de produire une descendance viable.
   2) comme Esser ci-dessus.

4. ESSER-BOIDIN: (combined from 2 and 3)
   Populations, even when isolated geographically and/or ecologically,
   belong to the same species when they are able to interbreed and to
   produce viable offsprings provided that an absence of this interfertility is caused only by those genetic parameters operating in
   entire sexual cycle.

5. KEMP:
   1. Two different intra-fertile populations cannot be the same spe-

cies if all members of one population are intersterile with all members of the other.

2. Two or more morphologically distinct taxa from a small area cannot belong to different species if they are interfertile and produce viable spores.

3. Two or more interfertile but morphologically distinct taxa may belong to different species, if they grow in widely separated areas; e.g. mycorrhizal species on hosts which have a disjunct distribution.

6. SINGER:

A species in Agaricales, and probably all of so-called higher fungi (not including Ustilaginales and Ascomycetes) is the lowest and first genetically fixed step in evolution producing a sum of individuals and/or populations having a specific number of heritable characters (and therefore a hiatus separating them from other species) which determine their ecological niche and, eventually, a definite area of distribution, and not being interfertile with other species at all.

7. THIERS:

A species is a population of individuals which possess a group of constant, reproducible characters in common (morphological, physiological, ecologic, genetic, etc., or a combination of these) and forwhich a hiatus exists between this and other populations.

8. WATLING:

Although ideal, the genetic species has as many if not more disadvantages than the morphological species when dealing with the world flora. In the small number of species analysed by experimental techniques morphological divisions have, on the whole, been supported. But if one has to wait until experimental work is carried out on all mycological units the world's fungus flora will be depleted by man's activity! Whenever possible, experimental work should be incorporated into the total understanding of the taxa. As an interim solution a sympathetic hearing to a macro- and microspecies concept, based on <u>correlation of constant morphological characters</u> should be made.

FINAL DISCUSSION AND COMMON CONCEPT

BLAICH: I wanted to ask Esser-Boidin why a geographical isolation is excluded? Geographical isolation has the same effect on evolution as the genetic isolation. Sor for the fromation of new species it has the same consequences.

BOIDIN: Je pense aussi que l'isolement géographique ou écologique est un moyen de spéciation. Ce n'est pas la spéciation. S'il y a encore possibilité d'échanges de gènes, la spéciation est encore à venir.

KUEHNER: Le cas des homothalles ne me paraît pas suffisamment précisé dans le projet de définition Esser-Boidin. Or il semble que les mycéliums secondaires de formes homothalles soient incapables de diploidiser des mycéliums primaires de formes hétérothalles, même si, pour un Systématicien de formation traditionnelle, ces deux formes semblent appartenir à la même espèce.

ESSER: Premièrement, j'aimerais faire une définition pour toute la biologie. Et, ça veut dire que certainement les auto-compatibles et auto-fertilité sont inclus.

You will never have, in biology, a definition which fulfills all examples we have in nature. We only try to approximate the nature as much as we can. Taxonomy does not do anything to what we all want to know how life functions.

KEMP: In the definition of Esser-Boidin do we have to have "when they are able to interbreed" surely it is adequate to say "when they are able to produce viable offspring". But I am afraid I cannot agree with this definition for the reason shown in the diagram. The diagram is an attempt to show that there could be different types of speciation in different species. In the diagram:
  a) = genetical isolation complete before morphological differences arise.
  b) = geographical isolation so great that interbreeding impossible: in nature, morphological variation may be considerable but genetical isolation could be absent.
  c) = genetical isolation and morphological difference geographical isolation take place at the same time.

These are three extreme forms which speciation could take, there are clearly many intermediate between these.
  a) This probably represents the species of the geneticist; if they are interfertile they must be the same species and if inter-sterile they must belong to different species.
  b) This probably represents the species of the taxonomist; if they are morphologically distinct, regardless of their genetical isolation, then they are distinct species.
  c) This represents allopatric speciation of the angiosperm taxonomist. In fungi I think this is only likely to be found in mycorrhizal and pathogenic species.
  The definition of Esser-Boidin does not allow the existence of species of type (b).

To summarise.
Species may be genetically isolated but be morphologically indistinguishable. Species may be morphologically distinguishable but completely inter-fertile. Some species might be intermediate in their mechanism of speciation. A narrow species definition would seem to be impossible in this group and its application would surely prevent agreement on a species definition.

BLAICH: We should replace "geographical" by physical isolation.

Discussion

WATLING: If they are interfertile, then they are the same species. If geographically isolated and show genetic drift in morphological differences then this is, in fact, what is called subspecies.

KEMP: Although they might be inter-fertile when we get them into culture they may not interbreed in nature because they are too far apart geographically. In effect they may be genetically separated on a non-genetical basis (i.e. geographical).

WATLING: Yes, I accept it.

THIERS: I would like to ask a geneticist, in their definition of species concepts how are they going to accomodate those fungi in which we cannot determine their intersterility? How do we fit these fungi into this concept?

KEMP: To me in my definition (item 1); if they will not cross, then they must be different. I am not suggesting they should be given a binomial. It might be useful to indicate this inter-sterility by a prefix e.g. Psathyrella a-candolleana.

THIERS: And what if we do not know?

ROMAGNESI: J'aimerais demander à Messieurs Boidin et Esser, s'ils considèrent comme espèce différentes des populations interstériles de Drosophila candolleana qui ne présentent aucune différences morphologiques?

BOIDIN: Nous ne savons pas, à ce jour, quel est le mécanisme qui interdit l'interfertilité:
   a) Est-ce une différence au niveau protéique (et nucléoprotéique) déjà poussée mais peu sensible dans ses effets aux moyens d'observation du morphologiste actuel? D'où les deux conclusions opposées: le morphologiste refuse le résultat d'interstérilité (et par là condamne les tests d'interfertilité-interstérilité), ou il reconnaît humblement qu'il n'a peut-être pas les moyens nécessaires.
   b) Est-ce un barrage d'une autre nature, tel celui dressé à l'intérieur d'une espèce par l'incompatibilité hétérogénique du Professeur Esser?
Je ne peux répondre aujourd'hui dans le cas des Drosophila gr. can-

<u>dolleana</u>.

KUEHNER: Ce peut être un blocage interspécifique mais non visible morphologiquement. Alors il faudrait admettre que ce sont des espèces, même si on ne peut pas les reconnaître morphologiquement.

BOIDIN: Ecoutez, je ne suis pas gêné par cette affirmation. Le phénomène inverse est parfois plus déroutant pour le morphologiste; par exemple que toutes les sortes de chiens soient interfertiles, que des races humaines très facilement reconnaissables au premier coup d'oeil (ce qui n'est pas le cas de tous les champignons!) soient interfécondes n'est pas évident à priori. Le morphologiste les distinguent, ce ne sont pas pour autant des espèces distinctes. D'ailleurs, je rejoins très volontiers la définition de Mr. Thiers à la condition d'y ajouter que pour juger du "hiatus" qui "exists between this and other populations" le meilleur critère, lorsqu'il est employable, est justement la confrontation des monocaryons!

SINGER: I wanted to make an observation on these separated geographic areas. A complete isolation is not possible. If you study the possibilities of spores brought from one continent to the other, if you consider modern tourist traffic, you cannot talk about absolute separation.

ESSER: I shall admit that the common definition we have provided does not consider if there is no crossing possible. We are here together to find out and discuss in some way. Maybe we should link Dr. Thiers more classical one with ours. You finished: "which a hiatus exists between this and other populations" and you could then follow: "such as which cannot be overbridged", and then comes the genetic concept.

THIERS: I have absolutely no opposition.

ESSER: What do you think, Professor Boidin?

BOIDIN: D'accord, parfait.

ESSER: If you (Thiers) could start, and your definition describes the morphological versatility, which we do not, so with our defini-

Discussion

tion we could fulfill together the gap of missing morphological description.

WATLING: I think, Professor Esser, really actually is misunderstanding the problem; I do not think there is a real conflict between the different views. Roger Kemp and people working with laboratory material would agree entirely. Maybe the geneticist need to tidy up their divergent views but there is certainly no disagreement between the morphologist and those working in physiology. A morphologist has the problem of receiving material to identify from all over the world. I am sure that the flowering plant people, like other morphologists, appreciate the genetic base of this material but still have to build up a morphological classification in which this genetic material can be slotted in at the right place.

PETERSEN: I quite agree. I do not see a conflict. I think I must use some of my own terminology again. Unless I am mistaken, the Esser-Boidin proposition deals with a biological species, and I think the classical taxonomists would certainly want to conform, if possible, to a biological species concept. In the formulation of any document, whichever words come first are always read first, and if people read only half way, they only read the first words. I would think, therefore, that because the morphological taxonomist could agree very readily with the biological concept, the morphological concept should be put second rather than first.

BAS: I agree with Ron Petersen, and I think it is rather simple if you add Thiers' definition after the one of Esser and Boidin. If you start with saying like "and in case such and such information is missing", then you have a principle, a definition, and at the same time you have a way out when you do not have sufficient information.

ESSER: That is a very good idea.

KEMP: I think Prof. Thiers definition a very useful one, but when I come to the words "characters in common" this could mean that a combination of genetic characters alone could be sufficient to delimit a species. The characters "in common" need not be morphological. The hiatus might be geographical and not genetic so that the whole range of speciation mechanisms could be covered.

BLAICH: There is still one difficulty. Both definitions, whether they are genetic or morphological, are descriptive. The definition of Singer is functional: "is the lowest and first genetically fixed step in evolution", and this functional definition is, I think, far more flexible, then purely descriptive definitions, as the definition Esser-Boidin e.g.

CLEMENCON: It is functional, but not practical.

BLAICH: The function should be included, Nobody knows what species are good for. And here, this species concept contains the answer: "they are the first genetically fixed step in evolution".

CLEMENCON: This is ideal, but in practical lab work, we cannot use it.

BLAICH: If you ever reach this level of science, then this should be the aim.

SINGER: But it comes to conditions afterwards, which are very practical. This is only the first statement.

THIERS: I would like to see that first sentence of Rolf's as the first sentence in our species concept definition.

SMITH: There seems to be one item that has not been emphasized quite enough. It is certain that two thirds of the species of plants on the face of the earth have not been analyzed enough genetically to allow one to define the species in genetical terms. To draw up a definition that puts genetic aspects first is unrealistic. When you come to teach a class you are forced to tell the students there are other aspects of the species concept. I would like to see our definition drawn up, so that this genetic feature for which we have no information on it, is secondary. I want to see it in there, but I do not like to see it emphasized as a main character.

PETERSEN: First: I agree with you. Second: I suppose that Esser and Boidin's statement becomes a faith statement for many: "This is what I believe a species should be". And I suppose, third, it provides a paradise toward which to move. When we do not have sufficient information, exhibiting the model gives me something toward which to proceed the next time.

SMITH: OK. Then let us start it out, "a species should be this"!

ESSER: You do not know if a species exists, if it "should".

SMITH: "A species should be", and then give your requirements.

ESSER: That is not logical. A species exists, then it must be "it is".

SINGER: I think "should be" is not acceptable.

WATLING: I think it is better to go from the generalized to the specialized. And I think the genetic is the general, but, unfortunately, because of lack of information, we have to go to the specialized. Hopefully, all the specialized cases will be confirmed in the future by the genetical observations.

CLEMENCON: I think so, too. We have sufficient information on genetics to say something like that.

ESSER: Yes, definitely.

CLEMENCON: And we hope, if we work and cannot in any way test this, we can go along in Dr. Petersen's sense: in the lack of our ability, we take morphological ... But I agree, now we can construct an ideal postulate.

BRESINSKY: A definition should be wide enough to include all possible examples, or cases, which may occur, all foreseeable possibilities. I think this is really a necessity for a good definition, otherwise it is useless. So, following this, I would be strongly opposed to anything like "even when isolated geographically or ecologically". I think this part of the definition, as we have put it on the blackboard, should be eliminated without any replacement. It makes the definition much to narrow.

KEMP: I came here with strong views that hybrids between basidiomycetes were impossible, but I have been mutated by Prof. Singer. I am now perfectly happy to accept morphologically distinct species which can interbreed in the laboratory but presumably they do not do so in nature, (e.g. mycorrhizal spp).

OBERWINKLER: I do not have all definitions of species, which ever have been published, in my mind, but I think the definition given by Dr. Esser is a very old one which has often be debated and not accepted, because it defines, as far as I can see, the species as a unit which cannot interbreed.

ESSER: This definition, yes, is very old. But this old definition does not take care of our present knowledge, that in many cases there is no interbreeding. This interbreeding is favored by very simple genetic mechanisms, such as more and more does not cross, or other cases of incompatibility. In Achlya sex organs are only formed when the strains come together. Unless people did not know the story, they would consider a female Achlya and a male Achlya different species. I do want to take care of all this knowledge which has accumulated now, by the studies of the morphologists, of all biologists which contribute to our plants. In higher plants, in fungi and all sorts of algae, you know, all the stages in the brown and red algae.

BLAICH: Can we restrict to the fungal species concept?

SINGER: Yes! Let us not fool ourselves thinking that in this Symposium we can go beyond fungi.

KEMP: I feel that Prof. Esser is making a special case of genetical variability and outbreeding mechanisms.

CLEMENCON: I have before me a paper signed Esser-Boidin-Thiers that comes close to a
COMMON CONCEPT:
POPULATIONS BELONG TO THE SAME SPECIES WHEN THEY ARE ABLE TO INTERBREED AND TO PRODUCE VIABLE OFFSPRING, PROVIDED THAT AN ABSENCE OF THIS INTERFERTILITY IS CAUSED ONLY BY THOSE GENETIC PARAMETERS OPERATING IN THE ENTIRE SEXUAL CYCLE. FOR TAXA, FOR WHICH THE INFORMATION NECESSARY, OR THE APPLICATION OF THE CRITERIA MENTIONED ABOVE IS MIS-

SING, THE FOLLOWING PRACTICAL DEFINITION IS TO BE APPLIED: A SPECIES IS A POPULATION WHICH POSSESS CONSTANT REPRODUCIBLE CHARACTERS (MORPHOLOGICAL ETC.) AND FOR WHICH A HIATUS EXISTS BETWEEN THIS AND OTHER POPULATIONS.

BRESINSKY: With this definition we run in some problems as far as viable offspring is concerned. There are so many cases where we have a reduced viability, and you have all percentages between fully viable offspring, 100%, 50%, 10%, and 1% viability. And you will never tell the line where to fix the definition. I think fungi are not well enough known yet. I have again to go back to the flowering plants where we have examples that the viability shows continuos variation.

ESSER: I would say, as long as you have viable offspring. And again, you will never find a definition where you have no exceptions.

PETERSEN: I am interested, under this species concept, in what happens to the various intersterility groups within, say, Psathyrella candolleana? Do they each demand a species epithet, or are they still part of one species?

KEMP: This definition clearly indicates that each breeding group in Psathyrella candolleana is a distinct species. Until morphological differences are found (or evolve) are not these microspecies which, at this stage, should be given a prefixed binomial rather than a full binomial. Naming these is clearly a problem. In the last sentence of the Esser-Boidin definition the words "Those parameters operating in the whole sexual cycle" to me means just affecting sexual reproduction. Are we now, in our definition, excluding species which are not reproducing sexually - parthenogenetic species?

PETERSEN: Can I now direct the same question to Dr. Esser? Under this species concept, would the various intersterility groups that can be identified within what we now call one species (i.e., Psathyrella candolleana) now receive individual species epithets?

ESSER: No.

PETERSEN: That is a different answer. (To Dr. Kemp:) Did you not just say they would?

KEMP: Yes, they should, but I do not think at this stage that it is useful to give them full binomials.

ESSER: On this they would not! Because of the ABC-system. If you have enough strains, and you find a way to fiddle around that you can eventually bring the genetic information from strain number 1, via strain 3, via 7, into strain 2.

KEMP: I do not think the ABC-system applies here. These strains of <u>Psathyrella</u> are completely lethal with each other, and they are completely genetically separate. So under this definition they would not be races - they are genetically completely separate, non-interbreedable newly formed species which lack known morphological differences. To call them races would not help.

ESSER: As I understood your scheme, do you not have what we call an ABC-system?

KEMP: No, not at all.

KUEHNER: Ayant suivi les travaux de Mr. J. Boidin et de son école, ainsi que ceux de Mlle. Lamoure, je suis devenu un très chaud partisan de l'utilisation, pour la délimitation des espèces, des tests interfertilité-interstérilité.
Je pense cependant que l'expérience que nous possédons actuellement n'est pas assez étendue pour qu'elle puisse nous permettre d'aborder la notion d'espèce par le biais de la génétique. Songez au nombre considérable d'espèces que nous ne savons pas cultiver et au nombre très faible des cas où l'on a réussi à obtenir des carpophores à partir de mycéliums secondaires issus de confrontations de mycéliums primaires compatibles et où l'on a essayé d'analyser la descendance. Pour moi le problème le plus urgent n'est pas d'essayer de définir l'espèce sur des critères génétiques; c'est bien plus de persuader les Systématiciens de formation traditionnelle d'utiliser la méthode expérimentale pour la délimitation des espèces dans chaque cas particulier.

ESSER: Monsieur, je ne suis pas d'accord avec vous, parce que la plupart des gens présente ici, have made the consence that his definition is the ideal one, which should be tried. There are not many taxa which are able to fulfill the requirements, and that is why we added a second part and have mixed our definition with that of Dr. Thiers. So we have then in this second part taken care of those taxa where for any reason (see genetic) requirements can not be fulfilled. I think that is a great advantage this decision has brought to us. We must have an ideal to look fore , at the time being, we must be aware of the incompleteness of human knowledge.

KUEHNER: Je ne cache pas que parmi les définitions de l'espèce proposées à l'issue de ce Symposium, celle du Dr. Singer avait ma préférence avant qu'il n'ait ajouté "and not being interfertile", cette adjonction n'étant qu'une concession à MM. Boidin et Esser, concession dont l'intérêt pratique n'est pas évident.

KEMP: I have read the last part of the Esser-Boidin definition many times and I am still unhappy. I do not think it is helpful to say that "provided that an absence of this interfertility is caused by those genetic parameters operating in the entire sexual cycle".

ESSER: Like more genes, less genes, etc.

SINGER: I have the impression that we have had as much approximation between very different views and that to-day we cannot force it any further. And, besides, I think, it is a good thing, that now we have everything down on paper. We should keep during the coming weeks and moths thinking about it, and then discuss it again.

KUEHNER: Quelle différence y a-t-il, dans la définition de Monsieur Thiers, entre une espèce et une variété?

CLEMENCON: Mais la variété est incluse.

KUEHNER: Ah oui, bien sur.

BAS: The second part of our definition actually applies to every taxonomic rank below the species, but I do not know how to avoid it without bringing in interbreeding again.

CLEMENCON: We should not try to define every taxonomic rank below the species.

PETERSEN: Now, I asked Roger Kemp a question, and I asked Karl Esser the same question, and I got two different answers. I think to myself, "Yes, I believe this" (unless, of course, this means intersterility groups within a morphological species, in which case I do not think they deserve species epithets). And if I believe that, then I am basing my species concept on morphology, and not on biology any more. I wind up very confused. I believe in that species definition up there (on the blackboard), except when it gets into Kemp's or a, and then it boggles my mind.
The second item is very small. In Harry Thiers' species concept I would like to see the word "cytological".

SMITH: When we get down to the rank of the species, and try to define it, we should try to do it within the context of the international rules of botanical nomenclature. And a form, and variety area, subspecies are all under the heading of species, so they are included in your definition of species. So we do not have to worry about it.

BLAICH: I think it is not necessary to consider heterogenic incompatibility. Either two strains are intersterile completely, and then, according to this definition, they are species. If they are not completely intersterile they belong to the same species. Not necessary to discuss heterogenic incompatibility. It is a step to complete intersterility.

WATLING: Thinking of Rolf's definition "A species in Agaricales, and probably in all of the so-called higher fungi, is the lowest and first genetically fixed step in evolution". That is the important bit, because, as I understand variety, subspecies and form, they are all interfertile. The variety differs from the type variety by only small differences, a form differing by ecological influences, and a subspecies, has some geographical implications.

BAS: I do not think it is true. I think that the same objection that has been made to Thiers' definition applies also here. If you say that a species is the lowest genetically fixed step in evolution, why not subspecies?

Discussion

WATLING: They are interfertile!

CLEMENCON: I got the impression that we are dealing with a very complex thing, and since the species concept is so complex we have to make a multidimensional image that is very hard to put into one statement or definition. We have, at our stage of knowledge, different statements that describe, from different view points and from different sides, the same thing. I think it is quite impossible to-day and for us, to make just one statement about the species concept.

SMITH: I want to point out the one element of fertility in our whole argument. We talk about things being interfertile and always interfertile, but if you were to make a million crosses you would probably find a few that would not cross. Well, then they are not completely interfertile. For any practical consideration that is OK, but for a definition, well, it creates an ambiguity. I think we have got to recognize it.

CLEMENCON: I think we should not use the word definition but characterization or something like that.

ESSER: Concept.

CLEMENCON: Concept, but not definition.

SMITH: Yes, that would please me a lot more.

Leçon Herbette

LA NOTION D'ESPECE CHEZ LES CHAMPIGNONS SUPERIEURS *)

R. Kühner

Introduction

Tout d'abord qu'est qu'un champignon supérieur?
Le Champignon de couche étant un champignon dit supérieur nous pouvons reconnaître sur lui ce qui caractérise les plus typiques des champignons supérieurs.

C'est la présence d'un appareil reproducteur massif, que chez les Champignon de couche et chez tous les Champignons auxquels il sera fait allusion dans cet exposé, on appelle carpophore. La partie que l'on consomme chez le champignon de couche correspond au carpophore. C'est bien un appareil reproducteur comme le montre une coupe pratiquée en travers d'une des lames qui rayonnent à la face inférieure du chapeau (Fig. 1). On y distingue, en effet, les spores, cellules qui, après s'être détachées, vont servir à la dissémination et à la reproduction du champignon.

Bien que le carpophore soit un appareil reproducteur massif, il n'est formé que des filaments, que l'on appelle des hyphes, et dont chacun n'est formé que d'une file de cellules. La figure citée montre, dans l'épaisseur d'une lame, ces hyphes qui, dans le cas présent, sont disposées parallèlement les unes par rapport aux autres; ailleurs elles peuvent se trouver enchevêtrées, formant un feutrage. C'est grâce à l'accolement des hyphes ou à leur enchevêtrement que l'appareil reproducteur des plus typiques des champignons dits supérieurs doit d'être plus ou moins massif et surtout relativement cohérent, bien que de structure fondamentalement filamenteuse.

Si nous examinons de plus près le mode de formation des spores chez le champignon de couche, nous voyons que les spores sont formées à

---

*) Adaptation d'une conférence publique de l'auteur, faite le 19 août 1976, à Lausanne, dans le cadre du Symposium international consacré à "La notion d'espèce chez les Hyménomycètes" organisé à l'Université de cette ville par M. le Professeur H. Clémençon, grâce à la fondation HERBETTE. Cette adaptation ne diffère de l'exposé oral de la conférence que par quelques modifications légères de forme et par une condensation de l'Historique.

l'extérieur de cellules particulières, dont chacune porte les spores à l'extrémité de petits diverticules en forme de cornes dits stérigmates. On appelle basides ces éléments qui produisent les spores à leur extérieur; aussi dit-on que le Champignon de couche est un Basidiomycète. On remarque que les basides sont disposées les unes à côté des autres, en une couche qui tapisse les lames. Cette couche a été appelée hyménium et l'on appelle Hyménomycètes, tous les Basidiomycètes dont l'hyménium est exposé à l'air libre comme celui du champignon de couche.

Dans notre exposé, il ne sera question que d'Hyménomycètes et nous choisirons nos exemples uniquement dans l'ensemble d'Hyménomycètes que l'on appelle aujourd'hui Agaricales, ensemble qui comprend les Bolets et, avant tout, les Champignons qui, comme le Champignon de couche, présentent des lames à la face inférieure d'un chapeau. L'unique raison de ce choix est que, dans l'ensemble des Hyménomycètes, l'ordre des Agaricales est le mieux connu de la plupart des Mycologues amateurs et même des simples mycophages.

Pour le Naturaliste débutant, la notion d'espèce peut sembler s'imposer. Tout Mycologue connaît l'Amanite tue mouches et l'Amanite phalloïde; ce sont deux espèces. Les scientifiques désignent chaque espèce par deux noms latins; c'est ainsi que les deux Amanites citées à l'instant s'appellent respectivement Amanita muscaria et Amanita phalloides. Pourquoi deux noms? C'est que les espèces sont si nombreuses qu'il faut les classer pour s'y reconnaître: muscaria et phalloides sont deux espèces classées dans un même genre Amanita.

Dans certains cas on distingue plusieurs variétés dans une espèce. Par exemple, dans l'espèce Inocybe geophylla, on distingue la variété type, dont chapeau et stipe sont blancs dès l'origine, et la variété lilacina, dont chapeau et stipe sont violacés au début. Pourquoi ne fait-on pas deux espèces de ces deux champignons puisqu'ils sont si facilement distinguables par leur couleur? C'est parce qu'ils semblent ne différer par rien d'autre que la couleur. Deux espèces diffèrent par plus d'un caractère, comme le montre l'exemple des deux Amanites citées plus haut. Les personnes non initiées aux Sciences naturelles utilisent rarement le mot espèce; elles appellent couramment variétés, à la fois nos variétés et nos espèces, voire nos genres.

Leçon Herbette   411

Les Mycologues ayant déjà une certaine expérience savent que bien des espèces sont loin d'être aussi faciles à distinguer les unes des autres que les deux Amanites citées à l'instant; ils savent que, dans nombre de cas, la délimitation des espèces est chose délicate et qui peut être sujette à discussion. Quels moyens utiliser pour délimiter les espèces? Tel est le sujet de cet exposé.

D é f i n i t i o n   e t   r e c o n n a i s s a n c e   d e s
e s p è c e s   n e   c o n s t i t u e n t   q u ' u n e   p a r -
t i e   d e   l a   S y s t é m a t i q u e   o u   T a x i n o m i e
Prise au sens large, la Systématique ou Taxinomie s'occupe de la classification des organismes, par exemple du rattachement de plusieurs variétés à une même espèce, du groupement de plusieurs espèces en genres, de plusieurs genres en familles, etc... Chacune de ces unités systématiques, variété, espèce, genre, etc. constitue ce qu'on appelle un taxon.

Chacun sait qu'en ce qui concerne les Agaricales, la publication de l'ouvrage que Fries a intitulé "Epicrisis sistematis mycologici" marque une date essentielle dans l'histoire de la systématique. Cet ouvrage, publié de 1836 à 1838, aeu, en 1874, une seconde édition assez légèrement remaniée, mais présentée sous un titre nouveau: "Hymenomycetes Europaei"; il s'agit de la dernière expression d'ensemble de l'oeuvre du Maître suédois; rien que parmi les Hyménomycètes qui, comme le champignon de couche, présentent les lamelles, Hyménomycètes que Fries appelait Agaricini, cet auteur y décrivait quelque 1800 espèces, qu'il classait en 20 genres, dont l'énorme genre Agaricus, lui même subdivisé en 35 sous-genres.

Nous qui ne saurions nous passer du microscope ne pouvons qu'être émerveillés en parcourant l'oeuvre de Fries, entièrement exécutée sans cet instrument. Elle nous prouve, ce que savent bien les Mycologues de terrain et les autres, qu'une foule d'espèces peuvent être reconnues sans microscope, mais, en nombre de points, elle matérialise les difficultés de la transmission écrite de la connaissance.

Même des descriptions détaillées, comme celles publiées en 1833 par Secrétan, dans l'ouvrage intitulé "Mycographie suisse. Description des champignons qui croissent en Suisse", laissent souvent perplexe,

en partie parce que nous les comprenons mal, en partie parce qu'elles
sont malgré tout incomplètes. Dans la manière de décrire l'aspect
d'un carpophore, il reste certainement de grands progrès à accomplir
puisque bien des espèces peuvent être reconnues au premier coup
d'oeil, alors que nous ne savons pas toujours dire pourquoi.

Un demi siècle après Secrétan, en 1889, c'est-à-dire juste après la
période friesienne, V. Fayod publiait son "Prodrome d'une Histoire
naturelle des Agaricinés", dans lequel l'auteur se préoccupe uniquement de la classification des espèces, et, en aucune façon, de leur
définition. A première vue, nous n'aurions donc pas dû mentionner ce
mémoire dans le cadre de notre exposé; si nous l'avons fait c'est
parce qu'en utilisant pour la première fois les caractères microscopiques les plus divers pour réaliser sa classification, Fayod a montré, en même temps, l'intérêt que présentent, pour les descripteurs
d'espèces, une foule de caractères microscopiques trop longtemps négligés. En d'autres termes, Fayod nous a enseigné la manière de décrire les caractères microscopiques d'un champignon comme Fries nous
avait appris à décrire ses caractères sensibles à l'oeil nu ou à la
loupe.

Il est hors de question d'essayer de citer, dans le cadre limité de
cet exposé, les nombreux Mycologues qui ont fait progresser nos connaissances de lecteur saisira de mieux en mieux au fil de notre exposé, nous n'en citeron qu'un seul: Jules Favre.

Les importants mémoires publiés par Favre de 1936 à 1960 sont d'ailleurs remarquables à divers points de vue.

Ils sont remarquables tout d'abord par la qualité des descriptions et
des illustrations, qui n'a guère été dépassée. Favre est l'un des
très rares Mycologues ayant su allier avec bonheur la concision des
descriptions avec leur précision. Concernant l'illustration de ses
mémoires, on ne saurait omettre de citer Madame Favre, qui fut la fidèle collaboratrice de son mari, au service de qui elle a mis son
beau talent d'aquarelliste.

L'oeuvre de Favre est encore remarquable par son originalité; cet auteur a en effet étudié les champignons de groupements végétaux particuliers: c'est ainsi qu'il a traité, en 1948, des associations fongiques des hauts marais jurassiens, milieux relativement peu explorés par la plupart des Mycologues, et, en 1955, des champignons qui poussent dans les hautes montagnes des Grisons, au-dessus de la forêt, c'est-à-dire dans la zone dite alpine, dont il a été le premier à inventorier les richesses fongiques de façon systématique. Nous aurons à revenir, au cours de cet exposé, sur ce dernier mémoire de 212 p. et 11 pl., qu'il a intitulé: "Les Champignons supérieurs de la zone alpine du Parc national suisse", et qui a été publié dans le cadre consacré aux "Résultats des recherches scientifiques entreprises au Parc national suisse".

D i v e r s i t é   d e s   c a r a c t è r e s   u t i l i s é s
p o u r   d é f i n i r   l e s   e s p è c e s

Pour définir, comme d'ailleurs pour classer les espèces, ont toujours joué un rôle essentiel les caractères des champignons que l'on peut représenter par le dessin ou par la photographie, ce qu'on appelle des caractères morphologiques. Les Icones publiées depuis les premiers balbutiements de la Mycologie en sont le témoignage le plus direct; la découverte du microscope et ses perfectionnements n'ont fait qu'étendre de plus en plus la gamme des caractères morphologiques connus. Longtemps limitées au carpophore, les recherches morphologiques se sont étendues dans la période moderne au mycélium, grâce à la poccibilité d'obtenir des cultures pures pour d'assez nombreuses espèces.

Récemment ont été faites des tentatives pour ajouter à cette panoplie de caractères morphologiques des caractères chimiques; on a par exemple montré que le stock d'enzymes peut varier qualitativement d'une espèce à une autre, même lorsque celles-ci sont voisines

Si pour définir ou pour déterminer une espèce on fait avant tout appel à des caractères du champignon lui-même, il ne faut pas oublier que, dans plusieurs cas, l'environnement du champignon fournit des caractères distinctifs que l'on aurait tort de négliger.

Afin de ne pas alourdir cet exposé, nous ne traiterons dans ce qui suit que de quelques caractères écologiques et de quelques particularités microscopiques.

Importance des caractères écologiques pour la délimitation des espèces

Chacun sait que certaines espèces de Bolets ne se trouvent que dans le voisinage de Mélèzes, d'autres qu'à proximité de Pins à 5 aiguilles. Comme ces essences n'existent pas en Scandinavie, Fries n'avait pas vu les Bolets en question dans la nature.

Entre 1830 et 1840 ont été décrits trois Bolets du Mélèze; en particulier le Bolet que l'on a pris depuis longtemps l'habitude d'appeler viscidus a été décrit dès 1833 par Secrétan sous l'étiquette aeruginascens, qui a la priorité.

Les Bolets qui ne poussent que sous des Pins à 5 aiguilles sont des espèces proches de B. granulatus. Le premier décrit de ces Bolets est un Bolet sans anneau, comme granulatus, dont le distingue immédiatement son chapeau blanc ou ivoire au début qui lui a valu le nom de Boletus albus que lui donna Peck lorsqu'il le décrivit d'Amérique du nord en 1873. Il est répandu en Europe dans les plantations des régions basses d'un Pin à 5 aiguilles importé d'Amérique dans la première moitié du siècle dernier, Pinus strobus, reconnaissable en particulier à ses cones allongés et pendants. C'est d'ailleurs de telles plantations d'Allemagne qu'il a été décrit pour la première fois sous l'étiquette Boletus placidus qui a la priorité sur Boletus albus car publiée en 1861, c'est-à-dire 12 ans avant la publication de B. albus. Quoi qu'il en soit on a découvert beaucoup plus tard B. placidus sous un autre Pin à 5 aiguilles, Pinus cembra, l'Arole, arbre indigène en Europe.

Deux autres Bolets n'ont été rencontrés que sous ce dernier Pin: il s'agit de B. plorans, sans anneau, et de B. sibiricus, qui possède un anneau. Ces Bolets spéciaux à l'Arole n'ont été découverts qu'à une époque relativement récente. C'est de l'arc alpin que Rolland a décrit pour la première fois B. plorans, mais cet auteur le croyait lié au Mélèze; c'est que, comme l'a fait remarquer Singer, le Mélèze se trouve mêlé à l'Arole dans la station explorée par Rolland, comme c'est d'ailleurs souvent le cas dans les Alpes, à la limite supérieure de la forêt.

Après avoir été découvert dans nos Alpes, B. plorans a été retrouvé en Asie centrale, dans l'Altaï, par Singer. C'est que Pinus cembra existe, non seulement dans les hautes montagnes d'Europe, où il n'occupe qu'une surface d'ensemble réduite, mais encore en Asie occidentale et centrale, où il occupe des surfaces énormes, formant une aire continue, même en régions basses (Fig. 2).

En fait l'Arole asiatique est différent de l'Arole européen, mais il l'est si peu qu'on ne saurait douter que tous deux dérivent d'un ancêtre commun qui a dû occuper une aire englobant les deux aires actuelles: l'aire européenne et l'aire asiatique. Lorsque, par suite de changements climatiques, cette aire s'est trouvée coupée en deux aires séparées l'une de l'autre, les descendants de cet ancêtre commun ont évolué dans des directions différentes en Europe et en Asie; ils ont pu le faire du fait que l'énorme distance séparant l'aire européenne de l'aire asiatique interdisait la pollinisation entre les Aroles d'Europe et ceux d'Asie. Certains auteurs considèrent que notre Arole est une espèce différente de l'Arole asiatique, qu'ils appellent Pinus sibiricus, mais d'autres pensent que les différences entre ces deux aroles sont trop faibles pour qu'il soit raisonnable de les considérer comme deux espèces distinctes; ils disent que Pinus cembra présente deux sous-espèces: la sous-espèce type et la sous-espèce sibiricus. Si l'évolution séparée se poursuivait et si l'homme pouvait la suivre assez longtemps il est probable qu'il constaterait une accentuation des différences telle que tous les botanistes seraient alors d'accord pour admettre que l'Arole européen et l'Arole asiatique sont deux espèces distinctes. Pour l'instant nous avons l'impression d'assister à la naissance progressive de deux espèces à partir d'une seule.

Pour nous, Mycologues, il est intéressant de noter que, selon Singer, le B. plorans qui pousse sous la sous-espèce sibiricus de l'Arole est légèrement différent du plorans de l'Arole d'Europe, notamment par sa chair bleuissant régulièrement à la coupure. Singer fait du plorans asiatique une sous-espèce du plorans d'Europe. Aux évolutions divergentes de Pinus cembra correspondraient des évolutions divergentes de Boletus plorans, toutes deux facilitées par l'isolement géographique.

Selon Singer Boletus sibiricus comprendrait également deux sous-espèces, correspondant aux deux sous-espèces d'Aroles; son nom spécifique rappelle que, contrairement à B. plorans, B. sibiricus a été découvert en Asie avant d'avoir été repéré en Europe.

Si les Bolets que nous venons de citer accompagnent fidèlement des essences déterminées, c'est qu'ils vivent en association, on dit en symbiose, avec les arbres en question, leur appareil végétatif, appelé mycélium, formant avec les racines de l'arbre associé des complexes dits mycorhizes, qui sont bénéfiques, à la fois au champignon et à l'arbre. Il est fort possible que les premiers Boletus placidus reconnus en Allemagne, dans des plantations de Pinus strobus, aient été involontairement importés d'Amérique avec de jeunes plants de ce Pin, qui présentaient déjà des racines mycorhizées par ce Bolet.

Favre a été frappé par le fait qu'en zone alpine les espèces les plus nombreuses, notamment celles qui appartiennent aux genres Cortinarius et Inocybe, ne viennent pas dans les prairies; elle viennent dans les tapis d'arbrisseaux nains que forment les Dryades et certains Saules, certainement parce que la plupart d'entre elles forment des mycorhizes avec ces plantes ligneuses naines, dont les tapis sont, en zone alpine, l'équivalent des forêts situées à des altitudes plus faibles.

Parmi les champignons capables de faire des mycorhizes, il y en a beaucoup qui n'ont pas une spécificité de partenaire aussi étroite que celle évoquée plus haut pour certains Bolets, c'est-à-dire que bien des champignons ont la possibilité de faire des mycorhizes avec des arbres d'essences différentes. D'autre part, parmi les champignons susceptibles de faire des mycorhizes avec une ou plusieurs essences, il y en a qui peuvent former leurs carpophores en l'absence de ces essences, c'est-à-dire en l'absence de symbiose mycorhizique.

Enfin il y a beaucoup de champignons qui ne forment jamais de mycorhizes; sont dans ce cas notamment les champignons qui poussent sur les excréments, comme nombre de Coprins; sont également dans ce cas les petits Marasmes dont les carpophores sont greffés sur les feuilles mortes; plusieurs de ces derniers ne poussent que sur des feuilles d'espèces déterminées, de buis (M. buxi), de houx (M. hudsoni) de lierre (M. epiphylloides); ces Marasmes nous montrent que, même en l'absence de relations mycorhiziques, les exigences de certains champignons vis-à-vis du milieu peuvent être très strictes. Les conditions requises par nombre d'autres champignons sont plus difficilement reconnaissables, ce qui souligne que, dans le domaine des rapports entre les champignons et leur environnement, il reste encore beaucoup à faire. Aux Mycologues de terrain, qui connaissent déjà bien un certain nombre d'espèces, nous ne pouvons que conseiller de s'intéresser à l'environnement des champignons.

Quelques caractères microscopiques utilisés pour définir les espèces

Caractères des spores.

La spore est l'élément microscopique auquel les systématiciens se sont tout d'abord intéressés. Ils s'y sont même intéressés avant l'utilisation du microscope; chacun sait l'importance que Fries accordait à la couleur de la sporée dans la définition des genres ou sous-genres.

La paroi sporique.

C'est très généralement la paroi de la spore qui est responsable de la couleur de la sporée; c'est aussi cette paroi qui conditionne les dimensions, la forme et l'ornementation de la spore, tous caractères dont on connaît l'importance systématique.

Chacun sait que l'on ne peut aborder la détermination de la plupart des Agaricales à spores brunes que sont les Inocybe sans savoir si les spores sont lisses ou gibbeuses (Fig. 3).

Pour les genres d'Agaricales qui renferment des espèces à spores blanches, il est souvent utile d'examiner les spores dans le réactif de Melzer. C'est en 1924 que Melzer s'est aperçu qu'en ajoutant de l'hydrate de chloral aux solutions aqueuses iodoiodurées classiques, on obtient un réactif qui colore en bleu-noir les ornements des spores des Russules, lesquels se détachant alors sur un fond beaucoup plus pâle, sont rendus particulièrement distincts; de tels ornements sont dits amyloïdes. L'utilisation de ce réactif a permis à Melzer de distinguer trois principaux types de spores de Russules, d'après leur ornementation: les spores aculéolées, qui sont ornées d'aiguillons isolés, les spores réticulées, qui diffèrent des précédentes par la présence de fines lignes unissant les aiguillons et formant un réseau, enfin les spores à crêtes, crêtes fines ou épaisses (Fig. 4). Melzer a montré les services que peut rendre une étude précise de l'ornementation des spores pour la détermination des Russules.

Dans l'ensemble des Agaricales à spores blanches autres que les Russulacées, le réactif de Melzer s'est révélé fort précieux, comme nous l'avons montré en 1934, après avoir examiné systématiquement dans ce réactif toutes les espèces contenues dans l'herbier de R. Maire. On trouve là des espèces à spores ornées, les unes à ornements amyloïdes comme ceux des Russulacées, les autres à ornements non amyloïdes; on y trouve aussi des espèces à spores lisses dont la paroi se colore en gris-bleu dans le Melzer (elle est amyloïde) et d'autres espèces à spores lisses, dont la paroi ne présente pas ce caractère. Il n'est pas rare que, dans un même ensemble d'espèces à spores lisses, un genre par exemple, se trouvent à la fois des espèces à spores amyloïdes et d'autres dont les spores ne présentent pas ce caractère; c'est ainsi que, dans l'ensemble très difficile des Clitocybe hygrophanes, les spores ne sont généralement pas amyloïdes, sauf chez C. syathiformis et un petit nombre d'espèces voisines; la détermination de ces derniers Clitocybe de trouve donc facilitée par l'emploi du réactif de Melzer.

Fayodia (Omphalia) bisphaerigera est l'une des rares espèces qui, dans les limites de la flore française, puisse être identifiée au seul examen de sa spore (Fig. 5); bien que celle-ci présente des ornements très développés, elle est lisse, car un mince plafond, dit ectospore, est tendu au-dessus des ornements; ce plafond est particulièrement reconnaissable dans le Melzer car il est amyloïde alors que les ornements ne le sont pas. Dans la flore française nous ne connaissons aucune autre Agaricale à spores blanches dont la paroi sporique présente ces caractères réunis.

Le contenu de la spore.

Ce contenu a été fort peu étudié. Signalons que dans quelques ensembles d'Agaricales à spores blanches, dominent les spores à un noyau dans certaines espèces, les spores à deux noyaux dans d'autres. L'étude que nous avons faite à ce point de vue des Hygrophores du sous-genre Hygrocybe nous a convaincu de l'importance de la détermination du stock nucléaire de la spore pour la définition et par conséquent pour la reconnaissance des espèces.

Caractères des Cystides.

Après les spores ce sont les cystides qui ont attiré l'attention des systématiciens, car, en recherchant les spores sur h'hyménium, les mycologues ne pouvaient manquer de voir les cystides, ces éléments stériles étant, soit disséminés parmi les basides, soit placés à proximité de celles-ci, par exemple sur la tranche des lames.

Très tôt on s'est aperçu que les cystides présentent un grand intérêt systématique. Certaines espèces ont des cystides alors que d'autres n'en ont pas, et les espèces qui possèdent des cystides peuvent différer les unes des autres par la répartition de ces cystides ou par les caractères de celles-ci.

On sait, par exemple, que dans le genre Inocybe (Fig. 6) certaines espèces ne présentent de cystides que sur la tranche des lames, alors que d'autres en présentent sur leurs faces. On siat aussi que les cystides faciales des Inocybe sont généralement couronnées de cristaux d'oxalate de chaux, que leur paroi est souvent épaissie dans leur partie supérieure et qu'elles sont souvent ventrues ou en bouteille. Chez les Inocybe qui n'ont de cystides que sur la tranche des lames, les cystides sont d'une autre nature: elles ne sont pas oxalifères; leur paroi est généralement mince et, dans la plupart des cas, elles ne sont pas en forme de bouteille.

Dans le genre Mycena, la forme des cystides peut varier beaucoup d'une espèce à une autre (Fig.)). Dès 1889 Schroeter avait reconnu que le sommet des cystides, simplement atténué ou arrondi dans certaines espèces, présente, chez d'autres, des appendices plus ou moins nombreux et plus ou moins allongés. Il arrive que ces deux formes extrêmes de cystides s'observent dans des espèces qui se ressemblent passablement par l'aspect extérieur, parfois au point d'avoir été confondues. C'est surtout sur la tranche des lames que Schroeter avait repéré les cystides; c'est évidemment là qu'on les voit facilement sans faire de coupes, simplement sur un lambeau de lame que l'on examine pour y rechercher les spores; mais, grâce au fait que, dans quelques espèces, le contenu des cystides est coloré, il s'était aperçu que certaines Mycènes ont des cystides également sur les faces des lames.

Dans la plupart des Russulacées le contenu des cystides est remarquable. D'abord il est bourré de granules qui le rendent très réfringent, comme l'avaient déjà reconnu d'anciens auteurs (Fayod a figuré une telle cystide). Ensuite ce contenu présente généralement une réaction colorée caractéristique, qui a été découverte par Arnould et Goris dès 1907: si des coupes sont traitées par la vanilline et par l'acide sulfurique, il se colore en bleu plus ou moins foncé, tranchant sur la coloration rose ou rouge que prennent les cellules foncé, tranchant sur la coloration rose ou rouge que prennent les cellules voisines, les basides notamment. Arnould et Goris ont cependant rencontré quelques exceptions: c'est ainsi qu'ils ont été amenés à individualiser leur Russula pseudointegra, espèce qui, selon eux, ressemble à R. integra, mais qui en diffère par le fait que le contenu de ses cystides se colore seulement en rose par le réactif sulfovanillique. Depuis, on a montré que l'aldéhyde qu'est la vanilline peut être remplacée, parfois avantageusement, par d'autres aldéhydes. Quand le contenu des cystides se colore de façon très différente du contenu des cellules environnantes, par exemple en bleu par la sulfovanilline, on dit que ces cystides sont sulfoaldéhyde-positives; on dit qu'elles sont sulfoaldéhyde-négatives dans le cas contraire.

Caractères structuraux de la partie superficielle du chapeau.

La structure de la partie superficielle du chapeau, que Fayod appelait "cuticule" et que nous appelons volontiers "revêtement piléique", a été longtemps négligée par les systématiciens, absorbés par l'étude des spores et des cystides hyméniales. Dès 1889 Fayod écrivait: "Il est singulier et déplorable que l'on ait presque entièrement négligé jusqu'ici l'étude de la cuticule", et encore: "Comme la cuticule offre le plus souvent des caractères tant génériques que spécifiques excellents, il serait à désirer qu'on en tint compte à l'avenir". Comme l'a fait remarquer Melzer, cet appel resta longtemps sans écho; pour les Russules par exemple, c'est seulement vingt années plus tard que R. Maire utilisa des caractères anatomiques du revêtement piléique pour classer les espèces, comme nous allons le voir.

Dans plusieurs Agaricales qui présentent des cystides sur les lames on trouve, à la surface du chapeau, des éléments qui leur ressemblent parfois tellement qu'on est obligé de les appeler cystides bien qu'ils ne soient pas accompagnés de basides, puisque ces dernières manquent toujours à la surface du chapeau; on parle de piléocystides pour éviter toute confusion avec les cystides des lames.

Par exemple de très nombreuses espèces de Russules présentent des piléocystides, que l'on repère facilement parmi les hyphes grêles qui constituent le fond du revêtement (Fig. 8). Dès 1889 Fayod a donné un dessin de coupe verticale dans le revêtement du chapeau d'une Russule où l'on reconnaît sans difficulté les piléocystides. L'homologie des piléocystides avec les cystides des lames ne peut faire de doute chez les Russules. En effet, dans une foule d'espèces de ce genre le contenu des piléocystides est tout à fait comparable à cului des cystides des lames. On y retrouve les mêmes granulations réfringentes, qui manquent aux hyphes grêles entre lesquelles elles se glissent; simplement, le système vacuolaire étant plus également réparti dans les piléocystides, il est plus facile de voir que ces granulations ne se trouvent pas dans les vacuoles; on les voit dans le cytoplasme qui tapisse la paroi et dans celui qui forme les travées séparant les vacuoles les unes des autres. En outre, comme l'ont reconnu Arnould et Goris dès 1907, le contenu des piléocystides se colore en bleu par la sulfovanilline, comme celui des cystides des lames, dans une foule d'espèces de Russules. Dans ce genre l'homologie des piléocystides avec les cystides des lames est donc absolument évidente; tout au plus peut-on dire que si les cystides des lames y sont toujours unicellulaires, les piléocystides y sont fréquemment formées d'une file d'un très petit nombre de cellules.

En 1910 R. Maire regroupe dans une section Alutaceae quelques espèces, dont R. alutacea et R. speudointegra, qui diffèrent des Russules qui leur ressemblent, par l'absence, dans leur revêtement piléique, de cystides à contenu colorable en bleu par la sulfovanilline.

Si, dans certaines Alutaceae, comme R. aurata, les cellules du revêtement piléique sont d'une seule sorte, dans d'autres on trouve, dans ce revêtement, à côté de cellules banales, des hyphes remarquables que Melzer, qui les a découvertes en 1934, a nommé "hyphes primordiales". La Fig. 9 reproduit le dessin que cet auteur a publié des hyphes primordiales de R. pseudointegra. On y voit que l'hyphe primordiale est caractérisée par le fait que sa paroi est incrustée extérieurement, au moins sur une partie de sa longueur. Melzer a montré que cette incrustation, difficilement visible sur le vivant, se colore fortement par la fuchsine basique de Ziehl et que la coloration qu'elle prend résiste relativement bien à l'action brève d'une solution diluée d'acide chlorhydrique, qui décolore le reste de la préparation; des incrustations qui se comportent ainsi sont dites acido-résistantes.

En dehors des Russules, la connaissance de la structure de la surface du chapeau peut permettre de distinguer commodément des espèces voisines. Tout d'abord dans certaines espèces, les cellules qui forment la surface du chapeau sont allongées, ce qui fait dire que leur revêtement est filamenteux, alors que dans d'autres ces cellules sont courtes, parfois rondes; c'est ce second type de revêtement que l'on appelle celluleux. La figure 10 montre qu'il est impossible lorsque l'on connaît la structure de la surface de leur chapeau, de confondre Boletus griseus (=B. carpini) à revêtement celluleux, avec B. scaber, à revêtement filamenteux. Certains revêtements, typiquement celluleux lorsque vus par dessus, montrent, sur une coupe verticale du chapeau, des cellules dressées, disposées les unes à côté des autres comme les basides dans un hyménium, d'où le nom d'hyméniformes que l'on attribue aux revêtements qui présentent cette structure qu'illustre la Fig. 10 chez deux Marasmes. On y voit que Marasmius epiphylloides, voisin, comme le rappelle son nom, de M. epiphyllus, s'en distingue très nettement par la structure de la surface du chapeau, dont les cellules sont hérissées en brosse par des diverticules de leur partie supérieure.

Il est rare que la structure du revêtement piléique d'une espèce soit suffisamment originale pour assurer à elle seule, une détermination sûre. Voici cependant un exemple de ce cas. En France, comme en Suisse, Clitocybe hydrogramma est apparemment la seule Agaricale présentant, dans son revêtement piléique (Fig. 11), à base de cellules cylindriques, grêles, des cellules enflées en ampoules, dont chacune renferme une inclusion muriforme, de nature indéterminée. Dès 1889 Fayod avait repéré et figuré de telles ampoules chez un Clitocybe qu'il déterminait candicans, probablement par erreur, car ce que nous appelons candicans n'en présente pas.

Nous ne pouvons terminer ce bref survol relatif aux caractères microscopiques utilisables pour définir et par conséquent pour déterminer les espèces sans évoquer deux thèmes qui n'avaient pas attiré l'attention de Fayod, à savoir la localisation, à l'échelle cellulaire, des substances naturellement colorées, de ces substances que l'on ap-'elle pigments, et la présence ou l'absence de boucles aux hyphes et aux basides.

Localisation des pigments à l'échelle cellulaire.

La connaissance de la microtopographie de la pigmentation peut rendre d'importants services aux systématiciens.

C'est notamment le cas chez les Rhodophyllus. Dans ce genre deux types de localisation pigmentaire sont possibles (Fig. 12). La pigmentation peut être intracellulaire, les substances colorées étant initialement dissoutes dans le suc vacuolaire (dans lequel elles précipitent assez souvent par la suite) ou bien les substances colorées se trouvent au niceau de la paroi cellulaire, très souvent dans des incrustations de la surface extérieure de celle-ci. Les deux types de pigmentation peuvent coexister dans une même espèce, voire dans une même hyphe.

Si de nombreux Rhodophyllus du sous-genre Entoloma présentent une pigmentation intracellulaire, au moins dans la partie supérieure de leur trame piléique, pigmentation accompagnée chez certains d'incrustations pigmentaires de la paroi, quelques espèces de ce sous-genre, même des espèces de coloration sombre, ne doivent celle-ci qu'à la paroi cellulaire, particulièrement à ses incrustations superficielles, le contenu des cellules vivantes n'étant pas coloré. R. sericeus illustre ce dernier cas. R. sericatus, qui peut ressembler beaucoup à R. sericeus, en diffère fort nettement par une pigmentation intracellulaire évidente.

Les boucles et la reproduction des Hyménomycètes.

Sur la Fig. 13 on voit, à droite, des hyphes sans boucles, à gauche des hyphes bouclées; on reconnaît que la boucle est une formation qui se trouve au niveau d'une cloison, mais seulement sur un côté de celle-ci.

Fréquence et répartition des boucles.

La fréquence et la répartition des boucles dans le carpophore constituent des caractéristiques importantes de certaines espèces que l'on ne peut négliger lorsque l'on étudie les Rhodophyllus. Chez les Entolomes les plus typiques, dont fait partie R. sericatus, les boucles sont abondantes dans toutes les parties du carpophore, alors que chez R. sericeus, elles n'abondent qu'au pied des basides et dans la région sous-hyméniale; elles manquent à nombre de cloisons des autres hyphes et sont pratiquement absentes aux hyphes du stipe, même aux hyphes grêles, où elles seraient pourtant faciles à repérer.

En étudiant la partie de l'herbier de J. Favre consacrée à la zone alpine, nous nous sommes aperçu que cet auteur a confondu, sous la dénomination spécifique Rhodophyllus sericeus, à la fois notre sericeus (qu'il décrit sous le nom de f. luridofuscus) et des champignons ayant les caractéristiques microscopiques indiquées plus haut pour sericatus, champignons qu'il rapporte à deux formes de sericeus, ses f. flexipes et nanus. Nous ne nous étonnons pas que Favre n'ait pas remarqué la pigmentation intracellulaire de ces deux dernières formes, car si cette pigmentation est absolument évidente sur matériel d'herbier, elle n'est bien sensible sur le vivant qu'en réalisant une plasmolyse. Mais il est intéressant de noter qu'en observation à sa f. luridofuscus (c'est-à-dire notre sericeus) cet auteur a écrit: "Je n'ai pu arriver à constater de boucles aux hyphes de cette forme et, si ce fait se révélait constant il faudrait probablement la considérer comme espèce propre, car les f. nanus et flexipes en montrent assez abondamment".

Si nous avons tenu à souligner la facilité avec laquelle sericeus peut être confondu avec des formes proches de sericatus, c'est évidemment pour inciter les agaricologues qui étudient les Entolomes à ne pas négliger, comme on l'a trop souvent fait, de rechercher si l'espèce étudiée présent ou non une pigmentation intracellulaire, et de préciser la fréquence et la répartition des boucles. Mais c'est aussi pour souligner à quel point Favre avait un sens aigu de l'espèce, puisqu'il soupçonnait la confusion qu'il faisait, décrivant prudemment, de façon séparée, les formes qu'il rapportait à tort à la même espèce; R. sericeus.

Les Rhodophylles dont il vient d'être question nous ont appris que, chez certaines espèces d'Agaricales, on trouve une boucle pratiquement à chaque cloison, alors que chez d'autres les boucles sont plus ou moins dispersées ou n'existent que dans certaines parties du carpophore, par exemple dans la région hyménium-sous-hyménium, pouvant manquer totalement dans d'autres parties, le stipe par exemple. Mais on peut également rencontrer des champignons entièrement dépourvus de boucles dans toutes leurs parties, non seulement carpophore, mais parfois aussi mycélium. L'absence de boucles dans toutes les parties du champignon est dans certains cas un bon caractère d'espèce, mais dans d'autres cas elle n'est que le résultat d'un accident survenu dans la reproduction normale de l'espèce. Pour le comprendre il est nécessaire de connaître ....

Les grandes lignes de la reproduction des Hyménomycètes (Fig. 14).

La baside est le siège des deux points cardinaux du cycle nucléaire. Comme la cellule sous-hyméniale qui la porte, la baside renferme à l'origine deux noyaux à n chromosomes; alors que les deux noyaux de la cellule sous-hyméniale restent distincts, les deux noyaux de la baside fusionnent en un seul noyau à 2 n chromosomes. Cette fusion nucléaire est l'un des points cardinaux du cycle. Le noyau de fusion se divise de manière à donner les noyaux des spores; c'est au cours des deux premières divisions que le nombre des chromosomes repasse de 2 n à n; c'est ce qu'on appelle la réduction chromatique ou méiose, qui est l'autre point cardinal du cycle nucléaire. Chaque noyau de spore ne renferme donc que n chromosomes.

Dans la plupart des espèces les deux noyaux qui fusionnent dans la jeune baside sont les descendants de noyaux de deux spores différentes, pas n'importe lesquelles, mais de deux spores convenables, dites compatibles; sur la fig. 14 une spore à noyau noir est compatible avec une spore à noyau blanc, mais deux spores à noyaux identiques sont incompatibles. Dans de telles espèces puisqu'il y a plusieurs sortes de spores, il y a plusieurs sortes de mycéliums ou thalles résultant de leur germination, ici des mycéliums ou thalles à noyaux noirs et des mycéliums ou thalles à noyaux blancs; c'est pourquoi on dit que de telles espèces sont hétérothalles.

Le mycélium issu directement de la germination de la spore est dit

mycélium primaire. Il y a en effet un autre type de mycélium, celui
que l'on appelle mycélium secondaire, qui résulte de l'union d'hyphes
de deux mycéliums primaires compatibles. Au point de vue nucléaire le
mycélium secondaire diffère des deux mycéliums primaires qui lui ont
donné naissance par la présence dans chacune de ses cellules, de deux
noyaux compatibles. Chez les espèces qui présentent des boucles au
niveau du mycélium, celles-ci se trouvent uniquement sur les mycéliums secondaires; les mycéliums primaires en sont constamment dépourvus.

La signification de l'absence totale de boucles.

Si dans nombre d'espèces hétérothalles le mycélium secondaire peut
seul produire des carpophores, dans d'autres espèces hétérothalles
certains mycéliums primaires sont susceptibles d'en porter également.
Les carpophores issus de mycéliums primaires diffèrent des autres par
leur comportement nucléaire: étant issus d'une seule spore, leur baside ne renferme dès son origine qu'un seul noyau, comme la cellule
qui la porte; il n'y a donc pas de fusion nucléaire, plus rien qui
rappelle les cycles nucléaires normaux des organismes sexués; on dit
que de tels carpophores sont parthénogénétiques. Ces carpophores sont
naturellement dépourvus de boucles puisqu'ils sont nés de mycéliums
primaires et que ces derniers sont constamment sans boucles, même
chez les espèces dont les mycéliums secondaires en présentent.

Hygrophorus conicus est une espèce qui produit fréquemment des carpophores parthénogénétiques dans la nature; à côté de carpophores à
boucles abondantes, dont la baside renferme à l'origine deux noyaux
qui fusionnent, elle présente des carpophores totalement dépourvus de
boucles et dont la baside ne renferme à l'origine qu'un seul noyau.
Chez cette espèce, l'absence de boucle ne se présente donc que dans
certains carpophores, les carpophores parthénogénétiques.

L'absence de boucles est dans certains cas un caractère spécifique.
On connaît en effet des champignons totalement dépourvus de boucles,
même dans les carpophores non parthénogénétiques; se comportent ainsi,
presque sans exception, les Psalliotes (Fig. 1) et les Russulacées.

Le nombre de spores par baside.

C'est le dernier caractère microscopique auquel il sera fait allusion dans cet exposé.

On sait que dans la majorité des carpophores d'Agaricales rencontrés dans la nature chaque baside porte quatre spores. Il arrive cependant que l'on rencontre des carpophores dont les basides ne portent que deux spores.

Dans plusieurs espèces, les carpophores bisporiques sont des carpophores parthénogénétiques. Il en est ainsi chez Hygrophorus conicus; alors que ses carpophores normaux, bouclés, ont 4 spores par basides, ses carpophores parthénogénétiques, non bouclés, n'ont que 2 spores par baside. Chez cette espèce la bisporie de certains carpophores n'est donc pas un caractère d'espèce; elle n'est qu'un accident lié à cet autre accident qu'est le développement parthénogénétique. C'est donc à tort que J.-E. Lange a caractérisé H. conicus par la bisporie et qu'il a, en conséquence, considéré comme appartenant à une espèce différente, son H. pseudoconicus, les carpophores tétrasporiques, qui ne sont en réalité que les carpophores normaux de H. conicus.

Dans d'autres espèces la bisporie n'est pas un accident; c'est un bon caractère spécifique, car on l'observe même chez des carpophores non parthénogénétiques. Fayodia (Omphalia) bisphaerigera illustre ce cas de façon spectaculaire, car ses carpophores bisporiques sont bouclés. La bisporie est également un bon caractère spécifique chez Agaricus (Psalliota) bisporus, le champignon de couche cultivé (Fig. 1), mais cette espèce étant dépourvue de boucles, on n'a pu le vérifier qu'en constatant que ses basides bisporiques renferment à l'origine deux noyaux qui fusionnent.

La délimitation des espèces par les méthodes classiques et ses difficultés.

Nous venons de voir comment, grâce à l'observation de caractères microscopiques variés, diverses espèces peuvent être mieux définies et par suite plus facilement reconnaissables.

L'intérêt des caractères microscopiques ne doit cependant pas faire sousestimer l'importance des autres caractères, d'autant que, comme l'a très justement fait remarquer R. Maire, les caractères microscopiques risquent de nous sembler plus constants parce qu'on les observe moins souvent. C'est donc en considérant l'ensemble des caractères de tous ordres que l'on aura le plus de chances de bien cerner les limites de chaque espèce.

Par exemple Russula laurocerasi, qui était autrefois confondue avec R. foetens, s'en distingue à la fois par son odeur d'amandes amères et par l'ornementation crêtée de ses spores, comme l'ont fait remarquer Melzer et Zvara, qui ont nommé cette espèce.

Plus on fait de récoltes d'une espèce et plus on les étudie attentivement, plus grande apparaît souvent l'amplitude de variation de ses caractères, plus grande aussi peut devenir la tentation de découper en plusieurs espèces ce que l'on considérait à l'origine comme une seule espèce. C'est ainsi que Romagnesi a été amené à séparer de R. laurocerasi sa R. illota, dont l'arête des lames se pique de brun de façon tout à fait remarquable, et qu'il soupçonne que dans ce qui reste de laurocerasi après séparation de illota, il y a encore deux espèces.

On peut se demander, non sans quelque inquiétude, jusqu'où ira le découpage de certaines espèces, car deux récoltes sont rarement tout à fait identiques. A la limite ne risque-t-on pas de considérer comme différences spécifiques des différences individuelles, c'est-à-dire des différences du même ordre que celles qui séparent deux personnes de l'espèce humaine, voire même des différences dues au fait que les carpophores se sont développés dans des conditions ambiantes différentes, différences alors non héréditaires?

Secrétan a été accusé d'avoir poussé trop loin l'analyse de certaines espèces. Pour éviter ce risque il est nécessaire, avant de proposer comme espèce nouvelle un produit du découpage d'une espèce décrite antérieurement, de vérifier la constance de ses caractères distinctifs sur de nombreuses récoltes de stations variées; c'est ce qu'a fait Romagnesi avant de proposer sa Russula illota.

Mais s'il faut être très prudent dans le découpage de ce qui était antérieurement considéré comme une espèce, il faut également se garder de la tendance inverse, tendance que l'on peut appeler syncrétique, et qui pousse à réduire au rang de variétés des taxons qui ont en réalité valeur d'espèces. Cette tentation est très forte lorsque ces taxons présentent en commun de très nombreux caractères.

C'est ce qui arrive pour les Russules de la section que Melzer et Zvara ont appelée Viridantinae, pour rappeler qu'elle rassemble les champignons de ce genre dont la chair se colore en verdâtre en présence de sulfate de fer. Les Viridantinae ont en commun une autre réaction colorée, également découverte par Melzer et Zvara: la chair se colore en rouge cuivré par l'eau anilinée. Sans le secours de ces réactifs les Viridantinae se reconnaissent facilement à l'odeur de crustacés cuits qu'elles développent dans la vieillesse ou par la dessiccation. Le microscope n'est d'aucun secours pour reconnaître cet ensemble, car les caractères qu'il révèle se retrouvent dans nombre de Russules d'autres sections: absence de crêtes ou de réseau bien développés sur les spores, qui sont plus ou moins purement verruqueuses; absence d'hyphes primordiales incrustées dans le revêtement piléique, qui présente presque toujours des cystides plus ou moins différenciés.

Si l'on rappelle que les Viridantinae se trouvent généralement en régions siliceuses, on conviendra que cette section est particulièrement homogène, ceci bien que la couleur du chapeau puisse varier beaucoup d'une récolte à une autre, par exemple de rouge-pourpre à brun-vineux, brun, ocre ou même vert-olive. Impressionné par la masse de caractères communs aux Viridantinae, on est naturellement tenté de considérer qu'il n'y a là qu'une seule espèce: Russula xerampelina, avec de nombreuses variétés de coloration.

Romagnesi pense que cette attitude est déraisonnable. Il fait remarquer que, dans l'ensemble des Viridantinae, la couleur de la sporée varie de crême à ocre et même jaune d'oeuf, alors qu'en dehors de cette section, aucune espèce de Russule ne montre une telle amplitude de variation de la couleur de la sporée. Il note d'autre part que toutes les Viridantinae n'ont pas la même écologie. Celles de la zone silvatique viennent sous des essences variées et certaines des soi disant variétés de coloration ne viennent que sous des essences déterminées; c'est ainsi que sous les conifères, sur sol acide, se rencontre une Viridantinae au chapeau d'un beau rouge-pourpre, à stipe également rouge, que Peltereau appelait erythropoda pour cette raison, et qui est sans doute la xerampelina, telle que décrite par Fries. Une Viridantinae ne vient qu'en dehors des forêts, par exemple dans les pâturages, d'où le nom de R. pascua que Moeller et Schaeffer ont choisi pour elle. Moeller l'a découverte au nord de la forêt boréale, et Favre l'a signalée au-dessus de la forêt de l'arc alpin, où nous l'avons nous-même rencontrée. Ses carpophores, de petite taille, ont une couleur initiale très constante, le chapeau rose carminé ou rose-pourpre tranchant sur le stipe blanc. Nous avons fait remarquer que ses piléocystides diffèrent de celles de beaucoup d'autres Viridantinae par leur contenu dépourvu de granulations réfringentes et sulfoaldéhyde-négatif. Il s'agit évidemment d'une bonne espèce et non d'une simple variété de xerampelina.

Il est donc certain que l'ensemble des Viridantinae ne doit pas être considéré comme correspondant à une seule espèce, avec de nombreuses variétés de coloration. Romagnesi pense qu'il correspond peut-être à l'ensemble de quelque 10 espèces, qu'il se sent toutefois incapable de débrouiller de façon satisfaisante à l'heure actuelle, après une trentaine d'années de recherches.

Il n'est pas rare en effet, qu'après avoir accumulé de nombreuses fiches descriptives correspondant à autant de récoltes d'un même ensemble, on ait l'impression que cet ensemble comprend plusieurs espèces, mais que l'on n'arrive pas à délimiter celles-ci avec certitude, soit à cause de l'existence d'une gamme graduée d'intermédiaires semblant unir deux types extrêmes pourtant sensiblement différents, soit parce qu'il y a plusieurs manières de grouper les fiches, suivant que l'on accorde plus d'importance à tel caractère plutôt qu'à tel autre de sorte qu'en présence des mêmes fiches descriptives des Mycologues différents pourront aboutir à des groupements différents.

C'est seulement dans la période moderne qu'a été mise au point une méthode expérimentale qui permet de grouper les fiches sans tenir compte de leur contenu, c'est-à-dire de façon indépendante des tendances personnelles de chaque Mycologue. La comparaison des fiches permet ensuite de dégager les caractères constants de chaque groupe et, éventuellement les caractères qui distinguent un groupe d'un autre.

Leçon Herbette   431

A la base de cette méthode, qui n'est utilisable qu'avec des champignons hétérothalles, se trouve la confrontation de mycéliums primaires, c'est-à-dire de mycéliums dont chacun a pour origine une seule spore et représente ce qu'on appelle, pour cette raison, une culture "monosperme". C'est cette méthode que nous allons décrire, illustrer et discuter brièvement dans la dernière partie de cet exposé.

L'aide apportée à la délimitation des espèces par la confrontation de cultures monospermes.

L'interfertilité et le postulat de Vandendries.

Nous avons vu que, chez une espèce hétérothalle, pour obtenir un mycélium secondaire susceptible de porter des carpophores normaux, on peut mettre en présence deux mycéliums primaires issus de spores d'un même carpophore, à condition que ces mycéliums primaires soient compatibles. Mais on peut naturellement aboutir au même résultat en confrontant deux mycéliums primaires issus de spores émanant de deux carpophores différents A et B de la même espèce; bien entendu il faut que ces deux mycéliums primaires soient compatibles, mais l'expérience a montré qu'ils ont beaucoup plus de chances de l'être s'ils proviennent de carpophores A et B nés loin l'un de l'autre que s'ils proviennent du même carpophore. Lorsque l'on obtient un mycélium secondaire par confrontation d'un mycélium primaire d'une souche A avec un mycélium primaire d'une souche B, on dit que les souches A et B sont interfertiles.

Dans l'optique de la délimitation des espèces, Vandendries a prétendu, dès 1923, que si les mycéliums primaires de deux carpophores sauvages sont "toujours et indéfiniment fertiles entre eux, ces deux carpophores appartiennent à la même espèce". Ayant connu assez intimement Vandendries nous pouvons affirmer qu'il ne formulait là qu'un postulat car ce biologiste ignorait tout de la Mycologie systématique.

Voici un exemple de l'application du postulat de cet auteur. On rencontre çà et là une petite Mycène entièrement blanche, qui déroute à première vue, mais que l'on est tenté de rapprocher de Mycena rosella dès que l'on en fait une étude microscopique détaillée. La souche blanche est parfaitement interfertile avec la souche typique de Mycena rosella, d'où on est tenté de tirer la conclusion que ces deux souches appartiennent à la même espèce. Nous pensons que c'est effectivement le cas car nous n'avons pas pu trouver d'autres différences entre les deux souches que la coloration du carpophore. Nous dirons donc que Mycena rosella présente deux variétés: une variété rose, la variété type, et une variété blanche.

On ne connaît jusqu'à présent que peu de cas auxquels le postulat de Vandendries semble ne pas s'appliquer. A l'occasion d'entretiens privés avec des personnalités présentes au Symposium de Lausanne nous avons appris que deux chercheurs travaillant sur des Pleurotacées, l'un sur des Omphalotus, l'autre sur un ensemble différent, ont constaté des interfertilités entre souches rapportées à des champignons considérés depuis longtemps par les systématiciens de formation traditionnelle comme espèces. Peut-être certaines de ces soi-disant espèces ne sont elles que des variétés. De toute façon, remarquons qu'avant de tirer une conclusion systématique de l'interfertilité constatée, il serait nécessaire d'obtenir des carpophores à partir des mycéliums secondaires issus de la confrontation de mycéliums primaires des espèces en question, de constater que ces carpophores sont normaux en tous points et de retrouver les types parentaux dans leur descendance. Dans cet ordre d'idées on ne peut en effet oublier le cas du Cheval et de l'Ane qui, comme chacun sait, s'hybrident parfaitement en apparence, mais dont les hybrides, le Mulet par exemple, sont stériles, ce qui confirme qu'il s'agit de deux espèces distinctes, génétiquement séparées.

Les cas connus d'interfertilité entre espèces différentes semblent d'ailleurs exceptionnels chez les Hyménomycètes, comme l'a fait remarquer, au Symposium de Lausanne, J. Boidin, à l'occasion de la présentation d'un mémoire dans lequel il a rassemblé toutes les observations d'auteurs variés figurant dans la littérature mondiale et tous les résultats acquis par divers chercheurs de l'Université de Lyon, dont lui-même.

Ceci peut évidemment choquer un naturaliste familiarisé avec l'étude des plantes à fleurs. En effet, dans les Flores consacrées à ces végétaux, on trouve souvent mention d'hybrides entre espèces, même entre espèces nettement différentes. Nous nous bornerons à rappeler ici un des exemples les mieux étudiés. Tout botaniste sait que les deux Benoites que sont Geum urbanum et G. rivale sont si différentes par l'aspect de leurs fleurs que, même une personne non initiée à la Botanique, ne risque de les confondre. Or l'expérience a montré que l'on peut hybrider sans difficultés ces deux espèces et que leurs hybrides sont parfaitement fertiles, comme on peut le constater dans les cultures. Dans les hybrides de ses deux Geum et dans leur descendance, on retrouve les caractères des deux espèces parentes, mais souvent recombinés autrement et de façon fort variées. Si ces types intermédiaires ne se rencontrent guère dans la nature c'est d'abord que ces deux Benoites ont peu de chances de se polliniser mutuellement car elles ne vivent pas côte à côte, G. rivale étant une plante d'endroits mouillés, contrairement à G. urbanum; c'est ensuite que, dans la nature, les types issus d'éventuels croisements entre les deux espèces sont éliminés par la concurrence de la végétation environnante.

L'interstérilité et la délimitation des espèces.

On dit que deux souches sont interstériles lorsqu'aucun mycélium primaire de l'une ne donne de mycélium secondaire par confrontation avec quelque mycélium primaire que ce soit de l'autre.

L'interstérilité peut se manifester entre espèces très voisines l'une de l'autre, par exemple entre ces petits Collybia qui poussent au printemps sur les cones, et pour lesquels Singer a créé son genre Strobilurus. Elle a été constatée entre S. tenacellus et S. stephanocystis, deux espèces qui viennent sur les cones de pins enfouis et qu'on ne distingue guère macroscopiquement que par une différence légère dans la coloration du chapeau. Elle confirme qu'il s'agit de deux bonnes espèces, ce dont un systématicien traditionnel ne saurait douter tant sont nets les caractères microscopiques qui les distinguent. S. tenacellus a des cystides à sommet aigu et des spores un peu arquées (Fig. 15). S. stephanocystis a des cystides à sommet largement arrondi, voire parfois capité et des spores non arquées (Fig. 16).

L'interstérilité constatée au laboratoire entre ces deux espèces suggère qu'elles ne peuvent s'hybrider dans la nature. En fait on ne

trouve jamais, dans la nature, de carpophores où les caractères des
cystides et des spores sont combinés autrement qu'il vient d'être
indiqué, et ceci bien que les deux espèces croissent souvent ensemble
dans les mêmes stations, contrairement aux deux Benoites dont le cas
a été évoqué plus haut.

En 1952 Morten-Lange a montré, dans un travail classique sur les Coprins de la section Setulosi, le parti que peut tirer le systématicien des tests d'interfertilité et d'interstérilité pour la délimitation des espèces dans un groupe difficile.

Plutôt que d'analyser ce travail très connu, nous préférons, pour illustrer notre exposé, utiliser deux mémoires très récents (1974 et
1975) consacrés par D. Lamoure à deux groupes d'un genre également
difficile, le genre Omphalina, mémoires parus dans les "Travaux
scientifiques du Parc national de la Vanoise", V, p. 149-164 et VI,
p. 153-166.

Application de la méthode des confrontations de cultures monospermes
a l'analyse taxinomique de deux ensembles d'Omphales.

Les Omphales étudiées par D. Lamoure gravitent autour de deux espèces
relativement répandues dès la plaine: O. pyxidata et O. velutipes.

Il s'agit d'Omphales banales par leur petite taille, l'absence de
cystides sur les lames, les basides à 4 spores lisses et par la présence de boucles.

Dans toutes la pigmentation est incrustante, mais elle n'a pas la
même teinte dans les deux ensembles. Dans l'ensemble comprenant O.
velutipes, chapeau et stipe sont gris-brun à noirâtres, les lames
elles-mêmes sont grises. Dans l'ensemble comprenant O. pyxidata, les
lames ne sont jamais grises et le chapeau est plus ou moins roussâ-
tre. Ces deux ensembles sont répandus en zone alpine, où D. Lamoure
en a étudié systématiquement de très nombreuses récoltes.

Elle a pu obtenir des mycéliums primaires pour 24 récoltes d'Omphales
rousses et pour 90 récoltes d'Omphales gris-noir. Confrontant ces mycéliums primaires elle est arrivée aux conclusions suivantes. Les 24
récoltes d'Omphales rousses se répartissent en 5 groupes tels que

toutes les récoltes d'un groupe sont interfertiles, mais interstériles avec toutes les récoltes des autres groupes. Sur les mêmes critères les 90 récoltes d'Omphales gris-noir se répartissent en 5 groupes également. D. Lamoure pense que chacun de ces 10 groupes interstériles d'Omphales correspond à une espèce. Elle le pense notamment parce que, dans chacun de ces deux ensembles d'Omphales, chaque groupe se distingue plus ou moins nettement des groupes interstériles avec lui par des caractères des spores. Pour illustrer ce fait il est indispensable d'examiner séparément les deux ensembles distingués par la couleur.

- Les Omphales gris-noir.

Comme on peut le voir sur la Fig. 17, les groupes interstériles diffèrent souvent les uns des autres par la forme et (ou) par les dimensions des spores. Si l'on ajoute que dans 2 des groupes interstériles dominent les spores à un seul noyau, alors que dans les 3 autres ce sont les spores à deux noyaux qui dominent, on peut difficilement échapper à l'idée que, conformément à l'opinion de D. Lamoure, chacun de ces groupes correspond à une espèce.

En systématicien traditionnel, ne faisant pas appel à des confrontations de mycéliums primaires. Favre avait individualisé, dans la zone alpine des Grisons, 2 des 5 espèces d'Omphales gris-noir reconnues depuis par D. Lamoure: O. velutipes, qu'il appelait O. umbratilis, var. minor, et O. obatra, espèce nouvelle. Sur la Fig. 17 on voit que les spores de ces deux espèces (obatra, Fig. c et velutipes, Fig. d) sont assez banales par leurs dimensions et par leur forme. Il est certain que si Favre avait examiné au microscope les autres espèces individualisées par D. Lamoure et dont les spores sont beaucoup plus remarquables (trigonospora, Fig. e et sphaerospora, Fig. f), il n'aurait pas hésité à les considérer comme de bonnes espèces; en analysant les Omphales rousses, nous verrons en effet que cet auteur tenait le plus grand compte de la forme des spores dans la distinction des espèces. Il est possible que Favre ait rencontré toutes les espèces gris-noir reconnues par D. Lamoure, mais sans avoir le temps d'en faire l'étude microscopique qui lui aurait permis de les individualiser avant elle, car, comme le fait remarquer D. Lamoure, il est délicat, voire impossible, de distinguer plusieurs des Omphales gris-noir en ne les examinant qu'à l'oeil nu ou même à la loupe.

- Les Omphales rousses.

La distinction des 5 espèces de cet ensemble qui s'élèvent en zone alpine peut être encore plus délicate que la distinction des espèces d'Omphales gris-noir, d'autant que les spores à un seul noyau dominent dans toutes.

Cependant, de la zone alpine des Grisons, J. Favre en a figuré deux: O. pyxidata et un champignon qu'il subordonnait à pyxidata sous l'étiquette O. pyxidata, var. rivulicola, et dont il n'avait fait qu'une seule récolte. L'étiquette rivulicola a été choisie par Favre en raison de la station de récolte: tapis de mousses imbibées d'eau au bord des ruisseaux. Favre a fait remarquer que rivulicola ne s'écarte pas macroscopiquement de pyxidata, mais qu'elle s'en distingue nettement par les spores, comme on peut le voir sur le dessin qu'il a donné (Ici, Fig. 18); chez pyxidata les spores ont une tendance amygdaliforme, alors qu'elles sont régulièrement elliptiques, plus larges, chez rivulicola.

Favre écrivait de rivulicola: "C'est certainement une bonne variété de pyxidata, peut-être même une espèce vicariante de cette dernière". L'interstérilité constatée par D. Lamoure plaide fortement en faveur de la dernière interprétation "espèce vicariante". Le fait que, sans le secours du test d'interstérilité, Favre ait pu, sur une seule récolte, créer son taxon rivulicola, en suggérant qu'il pouvait constituer une espèce vicariante de pyxidata souligne, une fois de plus, combien était aigu, chez ce Maître, le sens de l'espèce.

D. Lamoure pense que ses 5 groupes interstériles d'Omphales rousses correspondent à autant d'espèces, non seulement à cause de différences dans la forme et (ou) dans les dimensions des spores, mais encore parce que certains groupes paraissent caractéristiques d'habitats particuliers. Alors que pyxidata est une espèce des endroits non mouillés, qui ne semble pas s'élever au-dessus de la partie moyenne de la zone alpine, le champignon que D. Lamoure a appelé chionophila n'a été trouvé par elle que dans les combes à neige, souvent parmi Salix herbacea. O. rivulicola ne vient, comme l'avait bien supposé Favre, que dans les mousses mouillées du bord des ruisselets et O. kuehneri Lamoure se développe dans les mouillettes alpines.

Favre pressentait que, même après exclusion de son rivulicola, son pyxidata était encore hétérogène; il écrivait en effet qu'ayant étudié 10 récoltes de pyxidata il a pu se rendre compte que cette espèce est assez variable: par l'habitat, qui va des prairies aux endroits marécageux, des tapis de Dryas aux tapis de Saules nains; par les dimensions des spores qui vont de 8-9.5 x 4.5-5 microns à 9.5-11.5 x 5.5-6.5 microns. Les récoltes des endroits narécageux correspondent, par l'habitat, à O. kuehneri, des mouillettes alpines; celles des tapis de Saules correspondent peut-être à O. chionophila, qui vient souvent avec Salix herbacea, dans les combes à neige. Il est donc possible que Favre ait étudié les diverses Omphales rousses considérées comme espèces par D. Lamoure, mais qu'il ait groupé la plupart d'entre elles sous l'étiquette pyxidata.

Cette conclusion apparaît encore plus vraisemblable lorsque l'on considère les dessins que D. Lamoure a consacrés aux spores des Omphales rousses, que reproduit notre Fig. 19: pyxidata est figuré en haut (Fig. a, b, c), rivulicola juste au-dessous (Fig. d, e, f); on reconnaît bien les différences de forme des spores indiquées par Favre entre ces deux taxons. Plus bas se trouvent les dessins de spores relatifs à kuehneri (Fig. h et i) et à chionophila (Fig. j et k); On voit que, si les spores des deux dernières espèces sont plus grosses que celles de pyxidata, elles se rapprochent des spores de cette espèce par leur forme différente de celle caractéristique de rivulicola, et l'on comprend alors facilement que Favre ait pu confondre O. kuehneri et chionophila avec pyxidata.

Si l'on ne perd pas de vue que D. Lamoure avoue qu'elle est dans l'incapacité d'indiquer des différences sûres dans la morphologie superficielle des carpophores de ses espèces d'Omphales rousses, on ne peut que conclure que le test d'interstérilité est ici un test d'une finesse extrême, qui sépare des unités systématiques vraiment très peu différentes les unes des autres, certaines si peu que J. Favre, Mycologue de formation traditionnnelle, mais dont nous avons souligné à plusieurs reprises la classe exceptionnelle au cours de cet exposé, les confondait dans l'espèce pyxidata.

Nous sommes tout naturellement amenés à la question que pose le titre du paragraphe qui suit:

Lorsqu'une interstérilité totale se manifeste entre deux groupes formés chacun de souches interfertiles, doit-on toujours considérer que ces groupes correspondent à deux espèces différentes?

Il n'est pas douteux qu'au vu des résultats de D. Lamoure sur les Omphales rousses, bien des Systématiciens seront tentés de ne considérer que comme sous-espèces, voire comme simples variétés, les taxons que cet auteur présent comme espèces. A cette manière de voir on peut faire deux sortes d'objections. Tout d'abord une objection de principe: pourquoi ne pas considérer comme espèces des unités systématiques qui ne sont pas capables de s'hybrider, qui sont donc génétiquement séparées? Ensuite pyxidata pris dans un sens trop large ne présente aucun intérêt pour l'écologiste, puisqu'on le trouve un peu partout, dans des habitats fort variés; ce qui intéressera l'écologiste ce sont les espèces au sens étroit de D. Lamoure, espèces qui, elle, sont caractéristiques d'habitats particuliers.

Le cas de Psathyrella candolleana est plus embarrassant encore. M.-C. Galland a reconnu qu'il s'agit d'un complexe comprenant plusieurs groupes interstériles. Or Romagnesi, qui avait fourni à cet auteur les sporées utilisées pour ses recherches, écrit qu'il est incapable de distinguer ces groupes par des caractères morphologiques du carpophore, qu'ils soient macro- ou microscopiques. Autrement dit, la spore elle-même qui fournit des caractères distinctifs entre groupes interstériles d'Omphales rousses, n'est plus ici d'aucun secours.

On n'a pas encore recherché si les groupes interstériles de P. candolleana diffèrent les uns des autres par des caractères chimiques ou physiologiques. Si c'est le cas, ce qui est plus que probable, de tels groupes doivent-ils être encore considérés comme espèces?

Bien des Agaricologues de formation traditionnelle diront probablement non, car pour tous les organismes à morphologie complexe, l'espèce a été définie par des caractères morphologiques dès l'origine de la Systématique. Mais nul ne saurait oublier que, pour des organismes à morphologie très pauvre, comme les Levures ou les Bactéries, les Systématiciens considèrent comme caractères spécifiques des caractères non morphologiques, des caractères biochimiques, des caractères du fonctionnement cellulaire. L'Agaricologue de formation traditionnelle est tenté de dire qu'ils le font faute de mieux, à défaut de caractères morphologiques, mais il ne doit pas oublier que, dans les espèces morphologiquement différenciées, ce sont en définitive des différences dans le fonctionnement cellulaire qui sont à l'origine des différences morphologiques.

Conclusion

S'il est tentant de donner de la notion d'espèce chez les Hyménomycètes une définition basée sur les phénomènes d'interfertilité et d'interstérilité, il ne faut pas oublier que nous connaissons déjà quelques cas embarrassants (Pleurotacées et Psathyrella candolleana par exemple) et qu'il n'est pas certain que ces cas qui, dans l'état actuel de nos connaissances, semblent exceptionnels, resteront aussi isolés quand nous aurons une expérience plus étendue des phénomènes d'interfertilité et d'interstérilité dans l'ensemble des Hyménomycètes. Il est bon de rappeler que nous ignorons le comportement de très nombreuses espèces, et en particulier de toutes celles dont nous ne savons pas faire germer les spores, qui seules donneraient commodément les mycéliums primaires nécessaires aux confrontations. Et ces espèces dont nous ne savons pas faire germer les spores sont légion; sont même dans ce cas des ensembles entiers, qui comptent parmi les plus vastes et les plus difficiles, comme les Rhodophyllacées, les Russules, les Cortinaires et les Inocybes.

Même lorsque sera surmontée la difficulté technique à laquelle nous venons de faire allusion, c'est-à-dire lorsque nous saurons faire germer les spores de toutes les espèces, il restera des espèces auxquelles les tests interfertilité-interstérilité ne pourront s'appliquer; il s'agit d'espèces qui, contrairement aux hétérothalles, que nous avons seules considérées jusqu'ici, effectuent leur cycle nucléaire complet, avec fusion nucléaire suivie de réduction chromatique, à partir d'une seule spore, espèces que l'on appelle homothalles.

Ces réserves étant faites nous dirons, en terminant, que les brillants résultats taxinomiques obtenus grâce à l'utilisation de ces tests, et ceci dans les ensembles les plus variés d'Hyménomycètes, doivent inciter tous les Mycologues qui en ont les moyens matériels, à les mettre en oeuvre systématiquement pour tenter de débrouiller les ensembles difficiles qui s'y prêtent.

Fig. 1 - Agaricus (Psalliota) bisporus. Portion de coupe transversale de lame. (d'après A.H.R. Buller).

Fig. 2 - Aire (hachurée) de l'Arole: Pinus cembra (d'après H. Walter).

Fig. 3 - Spores de divers Inocybe.

Fig. 4 - Ornementation des spores de Russules. Spores réticulées (a et b); spores crêtées (c et d) (d'après V. Melzer).

Fig. 5 - Structure de la paroi sporique de Fayodia (Omphalia) bisphaerigera. Afin de rendre la structure plus intelligible la substance des piliers ornementaux est figurée en pointillé sur ce schéma, alors qu'elle est en réalité homogène, même en microscopie électronique.

Fig. 6 - Cystides de divers Inocybe. A gauche une cystide faciale oxalifère, à paroi épaissie; au milieu et à droite cystides d'espèces qui ne possèdent de cystides que sur l'arête des lames.

Fig. 7 - Cystides de diverses espèces de Mycena.

Fig. 8 - Revêtement piléique avec piléocystide chez une Russule.

Fig. 9 - Hyphes primordiales du revêtement piléique de Russula pseudointegra.

Fig. 10 - Eléments du revêtement piléique, en coupe radiale du chapeau. En haut: Boletus scaber (à gauche) et B. griseus (= carpini) (à droite) (d'après Singer). En bas Marasmius epiphyllus (à gauche) et M. epiphylloides (à droite).

Fig. 11 - Eléments du revêtement piléique de Clitocybe hydrogramma.

Fig. 12 - Divers types de pigmentation. Pigmentation incrustante de la paroi (à gauche). Pigmentation intracellulaire, vacuolaire (au milieu). Pigmentation à la fois incrustante et vacuolaire (à droite). Le contenu vacuolaire, figuré en pointillé sur ces schémas, apparaît en réalité uniformément coloré.

Fig. 13 - Hyphes bouclées (à gauche) et non bouclées (à droite) (d'après A.H.R. Buller).

Fig. 14 - Cycle d'un Hyménomycète hétérothalle. I.: mycélium primaire. II.: mycélium secondaire.

Fig. 15 - Strobilurus tenacellus (d'après J. Favre).

Fig. 16 - Strobilurus stephanocystis (d'après J. Favre).

Fig. 17 - Spores d'Omphales gris-noir. O. obscurata (a et b) - O. obatra (c) - O. velutipes (d) - O. trigonospora (e) - O. sphaerospora (f) (d'après D. Lamoure).

Fig. 18 - Omphalia pyxidata (en haut) et O. rivulicola (en bas) (d'après J. Favre).

Fig. 19 - Spores d'Omphales rousses. O. pyxidata (a, b et c) - O. rivulicola (d, e et f) - O. pseudomuralis (g et l) - O. kuehneri (h et i) - O. chionophila (j et k) (d'après D. Lamoure).

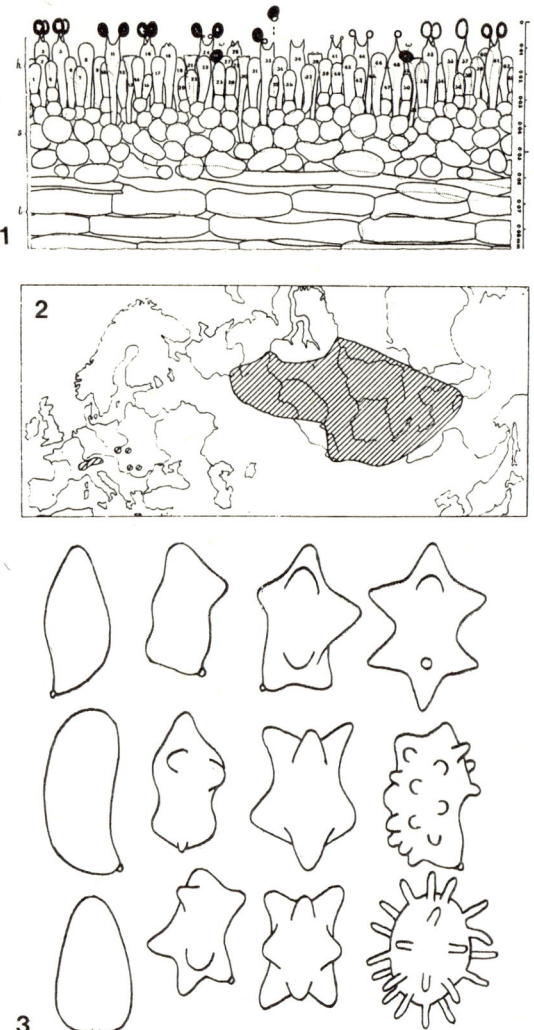

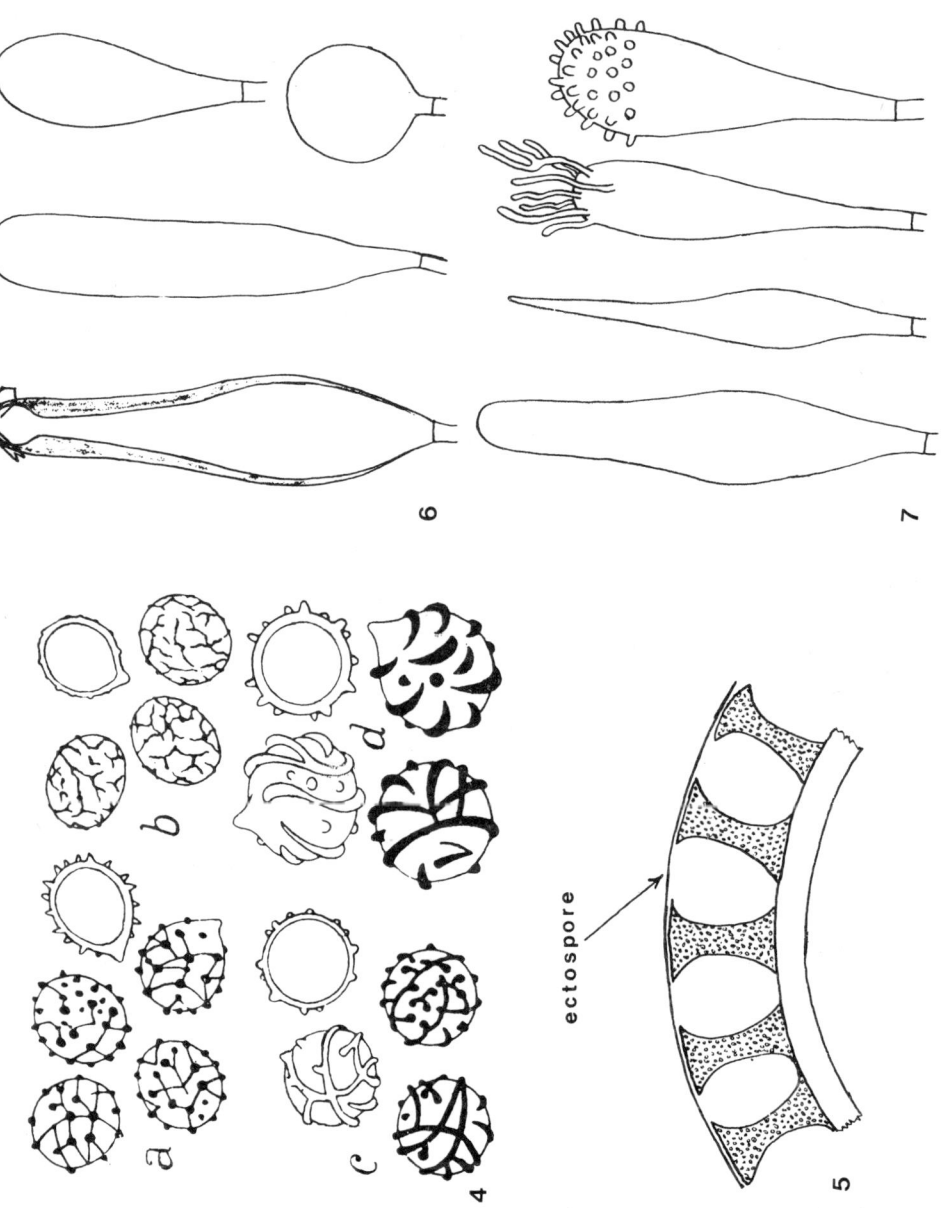

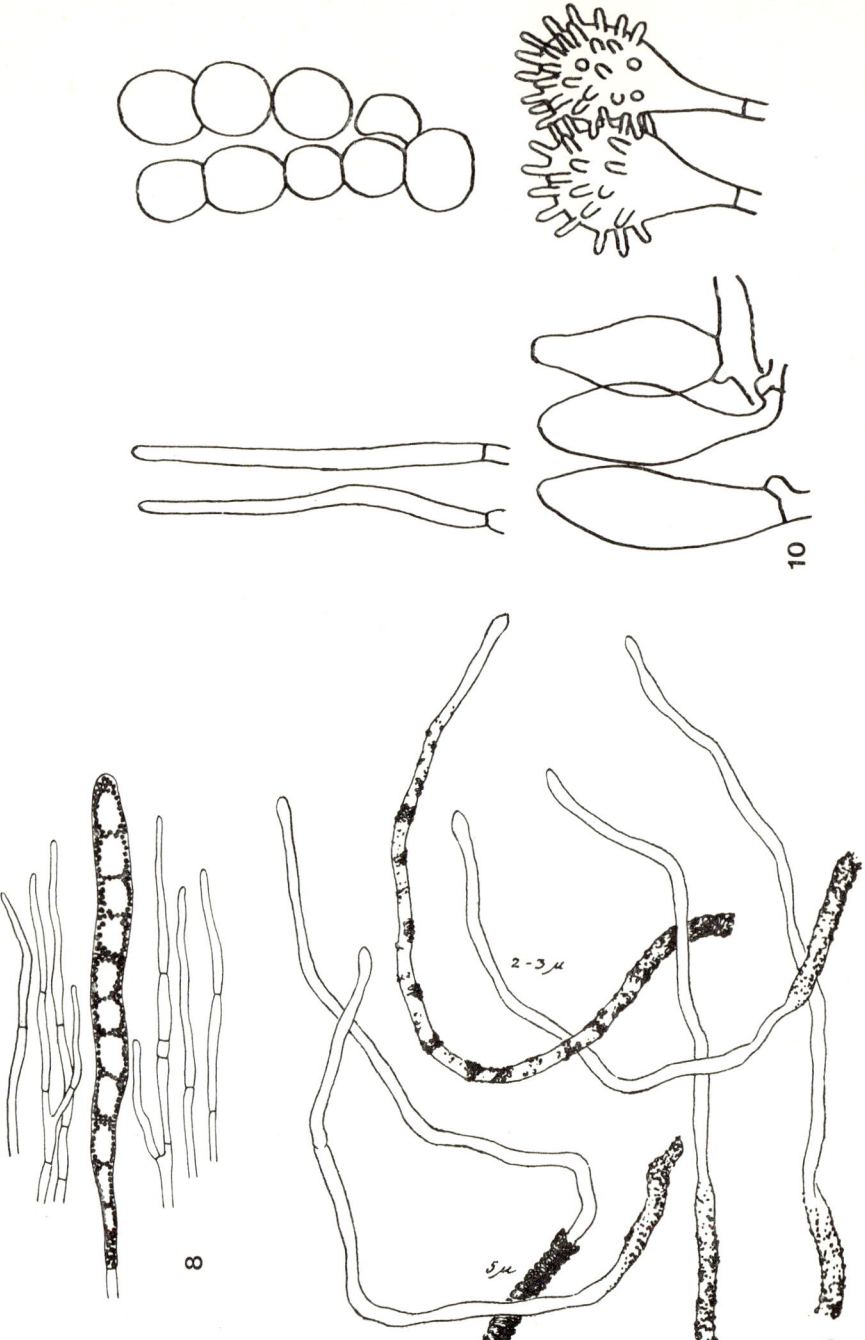

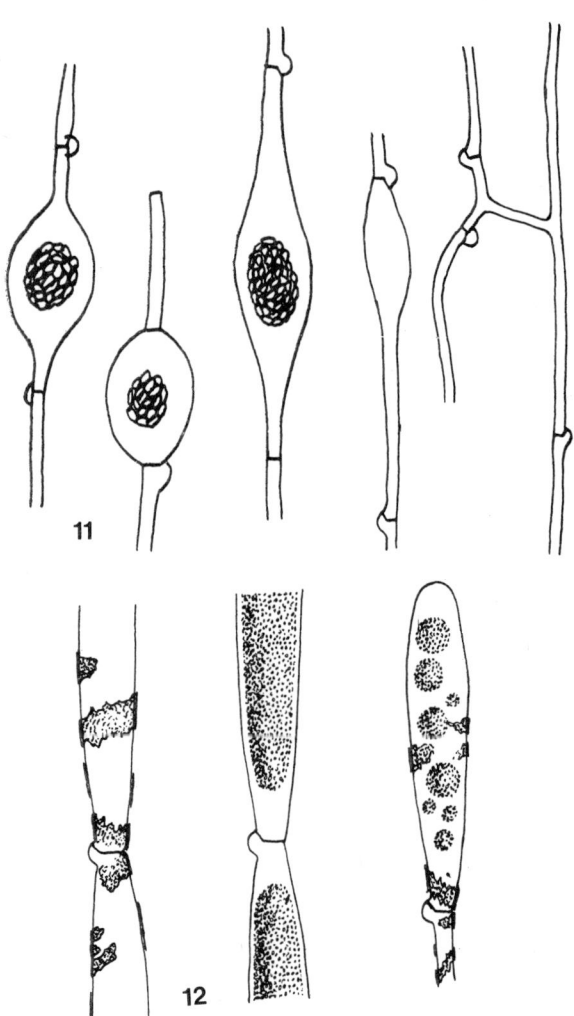

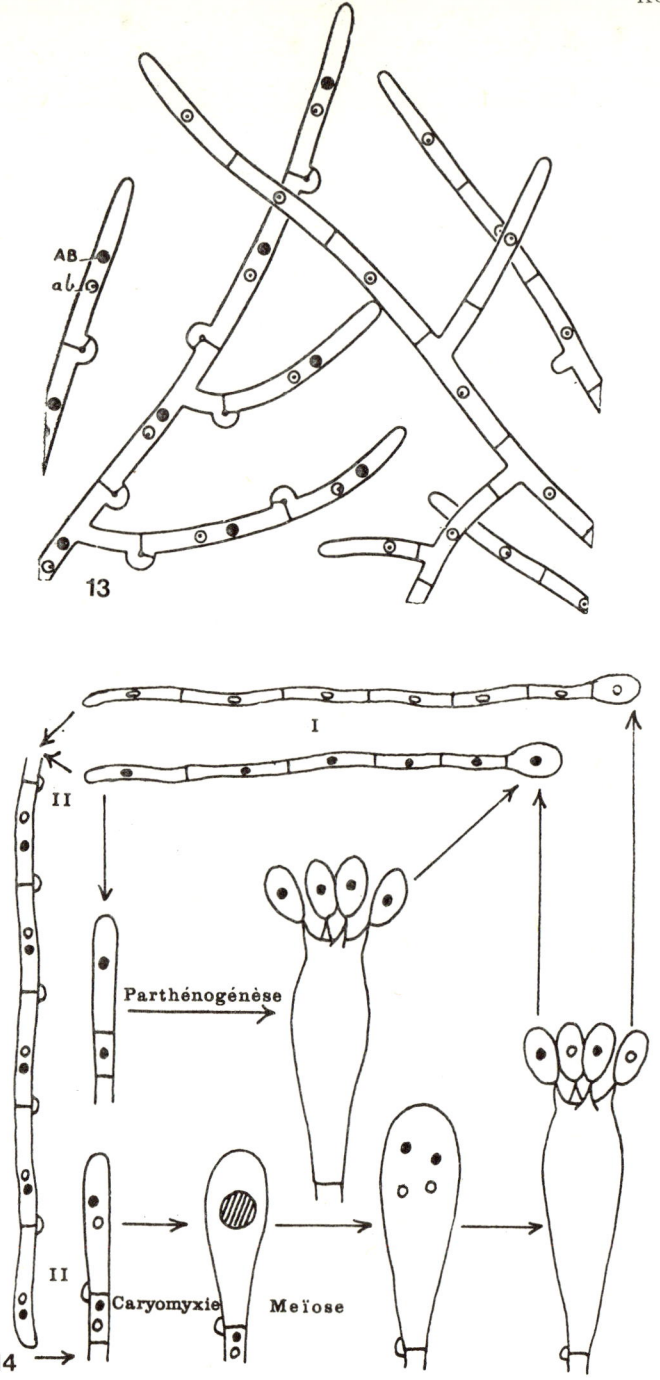

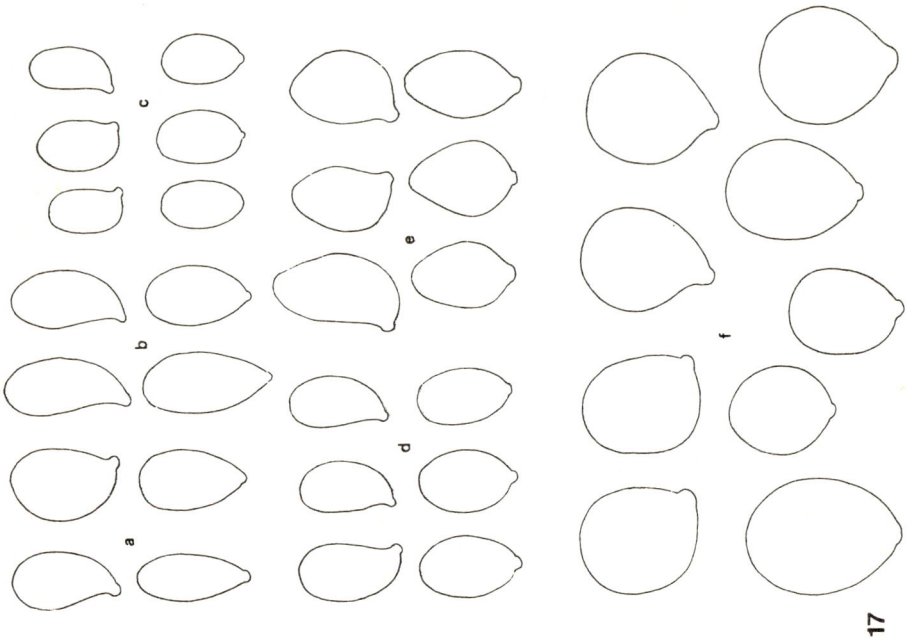

KÜHNER 8

19

GENERIC INDEX

Agaricus, 19
Agrocybe, 12, 13, 15, 16, 18, 20, 21, 23, 25, 26, 27, 29, 31, 32, 34, 35, 38, 39, 191, 204, 208
Aleurodiscus, 283, 288
Allomyces, 231
Amanita, 76, 79-90, 92-97, 99-102, 355, 369, 386
Amphinema, 341
Amylaria, 141
Amylostereum, 282, 283, 285, 294, 306, 308
Arcangeliella, 142
Aspergillus, 309
Asterostromella, 353
Athelia, 335-337
Athelidium, 337
Aureoboletus, 58
Auricularia, 51, 261, 282, 290, 292, 296, 297, 299, 302, 322, 366-367
Aurioporia, 300
Bolbitius, 12-18, 23, 25-27, 29, 31, 34, 38, 39, 92
Boletellus, 59
Boletinellus, 57
Boletinus, 57
Boletochaete, 58
Boletus, 52, 57, 61-65, 69
Bondarzewia, 140
Botryobasidium, 331, 333, 337
Bremia, 230
Candida, 309
Cantharellus, 94, 367-369, 375, 376
Cejpomyces, 334
Ceraceomyces, 338
Ceratobasidium, 334
Ceriomyces, 285
Chalciporus, 58
Chroogomphus, 118, 384, 390
Clavulinopsis, 369
Clitocybe, 105-116, 231, 353, 367
Collybia, 116
Confertobasidium, 337
Conocybe, 12-38, 39, 40, 41, 47, 51, 92, 94, 210, 385
Coprinus, 15, 28, 51, 119, 208, 209, 212, 231, 259-276, 290, 326, 353,
Coriolellus, 288
Coriolus, 309
Corticium, 283, 288, 333, 334, 337-340, 347
Cortinarius, 31, 98, 131, 357
Cristella, 339
Cylindrobasidium, 338

Cymatoderma, 281
Cystoderma, 386
Cyttarophyllum, 13, 14, 15, 40
Dacrymyces, 367
Daedaleopsis, 287
Dermocybe, 350
Dermoloma, 387
Descolea, 13, 16
Dichostereum, 287
Drosophila, see Psathyrella
Duportella, 283
Exobasidium, 334
Favolus, 194
Fayodia, 387
Fibulomyces, 336, 337
Fistulina, 226
Flammulina, 220, 263, 266
Fomes, 226, 283, 285, 289, 291, 292, 294, 298, 302
Fomitopsis, 53
Fuscoboletinus, 57
Galera, 12, 13, 29
Galerella, 29
Galerina, 12, 94
Galeropsis, 14, 40
Galzinia, 334, 348
Galziniella, 334-335
Gastrocybe, 13
Gerronema, 369, 378
Gloeocystidium, 288
Gloeodontia, 283
Gyrodon, 57, 76
Gyroporus, 58
Haematostereum, 305
Hebeloma, 16, 251
Hebelomina, 16
Heimiella, 59
Hirschioporus, 283, 285, 290, 292, 295, 298, 302
Hybogaster, 140
Hydnum, 377
Hygrocybe, 157-182, 227
Hygrophorus, 115, 157, 369
Hymenochaete, 334
Hyphoderma, 288, 298, 299, 334, 348
Hyphodontia, 341, 348
Hypholoma, 251
Inocybe, 16, 19
Laccaria, 385
Lactarius, 31, 100, 101, 123-135, 136-141, 204
Laurilia, 287
Laxitextum, 289, 302
Leccinum, 58, 61-65, 70, 211
Lentinus, 220
Lentodium, 8

Lenzites, 283, 287, 294
Leptocantharellus, 367
Leptosporomyces, 336, 337
Leucocortinarius, 16, 40
Leucoporus, 194, 278
Lopharia, 281, 322
Lyophyllum, 99, 116
Meiorganum, 59
Melzericium, 339-340
Merulius, 283, 295, 305-306
Monadelphus, 367
Multiclavula, 332
Mycena, 21, 40, 116, 131, 143, 204, 205, 207, 213, 289, 325, 387
Neurospora, 231, 271, 307
Omphalina (Omphalia), 102, 353, 356, 378, 385
Omphalotus, 366-367
Oudemansiella, 27, 186
Panus, 8
Paragyrodon, 57
Paullicorticium, 334
Penicillium, 213
Peniophora, 281-286, 293-295, 298-302, 306, 308, 309, 339, 340, 341, 356, 384
Phaeogyroporus, 58
Phaeophlebia, 283
Phellinus, 224, 351
Phlebia, 284, 296, 297, 339
Phlebiella, 339
Pholiotina, 12, 16, 23, 27, 28, 29, 30, 35
Platygloea, 334
Pleurotus, 194, 220, 224, 229-258, 352, 385
Pluteus, 205
Podoscypha, 278, 281, 287
Podospora, 117, 204, 210, 216, 242, 270-272, 275, 292
Polyporellus, 194, 351, 353, 357
Polyporus, 189-203, 208, 272, 2 281, 283-287, 291, 293
Porphyrellus, 59
Psathyrella, 8, 138, 208-210, 259, 261, 262, 269, 353-354, 357, 375, 387, 397, 403-404
Pseudohiatula, 119, 387
Psiloboletinus, 57
Psilocybe, 138, 253, 254
Pulveroboletus, 58
Punctularia, 283, 285
Pycnoporus, 226, 283, 287, 295, 300
Repetobasidium, 335
Rhizoctonia, 288
Rhodocybe, 116
Rhodophyllus, 100
Rozites, 118

Russula, 31, 387
Schizophyllum, 212
Schizosaccharomyces, 118
Scleroderma, 63
Scytinostroma, 283, 284, 287
Septobasidium, 334, 376
Septonia, 304
Serpula, 287
Setchelliogaster, 13
Sistotrema, 278, 280, 288, 289, 290, 302, 331, 333, 345-346, 348
Sistotremastrum, 333, 348
Sordaria, 117, 206, 292, 299
Stecchericium, 283, 289
Stereum, 282, 284, 294, 300, 304-305, 324
Strobilomyces, 59, 62
Subulicystidium, 334, 340, 341-343, 347
Suillus, 57, 62, 63, 65, 66-67, 68
Trametes, 217, 283, 286, 293
Trechispora, 288, 339
Tricholoma, 116
Tubariopsis, 13
Tylopilus, 59, 70
Typhula, 284, 285, 288, 300, 357
Tyromyces, 284, 287
Urnobasidium, 334-335
Vararia, 287, 288, 291, 301, 302, 353, 360
Waitea, 334
Xanthoconium, 58
Xenasma, 337, 339
Xenasmatella, 339-340
Xerocomus, 58

SUBJECT INDEX

albino, 368, 375, 377-378
amphithallie, 288, 302, 325
amyloidity
  spores, 64, 118-119, 386-387
  hyphae, 64, 118
anastomosis, 304-307, 310, 324
apomixis, 206-207

barrage, 196-199, 208, 210, 237, 290, 311, 323, 397
basidia, 19-21, 28, 105, 184, 185, 333-339, 341-342, 346, 348, 370
carotinoids, 367, 378
chemical characters, 31, 63, 65-67, 76-78, 85, 93-94, 95-96, 98-99, 105, 133, 134, 211, 215-221, 224, 307, 350-351, 358, 368-369
chimaeras, 268
clamp connection, 20, 25, 59, 85, 105, 113, 140, 205-206, 237, 293, 296, 299, 303, 325, 331
conidia, 28, 332
cristalloides, 186
crystals, 32, 33
cultivar, 386
culture, 20, 25, 27, 29, 31, 32, 46, 47-48, 65, 93, 102, 142-143, 186, 191, 204-205, 213, 217, 232, 233, 277, 329, 346, 352-353, 369, 375, 397
cystidia, 23-26, 28, 29, 59, 65, 85, 105, 113, 128, 137, 282, 335, 339-343

deme structure, 148-149
diversity, 76-77, 126, 138, 225, 328
ecology, ecological factors, 30, 35-36, 46, 49-50, 80-81, 112-113, 139, 141-143, 230, 255, 311, 351-352, 358, 382, 389
ecotype, 384-385
enzymes, 31, 32, 98, 118, 215-221, 224, 227, 239-242, 367
form, 90, 92, 287, 354, 360, 390
form-pool, 90
fruit body induction, 191, 204, 213, 214, 322, 324
  monocaryotic, 191, 204, 206, 208, 214, 288-89, 324-25
fusion (of hyphae), 197, 207-208, 209, 259, 264, 277, 304-307
gastroid genera, fruit bodies, 13, 14, 253, 254
genes, 191-193, 322, 371
genus concept, 331-343, 345, 347

homing, 259-273, 324
hologamodème, 311
hybridization, 114-115, 138, 196, 231, 237, 352-353, 401-402
imperfect forms, 284, 332, 345
incompatibility, 47, 52, 102, 117, 119, 120, 192-194, 196-198, 210-211, 230, 253, 272, 275, 280-307, 326, 352, 389, 406
interfertility, 277, 279-307, 310, 311, 322, 324, 353, 354, 358, 383, 397, 404-405, 407
intersterility, 237, 253, 260, 261, 277, 279-307, 311, 325, 353, 356-358, 383, 397, 403, 404-405, 406
macrospecies, 38, 148, 327, 383, 392
microspecies, 38, 47, 51, 92, 148, 327, 360, 384-385, 391-392
mutations, 118, 129, 138, 210, 254, 302, 379
mycelium, 31, 32, 126, 218, 240, 267-268, 284, 304
mycorrhiza, 59, 63, 71, 74, 76, 85, 98, 384, 390, 402
nuclei
  spores, 138-139, 162-182, 183-184
  hyphae, 184
oidia, 262
parasexuality, 211-212, 231
parthenogenesis (see also fruiting, monocaryotic), 289, 325
pigments, pigmentation, 17, 34, 60-61, 68, 69, 70, 71, 75, 80, 81, 95, 105, 107, 115, 127, 129, 133, 225-226, 233, 238-239, 350, 367, 376-377, 379
population, 190, 211, 325, 383, 403
proteins, 225, 227, 239, 307-308, 310
pseudovariety, 357, 360-361, 392
races, 21, 114, 224, 302, 383, 384, 404
recombination, somatic, 212
senescence, 204, 294-295
speciation, 119-120, 259-264, 269-271, 323, 389, 391, 395
SPECIES CONCEPT
  38, 121, 126, 134, 142, 144-148, 225, 229, 243-244, 273, 301, 332, 336, 345, 346, 349, 364-374, 381-387, 389, 393-407

species concept, continued

    biological, 45, 48, 49, 120, 147, 311, 364, 366, 370, 375, 392

    "bona fide", 48, 237, 243

    ecologic, 142, 358, 389

    genetic, 117, 120, 144, 189-190, 195-199, 207, 213, 323, 326, 329, 354, 400, 404

    macrospecies, 38, 148, 327, 383, 392

    microspecies, 38, 47, 51, 92, 148, 327, 360, 384-385, 391-392

    morphological, 45, 48, 71, 97-98, 102, 209-210, 326, 329, 336, 376, 379

    nomenclatural, 364, 370

    prae-species, 384, 385, 392

    sexual, 225, 352

    small species, 90, 92

    taxonomic, 50, 147, 311, 364, 370

    Grand, 367

    King, 190

    Linné, 189

    Möbius, 387

    Wettstein, 189, 384

species concepts proposed at the Symposium 1976:

    Bresinsky, 393

    Esser, 393

    Esser-Boidin, 393

    Kemp, 393

    Singer, 394

    Thiers, 394

    Watling, 395

    COMMON CONCEPT, 402

spore, 17-22, 64, 105, 109, 370

    amyloidity, 64, 110, 118-119

    colours, 59, 111, 117, 136, 138, 252

    cyanophily, 112

    dimensions, 22, 23, 36, 37, 48, 85, 88-90, 99-101, 109-111, 129-130, 136-137, 138, 235-236, 251, 253, 282, 331, 336

    nuclei, 162-182, 183-184

    ornamentation, 75, 109, 130-131, 137, 166, 339

    print, 16, 22, 82

    shape, 19, 22, 85-87, 99, 111, 136, 251, 331, 336, 340-342, 347

stirps, 392

subspecies, 45, 51-52, 74, 90, 278, 355, 358, 360-361, 383-385, 390-391, 397

types, nomenclatural, 123-125, 134, 150-155, 261, 371

variation, 76-77, 120, 126-129

variety, 45, 47, 74, 114, 287, 354, 360-361, 377, 384, 385, 390-391

virus, 253

INDEX OF NEW SPECIES AND NEW COMBINATIONS

New species:

    Conocybe mairei (Kühner) ex Watling: 41

    Subulicystidium meridense Oberw.: 343

    Subulicystidium naviculatum Oberw.: 343

New combinations:

    Ceraceomyces evolvens (Fr. ex Fr.) Oberw.: 343

    Phlebiella insperata (Jacks.) Oberw.: 343

    Phlebiella subflavido-grisea (Litsch.) Oberw.: 343

    Phlebiella tulasnelloidea (v.Höhn. & Litsch.) Oberw.: 343

BIBLIOTHECA MYCOLOGICA

A series of original papers and reprints of books on Fungi

The series include as well reprints as original papers which are too lengthy for inclusion in regular journals and too limited in scope to be published as a single book.

Texts will be reproduced directly from the typescript by offset, and plates will be printed in the conventional manner. To facilitate preparing of the typescript, the publisher has manuscript sheets available which will be sent on request free of costs to interested authors.

- **Authors will receive 50 copies of their book free of cost,**
- **no page charge will be billed the author,**
- **the number of pages and plates is not limited,**
- **publication of a paper of 300 to 400 pages will be within 6 - 8 weeks.**

The series is available on standing order basis with a special 20% discount on the list prices. For well established libraries a specieal standing order service is possible, covering only the original publications. Reprints will be offered shortly before publication and can be purchased if needed at the special 20% discount. Each issue is available separately at list price.

Contact the editor for further information: J. Cramer, P.O. Box 48, D-3306 Lehre, Germany

Recently published:

Volume 35: M.L. FARR, **An Annotated List of Spegazzini's Fungus Taxa.** 2 volumes. 1973. IV, 1662 pages. per vol. DM 125,--

Volume 38: C.D. MARR & D.E. STUNTZ, **Ramaria of Western Washington.** 1973. 232 pages, 109 figures on 34 plates. Soft-cloth.
DM 50,--

Volume 43: R.H. PETERSEN, **Ramaria subg. Lentoramaria** with Emphasis on North American Taxa. 1975. 174 pages, 18 figures, 15 plates (12 colored). DM 50,--

Volume 44: O.K. MILLER, Jr. & D.F. FARR, **An Index of the Common Fungi of North America** (Synonymy and Common Names). 1975. II, 206 pages. DM 37.50

Volume 45: I. NUSS, **Zur Ökologie der Porlinge.** 1975. 258 pages, 62 figures, 36 tables. DM 50,--

BIBLIOTHECA LICHENOLOGICA

A series of original papers and reprints of books on Lichens

The series include as well reprints as original papers which are too lengthy for inclusion in regular journals and too limited in scope to be published as a single book.

Texts will be reproduced directly from the typescript by offset, and plates will be printed in the conventional manner. To facilitate preparing of the typescript, the publisher has manuscript sheets available which will be sent on request free of costs to interested authors.

- **Authors will receive 50 copies of their book free of cost,**
- **no page charge will be billed the author,**
- **the number of pages and plates is not limited,**
- **publication of a paper of 300 to 400 pages will be within 6 - 8 weeks.**

The series is available on standing order basis with a special 20% discount on the list prices. For well established libraries a specieal standing order service is possible, covering only the original publications. Reprints will be offered shortly before publication and can be purchased if needed at the special 20% discount. Each issue is available separately at list price.

Contact the editor for further information: J. Cramer, P.O. Box 48, D-3306 Lehre, Germany

Recently published:

Volume 1: B. HANNEMANN, **Anhangsorgane der Flechten** (Appendix Organs of Lichens), ihre Struktur und ihre systematische Verteilung. 1973. IV, 196 Seiten, 4 Tabellen, 219 Abbildungen auf 67 Tafeln.   DM 50,--

Volume 2: D.D. AWASTHI, **A Monograph of the Lichen Genus Dirinaria.** 1975. IV, 116 pages, 59 figures, 16 maps.   DM 40,--

Volume 3: H. WUNDER, **Schwarzfrüchtige saxicole Sippen der Gattung Caloplaca** (Lichens, Teloschistaceae) in Mitteleuropa, dem Mittelmeergebiet und Vorderasien. 1973. II, 186 Seiten, 9 Tafeln.   DM 40,--

Volume 4: P. JÜRGING, **Epiphytische Flechten als Bioindikatoren der Luftverunreinigung** - dargestellt an Untersuchungen und Beobachtungen in Bayern. 1975. II, 164 Seiten, 39 Abbildungen, 4 Tabellen.   DM 50,--

CRAMER BOOKS ON BOTANY

The Genera of Fungi Sporulating in Pure Culture
by J.A. von ARX (Baarn, Netherlands)
2nd fully revised edition 1974. IV, 316 pages, 136 figures. Balacron bound.   DM 100,--

Index Hepaticarum - An Index to the Liverworts of the World
Founded by C.E.B. BONNER (†) continued by H. BISCHLER (Paris)
8 parts with together 3.000 pages are published from 1962 to 1976.   Price on request

Mangrove Vegetation
by V.J. CHAPMAN (Auckland, N.Z.)
1976. VIII, 448 pages, 298 figures. Balacron bound.   DM 150,--

British Ascomycetes
by R.W.G. DENNIS (Richmond, England)
Fully revised edition 1977. 500 pages, 44 coloured and 32 plain plates. Balacron bound.   DM 200,--

Chlamydomonas und die nächstverwandten Gattungen
von H. ETTL (Chrastavec, CSSR.)
3 volumes. 1970-1978. More than 2.000 pages, fully illustrated. Boards.   Price on request

Nova Scotian Boletes
by D.W. GRUND & K.A. HARRISON (Wolfville, Canada)
1976. IV, 284 pages, 80 figures, 68 plates. Balacron bound.   DM 60,--

Biographisch-bibliographisches Handbuch der Lichenologie
von V. GRUMMANN (†) - nach dem Tode des Verfassers für den Druck durchgesehen von O. KLEMENT
1974. 898 pages, 43 plates. Cloth bound.   DM 250,--

Diatomeenschalen im elektronenmikroskopischen Bild
herausgegeben von J.-G. HELMCKE, W. KRIEGER (+) & J. GERLOFF
10 volumes with together 1022 plates are published from 1962 to 1977.   per volume DM 150,--

Plant Chromosomes
by A. & D. LÖVE (San José, California)
1975. XVI, 184 pages, 29 figures. Balacron bound.   DM 36,--

Cytotaxonomical Atlas of the Pteridophyta
by A. & D. LÖVE and R.E.G. PICHI SERMOLLI
1977. XVIII, 398 pages. Balacron bound.   DM 150,--

Bestimmungsschlüssel europäischer Flechten
von J. POELT (Graz).
1969. 72, 758 pages, 9 plates. Boards.   DM 80,--

The Agaricales in Modern Taxonomy
by R. SINGER (Chicago, Illinois)
3rd edition 1975. VI, 912 pages, 84 plates. Cloth bound. DM 250,--

The Boleti of Northeastern North America
by W.H. SNELL & E.A. DICK (Providence, R.I.)
Quarto. 1970. XII, 116 pages, 87 plates (71 coloured).
Cloth bound.   DM 250,--

Prodrome of the European Plant Communities
edited by R. TÜXEN (Todenmann, Germany)
3 volumes are published from 1973 to 1976.   DM 112,--

**Catalogue of the Fossil and Recent Genera
and Species of Diatoms and their Synonyms**
by S.L. VANLANDINGHAM (San Francisco, California)
5 volumes with more than 3.000 pages are published from 1967
to 1976.   per volume DM 100,--

The Pyrenomycetous Fungi
by L.E. WEHMEYER (†)
1975. 262 pages. Balacron bound.   DM 80,--

A Revision of the Characeae
by R.D. WOOD & K. IMAHORI
2 volumes. 1964-65. XXIV, 1346 pages, 403 plates.   DM 300,--

Mucorales
von H. ZYCHA & R. SIEPMANN
1970. VIII, 356 pages, 155 figures. Cloth bound.   DM 100,--